SLOPE STABILIZATION AND EROSION CONTROL: A BIOENGINEERING APPROACH

SLOPE STABILIZATION AND EROSION CONTROL: A BIOENGINEERING APPROACH

Edited by

R. P. C. Morgan and R. J. Rickson

Silsoe College, Cranfield University, UK

Taylor & Francis

Taylor & Francis Group

LONDON AND NEW YORK

Published by Taylor & Francis
2 Park Square, Milton Park, Abingdon, Oxon, OX14 4RN
711 Third Avenue, New York, NY 10017

First edition 1995

First issued in paperback 2011

© 1995 Taylor & Francis

Typeset in 10/12 Palatino by Colset Pte. Ltd., Singapore

ISBN13: 978-0-419-15630-7 (hbk)

ISBN13: 978-0-415-51176-6 (pbk)

Publisher's Note
The publisher has gone to great lengths to ensure the quality of this reprint
but points out that some imperfections in the original may be apparent

CONTENTS

CONTRIBUTORS

N. J. Coppin
Environmental Consultancy Unit, Wardell Armstrong, Grove House Court, 11a King Street, Newcastle-under-Lyme ST5 1EH, UK

Nick Coppin is a Partner of Wardell Armstrong where he is Manager of the Environmental Consultancy Unit. After graduating as a botanist he took an MSc in environmental science and then followed a career as an environmental and landscape scientist in civil engineering projects. He has particular interests in restoration of non-metal mining sites and control of soil failures on cut and embanked slopes. He was the Coordinator and Joint Editor of the CIRIA project on the use of vegetation in civil engineering.

R. P. C. Morgan
Silsoe College, Cranfield University, Silsoe Campus, Silsoe, Bedford MK45 4DT, UK

Roy Morgan received his BA from the University of Southampton, his MA from the University of London and PhD from the University of Malaya. After a period as an Assistant Lecturer at the University of Malaya, he returned to the UK, taking up a post at Silsoe College, Cranfield University where he is currently Professor of Soil Erosion Control. His research interests are in the use of vegetation for erosion control and erosion modelling. He is Joint-Coordinator of the project to develop a European Soil Erosion Model. He has carried out research and consultancy work in many countries and is the author of two books on soil erosion and over 100 research papers. He is currently President of the European Society for Soil Conservation.

R. J. Rickson
Silsoe College, Cranfield University, Silsoe Campus, Silsoe, Bedford MK45 4DT, UK

Jane Rickson obtained her BSc degree from King's College, University of London and her MSc from Silsoe College, Cranfield University where she was appointed Lecturer in Soil Erosion Control. Her research and consultancy interests include the use of geotextiles for erosion control, mulching techniques, rainfall simulation, soil erosion risk assessment and the non-technical aspects of soil conservation. She has carried out research and consultancy work in a number of African and Asian countries, is Joint-Coordinator of the project on the development of a European Soil Erosion Model, and is author of several papers and confidential reports on geotextiles, mulching and related issues.

R. Stiles
School of Landscape, Department of Planning and Landscape, University of Manchester, Oxford Road, Manchester M13 9PL, UK

Richard Stiles is a Lecturer in the Department of Planning and Landscape at Manchester University where he is course leader for the Master of Landscape Design. He is an Associate Member of the Landscape Institute. After graduating as a botanist, he studied landscape architecture at graduate level before going on to work as a landscape architect in the UK and Germany. It was during an extended stay in Germany that he became interested in the technical use of vegetation as a material for landscape construction. Since returning to the UK to take up a post at Manchester University, he has followed up this interest with further research which has resulted in a number of articles, papers and contributions to several recent publications on bioengineering.

M. E. Styczen
Dansk Hydraulisk Institut, Agern Allé 5, DK-2970 Hørsholm, Denmark

Merete Styczen received her MSc and PhD degrees from the Royal Veterinary and Agricultural University, Copenhagen. She is currently employed as an agronomist with the Danish Hydraulic Institute where she is responsible for the development of the soil erosion and soil physics components of hydrological models. She has acted as an adviser on land development and soil conservation in many projects in Africa and Asia. Her research interests include soil erosion modelling and watershed management.

T. H. Wu
Department of Civil Engineering, Ohio State University, 470 Hitchcock Hall, 2070 Neil Avenue, Columbus, OH 43210-1275, USA

Tien H. Wu received his BSc degree from St John's University, Shanghai and his MSc and PhD degrees from the University of Illinois. He has been Professor of Civil Engineering at the Ohio State University since 1965. His research and consulting interests include reliability of geotechnical design and construction, stability of embankments and slopes, soil-reinforcement by natural and synthetic materials, and engineering properties of geotechnical materials. He is the author of some 60 papers on these topics and two books.

PREFACE

The inspiration for a text on the role of vegetation in slope stabilization and erosion control came when we were involved in a project for the Construction Industry Research and Information Association (CIRIA) to look at the use of vegetation in civil engineering. The results of that project were published in a multi-authored volume (Coppin and Richards, 1990), aimed as a guide to the practising civil engineer on the state-of-the-art of bioengineering. It seemed appropriate to us to complement this work with a more analytical treatment of the subject, aimed at both student and professional readership, written at a suitable level for undergraduates and postgraduate researchers and giving a more global view of the recent research on the engineering role of vegetation in the landscape. Given the interdisciplinary nature of bioengineering, the text endeavours to bridge the gap between the engineer and the vegetation specialist. It provides the engineer with a basic understanding of the principles and practices of vegetation growth and establishment, as well as explaining in detail how vegetation can be regarded as an engineering material. At the same time, the text aims to show the plant specialist how his or her skills can be applied to engineering problems and gives a background to the kinds of questions engineers need answers to before they can design vegetation-based systems for erosion control and slope stabilization.

In order to gain a global perspective and provide an up-to-date specialist coverage, we invited contributors to write certain chapters in their fields of expertise. Although we proposed a philosophy for the text and a synopsis for each chapter so as to give some continuity and cohesion, we did not attempt to impose strict editorial control on the content and the individual contributions represent a personal view of each of the authors.

We owe a special gratitude to Dr Andrés Arnalds who reviewed much of the material in Chapter 6, provided invaluable references on the work of the State Soil Conservation of Iceland and checked that the author's interpretations of articles available only in Icelandic were, in fact, correct.

We are also indebted for early ideas on Chapter 5 to the late Dr Clinton Armstrong of the University of Saskatchewan. He should have been a co-author of this chapter but unfortunately he died before any of the work came to be written.

Despite the lengthy gestation period for the text to emerge, we have enjoyed preparing the work. We hope that the readers will find the contribution useful and also enjoy reading it.

Roy Morgan
Jane Rickson

REFERENCES

Coppin, N. J. and Richards, I. G. (eds) (1990) *Use of Vegetation in Civil Engineering*, CIRIA/ Butterworths, London.

INTRODUCTION

1

R. J. Rickson and R. P. C. Morgan

Slope instability and erosion of the soil by water and wind are major environmental hazards. Although they are the result of natural geomorphological processes, they are both affected by and have consequences for human activity, often incurring economic and social damage. In nature, vegetation is one factor maintaining equilibrium in the landscape between the destructive forces of landscape instability and the constructive or regenerative forces of stability. The risk of slope failures and erosion is enhanced when the vegetation cover is removed. The question is whether the situation can be repaired if the vegetation cover is restored. This book aims at tackling this important issue by examining the mechanisms by which vegetation plays its protective role in the landscape.

The use of vegetation for slope stabilization and erosion control can be referred to as bioengineering. Bioengineering and biotechnical engineering are terms which are commonly found in the literature, but there is much confusion as to their precise definitions. In this book, **bioengineering** refers to the use of any form of vegetation, whether a single plant or a collection of plants, as an engineering material (i.e. one that has quantifiable characteristics and behaviour). **Biotechnical engineering** refers to techniques where vegetation is combined with inert structures such as crib walls, so combining the structural benefits of both the vegetative and non-vegetative components of the scheme.

Bioengineering is a classic example of where there is a significant gap between the 'art' (or application of the techniques proposed) and the 'science' (or the scientific quantification and hence objective justification of the practices). In Europe (especially in Germany, Switzerland and Austria) and in the United States of America, pioneers have been using bioengineering and biotechnical engineering techniques for many decades (Schiechtl, 1973, 1980). These relatively few, but significant case studies have illustrated the success of bioengineering, but we cannot continue to wait a further 50 years or so, whilst new schemes become established and fully matured, to evaluate the potential of these techniques. This book aims to state the potential of bioengineering and demonstrate the science behind it as a means of justifying the techniques involved to practitioners.

As such, the book is not intended as a 'stand-alone' practical handbook of how to apply the diverse techniques of bioengineering. Instead, it aims to describe and analyse the research base underlying bioengineering in order to provide a better understanding of the role of vegetation and how it can be regarded as an engineering material. It is intended, therefore, that the book will answer many of the questions that engineers raise when expressing their uncertainty about the potential of bioengineering techniques and go some way towards showing how vegetation can be incorporated as quantifiable inputs to landscape engineering design procedures.

The book was partially prompted by the increasing awareness of the environment, and the sustainability of landscape management practice. Traditional civil engineering techniques ('grey solutions', such as concreting of welded wire walls for slope stabilization) may not be sustainable in the long term due to high

Slope Stabilization and Erosion Control: A Bioengineering Approach. Edited by R. P. C. Morgan and R. J. Rickson.
Published in 1995 by E & FN Spon, 2-6 Boundary Row, London, SE1 8HN. ISBN 0 419 15630 5.

initial capital expenditure and (more importantly) increasing maintenance requirements over time. Carefully selected and implemented bioengineering techniques are bound to be more sustainable over time as vegetation is self-regenerating and able to respond dynamically and naturally to changing site conditions, ideally without compromising or losing the engineering properties of that selected vegetation. Indeed, there are examples where a grey solution to a landscaping problem has been wholly replaced with a more natural, environmentally sensitive vegetative approach. Schürholz (1992) outlines a scheme for river channelization of the River Enz using vegetation and natural geotextiles, which were shown to have significant advantages hydraulically, aesthetically and financially compared with the original, concrete-based channelization scheme.

Any attempt to answer the question of whether vegetation can be used to alleviate landscape instability will be of interest to a wide audience, for whom this book is intended. Prior to the publication of this book, the only major reviews of bioengineering are those of Schiechtl (1973, 1980), Gray and Leiser (1982) and Bache and MacAskill (1984). This means that there has not been a substantive publication for nearly a decade, during which time much state-of-art material concerning vegetation and its effect on slope stability and erosion processes has been published in diverse and in some cases obscure academic journals. These are often not easily accessible to non-academics, and the formal presentation of such work is not in a format that is readily usable by the practitioner in the field. At the other extreme, our knowledge is often confined to a few experts' experiences, whose work may not have received the widespread exposure it deserves.

This is one consequence of the multi- and interdisciplinary nature of the subject matter being addressed. There are few publications or journals whose subject matter ranges from the detailed physics of soil erosion processes (important when attempting to understand the nature of the problem being faced) through to the tech-

niques of vegetation establishment, for example. This book aims to encompass and integrate the diversity and complexity of the role and use vegetation for landscape protection and management.

There is increasing awareness by civil engineers of the potential role of vegetation in construction work, over and above the aesthetic qualities the vegetation may have. This awareness is reflected by the publication of books such as Coppin and Richard's *Use of Vegetation in Civil Engineering* (1990), initiated and supported by the United Kingdom's Construction Industry Research and Information Association (CIRIA). Geomorphologists will also find helpful the synthesis of the most recent research on the complex relationships between vegetation and erosion processes presented in this book. In this respect, the book will complement other recent expositions on the role of vegetation, notably those edited by Viles (1988) and Thornes (1990). Other users of the book may be involved with the expansion of the landscaping industry. The number of sites and applications where the techniques presented in this book could be utilized is growing rapidly, such as land reclamation of landfill and mine spoil. Such sites require environmentally sensitive solutions to reclamation, given the public's concerns over the ways we manage and restore our diverse and ever-increasing wastelands. Recreational sites such as golf courses and ski slopes also have to be designed and maintained to cope with the increasing pressure as leisure time expands.

Although the book deals primarily with the engineering and geomorphological roles of vegetation, the cost implications of using bioengineering are not ignored. The economic differentials between conventional, grey solutions and the use of vegetation may be significant in areas where the availability of products such as concrete, sheet piling, rip-rap and gabions is severely restricted, as in inaccessible areas of developing countries. Already, bioengineering techniques have been used in developing counties such as Nepal, where experience has shown the conventional methods of slope stabilization are prohibitively expensive on implementation and in maintenance, as well as being inappropriate to the local technology and expertise used to combat slope instability of the area.

The book is organized into sections covering firstly the principles behind the use of vegetation, and secondly, the practices which have been founded on these principles. Chapter 2 reviews the scientific research which has built up a quantified database on the interactions between vegetation and both surface erosion and deeper seated processes and leads to a discussion on the salient properties of vegetation for engineering purposes. Chapter 3 covers the main considerations of whether the vegetation will establish and develop into a form which meets these engineering needs. No matter how effective vegetation may be in controlling rainsplash erosion, for example, the vegetation will never reach the design requirements unless the correct growing conditions exist for that vegetation type to establish and develop successfully. Chapter 4 concentrates on the practice of using simulated vegetation, which may circumvent the problems of achieving the required vegetation characteristics in hostile areas, or when time is limited for the vegetation to establish and reach maturity at which it realizes its potential, as outlined in Chapter 2. Chapters 5 and 6 report on the practices used for the control of erosion by water and wind, based on bioengineering and biotechnical engineering principles. Many of these techniques have been adapted from agricultural engineering practice, again reflecting the multidisciplinary nature of the subject, and the fact that the detrimental impacts of erosion were first felt on

agricultural land. Hence the experience and expertise on using vegetation to control soil erosion originate from this discipline. This book aims to widen the audience to whom these proven techniques may be helpful. With increasing concern over sediment production from non-agricultural land uses, it is wise to adopt techniques already proven to be successful. The role of vegetation in slope stability is covered in Chapter 7, where particular emphasis is placed on how conventional approaches to modelling and calculating slope stability and instability can be modified and adapted to account for the role of vegetation.

REFERENCES

Bache, D. H. and MacAskill, I. A. (1984) *Vegetation in Civil and Landscape Engineering*. Granada, London.

Coppin, N. J. and Richards, I. G. (1990) *Use of Vegetation in Civil Engineering*. CIRIA/Butterworths, London.

Gray, D. H. and Leiser, A. T. (1982) *Biotechnical Slope Protection and Erosion Control*. Van Nostrand Reinhold, New York.

Schiechtl, H. M. (1973) *Sicherungsarbeiten im Landschaftsbau*. Callway, München.

Schiechtl, H. M. (1980) *Bioengineering for Land Reclamation and Conservation*. University of Alberta Press, Edmonton.

Schürholz, H. (1992) Use of woven coir geotextiles in Europe. Paper presented to UK Coir Geotextile Seminar, Organised by ITC, UNCTAD/GATT, Coir Board of India and SIDA.

Thornes, J. B. (1990) *Vegetation and Erosion*. Wiley, Chichester.

Viles, H. A. (1988) *Biogeomorphology*. Blackwell, Oxford.

ENGINEERING PROPERTIES OF VEGETATION

2

M. E. Styczen and R. P. C. Morgan

2.1 INTRODUCTION

Vegetation provides a protective layer or buffer between the atmosphere and the soil. Through the hydrological cycle, it affects the transfer of water from the atmosphere to the earth's surface, soil and underlying rock. It therefore influences the volume of water contained in rivers, lakes, the soil and groundwater reserves. The above-ground components of the vegetation, such as leaves and stems, partially absorb the energy of the erosive agents of water and wind, so that less is directed at the soil, whilst the below-ground components, comprising the rooting system, contribute to the mechanical strength of the soil.

Traditionally, the role of vegetation has been viewed rather simplistically, as seen by the somewhat superficial way it is dealt with in water erosion studies. The most commonly used approach has been to assign to it a coefficient, such as the C-factor in the Universal Soil Loss Equation (Wischmeier and Smith, 1978) which, for a certain stage of growth and plant density, describes the ratio of soil loss when vegetation is present to the amount lost on a bare soil. Values of this soil loss ratio are derived experimentally from field trials and, while they are true values for those situations, they cannot be readily used to predict the effect of the same or other vegetation in different climatic and pedological conditions.

Wischmeier (1975) tried to split the C-factor into C_I, C_{II} and C_{III} subfactors (Figure 2.1). C_I describes the effect of the presence of a plant canopy at some elevation above the soil. C_{II} is defined as the effect of a mulch or close-growing vegetation in direct contact with the soil surface. Root effects are not included. C_{III} represents the residual effects of land use on soil structure, organic matter content and soil density, the effects of tillage or lack of tillage on surface roughness and soil porosity, and the effects of roots, subsurface stems and biological activity in the soil. This approach has been used in erosion prediction models (Beasley, Huggins and Monke, 1980; Park, 1981; Park, Mitchell and Scarborough, 1982) but is limiting because at least two of the three subfactors may influence more than one erosion process. It is difficult therefore to give them a precise physical meaning.

The conflicting views, provided by field and laboratory experiments (Figure 2.2) on what level of vegetation cover is required to reduce the soil loss ratio from 1.0 to 0.5, illustrate the inadequacy of the above approach. In order to understand the role of vegetation in combating erosion it is necessary to:

1. understand the erosion processes;
2. consider how each of these processes may be affected by vegetation;
3. determine the salient properties of the vegetation which most affect these processes;
4. try to quantify the combined effect of vegetation on the processes acting together in different situations.

Slope Stabilization and Erosion Control: A Bioengineering Approach. Edited by R. P. C. Morgan and R. J. Rickson. Published in 1995 by E & FN Spon, 2-6 Boundary Row, London, SE1 8HN. ISBN 0 419 15630 5.

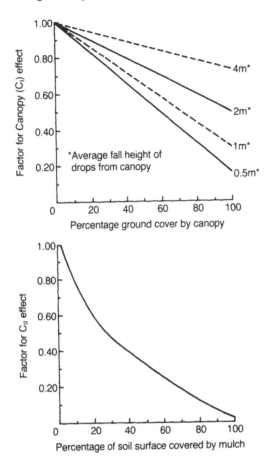

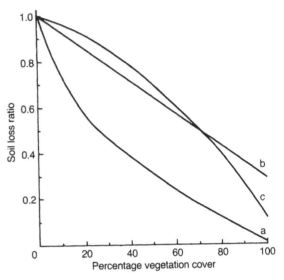

Figure 2.2 Examples of relationships between the soil loss ratio and percentage vegetation cover. a = ground level vegetation (Laflen and Colvin, 1981); b = vegetation canopy at 1 m above the ground (Wischmeier, 1975); c = oat straw mulch (Singer and Blackard, 1978).

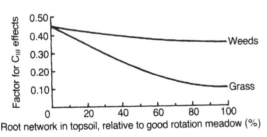

Figure 2.1 Soil loss ratios for subfactors of the C-factor in the Universal Soil Loss Equation (after Wischmeier, 1975). C_I describes the canopy effect, C_{II} the effect of plant residues and ground vegetation, and C_{III} the residual effects of previous land use. The graph shown here for the C_{III} effect applies to previously undisturbed land only and not to cropland or construction sites.

Such a detailed understanding is difficult to achieve. It is hampered by the fact that much previous research has concentrated on establishing C-factor values rather than on understanding how vegetation operates within the erosion system. Analysis is also hampered by the complexity of the interaction between vegetation, climate, soil properties and hydrology. Nevertheless, the relatively low rates of erosion observed in well-vegetated areas compared with the catastrophic rates which can arise when vegetation is cleared, demonstrate that vegetation performs a major engineering role in protecting the landscape. This chapter aims to explore that role by reviewing its hydrological, hydraulic and mechanical effects. These are summarized diagrammatically in Figure 2.3.

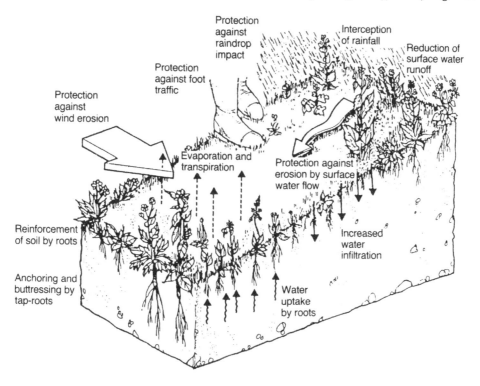

Figure 2.3 Engineering role of vegetation (after Coppin and Richards, 1990).

2.2 HYDROLOGICAL EFFECTS OF VEGETATION

2.2.1 EVAPOTRANSPIRATION

Evapotranspiration is the combined process of the removal of moisture from the earth's surface by evaporation and transpiration from the vegetation cover. Evapotranspiration from plant surfaces is compared to the equivalent evaporation from an open water body. The two rates are not the same because the energy balances of the surfaces are markedly different. For example, the albedo value, defined as the proportion of incoming short-wave radiation which is reflected, is, depending on the altitude of the sun, about 0.1 for water, but varies between 0.1 and 0.3 for a plant cover. The effect of vegetation is expressed by the E_t/E_o ratio, where E_t is the evapotranspiration rate for the

vegetation cover and E_o is the evaporation rate for open water. Table 2.1 gives some typical values for plant covers at different stages of growth and in different seasons (Withers and Vipond, 1974; Doorenbos and Pruitt, 1977).

The values of E_t/E_o ratios assume that evapotranspiration is not limited by the supply of water; in other words, it is taking place at the potential rate (E_p). Where high rates of evapotranspiration occur, however, the top layers of the soil rapidly dry out and the plants find it more difficult to extract water from the soil by suction through the roots. To prevent dehydration, plants reduce their transpiration so that actual evapotranspiration (E_a) becomes less than potential. The ratio of actual to potential evapotranspiration (E_a/E_p) depends upon the soil moisture deficit (SMD) which is defined as the difference between the reduced soil moisture level and that pertaining at field capacity. The

Table 2.1 E_t/E_o ratios for selected plant covers (after Withers and Vipond, 1974; Doorenbos and Pruitt, 1977)

Plant (crop) cover	E_t/E_o ratio
Wet (padi) rice	1.35
Wheat	0.59–0.61
Maize	0.67–0.70
Barley	0.56–0.60
Millet/sorghum	0.62
Potato	0.70–0.80
Beans	0.62–0.69
Groundnut	0.50–0.87
Cabbage/Brussels sprouts	0.45–0.70
Banana	0.70–0.77
Tea	0.85–1.00
Coffee	0.50–1.00
Cocoa	1.00
Sugar cane	0.68–0.80
Sugar beet	0.73–0.75
Rubber	0.90
Oil palm	1.20
Cotton	0.63–0.69
Cultivated grass	0.85–0.87
Prairie/savanna grass	0.80–0.95
Forest/woodland	0.90–1.00

but when SMD > C, a further 25 mm of moisture can be extracted at a reduced rate until, at SMD > 3C, extraction becomes minimal $(E_a = 0.1\,E_p)$.

Although the ability of vegetation to reduce soil moisture is recognized qualitatively, it is hard to quantify. Reduced soil moisture increases soil suction which affects both hydraulic conductivity and pore-water pressure. Only limited information is available, however, on differences in the hydraulic conductivity of soils with and without a vegetation cover and the effect of vegetation on slope stability through soil moisture depletion is difficult to separate from that of soil reinforcement by the rooting system. Nevertheless, through modification of the soil moisture content, vegetation affects the frequency at which the soil becomes saturated which, in turn, controls the likelihood of runoff generation or mass soil failure. The strength of this effect depends upon the local soil and climatic conditions and the vegetation type. It will also show, often substantial, seasonal variation, being greatest in summer and lowest in winter or whenever the vegetation is dormant.

amount of soil moisture which can be extracted by a plant cover when water is not limiting is defined by a root constant (C); typical values are given in Table 2.2 (Grindley, 1969). Actual evapotranspiration taking place as a soil dries out can be estimated using the model of Penman (1949) whereby actual evapotranspiration equals potential $(E_a = E_p)$ as long as SMD < C

2.2.2   INTERCEPTION

On contact with the canopy of a vegetation cover, the rainfall is divided into two parts. These are (1) direct throughfall, that which reaches the ground after passing through gaps in the canopy, and (2) interception, that which strikes the vegetation cover. If it is assumed that

Table 2.2 Values of the root constant (C) for use in estimating evapotranspiration (after Grindley, 1969)

Vegetation	Maximum SMD (mm)[a]	Root constant, C (mm)
Cereals	200	140
Temporary grass	100	56
Permanent grass	125	75
Rough grazing	50	13
Trees (mature stand)	125–250	75–200

[a] SMD = soil moisture deficit. The actual value of maximum SMD varies with the depth of roots, being higher for deep-rooted vegetation than for shallow-rooted types.

the rain falls vertically, the volume of rainfall intercepted (*IC*) can be calculated from the simple relationship:

$$IC = \text{RAIN} \cdot CC, \qquad (2.1)$$

where *CC* = percentage canopy cover.

Some of the intercepted rainfall is stored on the leaves and stems and is later returned to the atmosphere by evaporation. The remainder of the intercepted rainfall, termed 'temporarily intercepted throughfall' (TIF), reaches the ground either as stemflow (i.e. that running down the stems, branches or trunks of the vegetation) or as leaf drainage.

Interception storage

Observed interception storage (*IC*$_{store}$) varies widely, depending upon the type of vegetation and the intensity of the rain, but, during a storm, it increases exponentially to a maximum value (*IC*$_{max}$) in a relationship similar to that proposed by Merriam (1973):

$$IC_{store} = IC_{max}(1 - \exp R_{cum}/IC_{max}), \qquad (2.2)$$

where R_{cum} is the cumulative rainfall received since the start of the storm. Values of maximum interception storage are difficult to determine but probably range from 0.5 mm for deciduous forest in winter to 1 mm for coniferous forest, deciduous forest in summer and many agricultural crops, 1–2 mm for grasses and 2.5 mm for a multi-layered tropical rain forest (Table 2.3). Since storage often returns to the maximum value between storms, its cumulative effect over a year can be considerable and can account for 10–15% of the annual rainfall in cool-temperate hardwood forests, 15–25% in temperate broad-leaved forests, 20–25% for cereals and grass covers, and 25–30% in temperate coniferous and in tropical rain forests. Interception storage thus reduces the volume of rainfall reaching the ground surface by these amounts.

Stemflow

The amount of water shed by stemflow depends upon the angle of the stems of the plant to the ground surface (De Ploey, 1982; van Elewijck, 1989). For plants where the stem diameters are less than the median volume drop size of the rainfall, such as grasses, stemflow is at a maximum when the stem angles are between 50° and 70°. For plants with larger diameter stems, the situation is less clear. Van Elewijck (1988) recorded maximum stemflow on maize leaves at stem angles between 10° and 20° and on simulated branches at stem angles between 5° and 15° whereas Herwitz (1987) found that stemflow on branches (>4 cm diameter) of *Toona australis* and *Aleurites moluccana* increased linearly with stem angle to reach a maximum at a branch angle of 60°, the highest angle used in his experiments.

Very little information exists on volumes of stemflow. Measurements by Noble (1981) and Finney (1984) show stemflow volumes to be about 3–7% of storm rainfall for both Brussels sprouts with canopy covers of 40–50% and potatoes with 20–25% canopy. Higher values were observed for sugar beet at 42% of storm rainfall with 28% canopy cover (Finney, 1984). A figure of 55% was also recorded for sugar beet by Appelmans, van Hove and De Leenheer (1980). Values of 44% and 31% were recorded by Bui and Box (1992) in laboratory experiments under maize and sorghum respectively. High stemflow volumes can therefore be expected for plants with an architecture designed to concentrate water at their base and characterized by stems and leaves which converge towards the ground. De Ploey (1982) estimates that tussocky grasses may produce stemflow volumes that amount to 50–100% of the intercepted rainfall and Herwitz (1987) found that more than 80% of the impacting rain on tree branches inclined at 60° contributed to stemflow. Such concentrations of rainfall over relatively small areas can increase the effective rainfall intensity locally beneath tussocky grasses to 150–200% of that received at the top of the canopy (De Ploey, 1982). Even greater concentrations can occur in forests. Herwitz (1986) recorded an instance in the tropical rain forest in northern Queensland where stemflow fluxes measured during a rain-

Table 2.3 Interception storage capacity for different vegetation types (after Horton, 1919; Leyton, Reynolds and Thompson, 1967; Zinke, 1967; Rutter and Morton, 1977; Herwitz, 1985)

Vegetation type	Interception storage capacity, $IC_{max}(mm)$
Fescue grass	1.2
Molinia	0.2
Rye grass	2.5
Meadow grass, clover	2.0
Blue stem grass	2.3
Heather	1.5
Bracken	1.3
Tropical rain forest	0.8–2.5
Temperate deciduous forest (summer)	1.0
Temperate deciduous forest (winter)	0.5
Needle leaf forest (pines)	1.0
Needle leaf forest (spruce, firs)	1.5
Evergreen hardwood forest	0.8
Soya beans	0.7
Potatoes	0.9
Cabbage	0.5
Brussels sprouts	1.0
Sugar beet	0.6
Millet	0.3
Spring wheat	1.8
Winter wheat	3.0
Barley, rye, oats	1.2
Maize	0.8
Tobacco	1.8
Alfalfa	2.8
Apple	0.5

fall of 11.8 mm in 6 min gave local depth equivalents of between 83 and 1888 mm. These large quantities of water beneath plants can play an important role in the generation of runoff.

Based on the work of van Elewijck (1988), the volume of stemflow (SF) may be estimated as a function of the average angle of the plant stems to the ground (PA) using the following equations:

for stem diameters < median volume drop diameter:

$$SF = TIF \left(\cos PA \cdot \sin^2 PA \right); \quad (2.3)$$

for stem diameters > median volume drop diameter:

$$SF = TIF \cos PA. \quad (2.4)$$

In the above, sin PA expresses the effect of gravity and cos PA expresses the effect of the projected length of the leaves and stems on the plant.

Leaf drainage

The volume of leaf drainage is equal to the volume of temporarily intercepted throughfall less the volume of stemflow. Leaf drainage comprises raindrops that are shattered into small droplets immediately they strike the vegetation and large drops formed by the temporary storage and coalescence of raindrops on the leaf

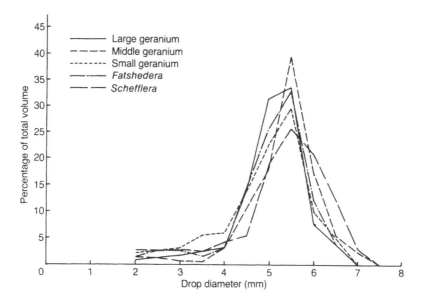

Figure 2.4 Drop-size distribution of leaf drainage (after Brandt, 1989).

and stem surfaces before they fall to the ground. Thus the rainfall beneath a plant canopy has higher proportions of small (<1 mm) and large (>5 mm) drops and fewer medium-sized drops compared with the original rainfall. In this way the canopy cover changes the drop-size distribution of the rain.

For plants with long leaves, like maize, the drops are mainly channelled along the centre vein and form leaf drips with diameters of 5–5.5 mm. For soya beans, the average size of the leaf drips is smaller, at about 4.5 mm, partly because more raindrops are rejected instantaneously by the leaves (Armstrong and Mitchell, 1987). Brandt (1989), in a review of previous literature combined with results of her own laboratory studies, concludes that leaf drainage has a normal drop-size distribution with a mean volume drop diameter of between 4.52 and 4.95 mm and a standard deviation of 0.79–1.30 mm (Figure 2.4).

Concentrations of water from leaf drip points can result in very high localized rainfall intensities, over 1000% greater than the intensity received at the canopy (Armstrong

and Mitchell, 1987). These can exceed infiltration rates and result in surface runoff. This effect would be most marked in calm conditions. In strong winds, movement of the leaves and branches, as well as the falling water drops, will help to spread the leaf drainage more uniformly.

Soil detachment by raindrop impact

Soil detachment by raindrop impact has been related to various properties of the rain; KE (kinetic energy), EI_{30} (kinetic energy times the maximum intensity of the storm, measured over 30 min) and I^2 (intensity squared) being the most commonly used parameters. Vegetation affects these properties by altering the mass of rainfall reaching the ground, its drop-size distribution and its local intensity.

The energy of the rainfall available for soil detachment under a vegetation cover is dependent upon the relative proportions of the rain falling as direct throughfall and as leaf drainage. The ability of stemflow to detach soil particles is normally ignored. Thus the kinetic energy of the rain can be expressed by the simple arithmetical relationship:

$$KE = [(DT/TV) \cdot KE(DT)] +$$
$$[(LD/TV) \cdot KE(LD)], \qquad (2.5)$$

where KE = the kinetic energy (J/m^2 mm) of the rain; DT = the volume of direct throughfall; LD = the volume of leaf drainage; and TV = the total volume of direct throughfall and leaf drainage.

The energy of the direct throughfall is assumed to be the same as that of the natural rainfall. A reasonable approximation of the drop-size distribution of steady rain in temperate mid-latitude climates is that described by Marshall and Palmer (1948):

$$N(\delta) = N_0 e^{-\Lambda\delta}, \qquad (2.6)$$

where $N(\delta)d\delta$ = the number of drops per unit volume with diameters between δ and $\delta + d\delta$; $\Lambda(I) = 41\,I^{-0.21}$, where Λ has units of cm^{-1} and I is the rainfall intensity (mm/h); and N_0 = approximately $0.08\,cm^{-4}$.

Other drop-size distributions have been presented by Carter *et al.* (1974) for Florida, Hudson (1963) for Zimbabwe, and Kowal and Kassam (1976) for northern Nigeria. In the case of the Marshall–Palmer distribution, the kinetic energy (J/m^2 mm) of a unit rain can be estimated from (Brandt, 1990):

$$KE(DT) = 8.95 + 8.44 \log I, \qquad (2.7)$$

where I = the intensity of the rain (mm/h).

If the drop-size distribution of the leaf drainage follows that described above, its energy may be calculated from (Brandt, 1990):

$$KE(LD) = (15.8 \cdot PH^{0.5}) - 5.87, \qquad (2.8)$$

where PH = the effective height (m) of the vegetation canopy.

For non-cohesive soils, the rainfall energy is not spent on detaching individual soil particles from the soil mass. It is primarily used for deformation of the surface and the lifting and moving of the already-discrete particles. In this case, splash erosion can be expected to be propor-tional to the kinetic energy of the rain (Free, 1960; Moss and Green, 1987), which is approx-imately proportional to $I^{1.14}$. Soil detachment (DET; g/m^2), in the sense of dislodgement of soil particles by raindrop impact, can then be estimated from the simple relationship:

$$DET = k \cdot KE^{1.0} \cdot e^{-ah}, \qquad (2.9)$$

where k = an index of the detachability of the soil (g/J); h = the depth (m) of the surface water layer, if any; and a = an experimental coeffi-cient varying between 1.0 and 3.0 in value, depending upon the soil texture (Torri, Sfalanga and Del Sette, 1987).

It follows from this analysis that the rate of soil detachment beneath a vegetation cover depends upon the percentage canopy area, which controls the volumes of direct throughfall and leaf drainage, and the height of the canopy, which determines the energy of the leaf drain-age. Numerous studies have shown that the energy of rain under vegetation can exceed that of an equivalent rainfall in open ground, both for trees (Chapman, 1948; Wiersum, Budirijanto and Rhomdoni, 1979; Maene and Chong, 1979; Mosley, 1982) and for lower-growing agricul-tural crops (Noble and Morgan, 1983; Morgan, 1985) with consequent increases in the rate of detachment (Finney, 1984; Wiersum, 1985). Field measurements with rainfall simulation showed that soil detachment under maize increased with percentage cover to double that recorded on bare soil when the canopy reached about 90% cover and was about 2 m above ground level (Morgan, 1985).

Recent research (Styczen and Høgh-Schmidt, 1988) has suggested that kinetic energy may not be the best parameter of the rain to explain soil detachment under vegetation or on cohesive soils. A different approach is proposed in which soil detachment is proportional to the sum of the squared momenta of the raindrops:

$$DET = A(2\hat{e})^{-1} Pr \sum^{\delta} N_\delta p_\delta^2, \qquad (2.10)$$

where A = a soil-dependent constant of propor-tionality; $\hat{e}$ = the average energy required to

Table 2.4 Values of squared momentum for different intensities of rain

Rainfall intensity, I (mm/h)	Squared momentum, M_R $((N\,s)^2/m^2\,s)$
5	2.66×10^{-7}
10	8.88×10^{-7}
20	2.86×10^{-6}
35	7.11×10^{-6}
50	1.25×10^{-5}
75	2.32×10^{-5}
100	3.56×10^{-5}
125	4.92×10^{-5}
150	6.38×10^{-5}
175	7.93×10^{-5}
200	9.55×10^{-5}
225	1.12×10^{-4}
250	1.30×10^{-4}

break the bonds between two micro-aggregates of soil, and the energy lost by heat in the process; Pr = the probability that the kinetic energy received by the detached micro-aggregate(s) is large enough to make it measurable as splash, i.e. to make the micro-aggregate jump a minimum distance; N_δ = the number of raindrops of size (diameter) δ; and p_δ = the drop momentum ($m_\delta \cdot v_\delta$).

A, $\hat{e}$ and Pr are related to soil properties, while N_δ and p_δ are rainfall properties; m and v refer respectively to the mass and velocity of the raindrop.

For the Marshall–Palmer drop-size distribution, $N_\delta p_\delta^2$ is proportional to $I^{1.63}$ for $0 < I < 100$ mm/h and $I^{1.43}$ for $100 < I < 250$ mm/h. Values for the squared momentum, $M_R = N_\delta p_\delta^2$, are listed in Table 2.4.

The squared momentum of the leaf drainage (M_{RC}) can be calculated in the following way (Styczen and Høgh-Schmidt, 1986), given that the amount of leaf drainage equals $CC \cdot I \cdot [1 - (SF + IC_{store})]$ and the number of drops equals $CC \cdot I \cdot [1 - (SF + IC_{store})/\text{vol}(\delta)]$ (equation 2.11) where $\text{vol}(\delta)$ = the volume of a drop with diameter (δ); ρ_w = the density of water; $v_{\delta H}$ = the velocity of the drop as a function of its diameter (δ) and fall height (H); and vol δ) $\rho_w^2 v_{\delta H}^2 = DH$, listed in Table 2.5.

$$M_{RC} = \frac{CC \cdot I \cdot [1 - (SF + IC_{store})]}{\text{vol}(\delta)} \cdot \rho_w^2$$
$$[\text{vol}(\delta)^2 \cdot v_{\delta H}^2]. \qquad (2.11)$$

When the sum of the squared momenta with and without a vegetation cover are known, the

Table 2.5 Values of the parameter DH ($\rho_a^2 \pi \delta^3 v_\delta^2/6$) (kg^2/m s) computed for different drop sizes (δ) and fall heights

Fall height (m)	Drop sizes (δ)			
	4.5 mm	5.0 mm	5.5 mm	6.0 mm
0.5	0.4180	0.5734	0.7633	0.9909
1.0	0.7942	1.1002	1.4787	1.9384
1.5	1.2120	1.6890	2.2836	2.9996
2.0	1.5720	2.1866	2.9508	3.8837
3.0	2.1291	2.9998	4.0757	5.4158
4.0	2.5706	3.6229	4.9526	6.6014
5.0	2.9029	4.1470	5.6452	7.4386
6.0	3.1459	4.4763	6.0883	8.0182
7.0	3.2949	4.6733	6.3533	8.4036
8.0	3.3907	4.7957	6.5331	8.6590
9.0	3.4554	4.8971	6.6696	8.8381
10.0	3.5125	4.9769	6.7768	8.9584
13.0	3.6530	5.1843	6.9936	9.2016
∞	3.8647	5.4080	7.2934	9.5310

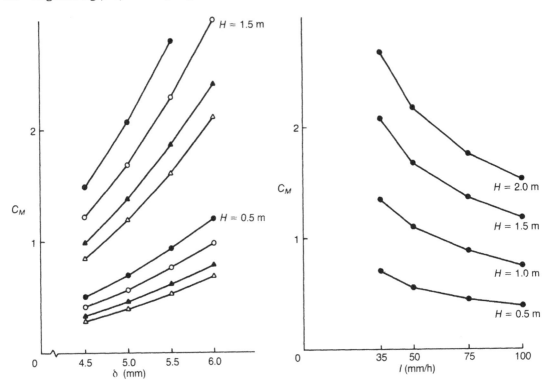

Figure 2.5 C_M as a function of the drop size of transformed rain (δ) for different rainfall intensities (I) and two canopy heights (H) (after Styczen and Høgh-Schmidt, 1988). The canopy cover is 100%. Storage and stemflow are estimated as 10% of the rainfall. ●, $I = 35$ mm/h; ○, $I = 50$ mm/h; ▲, $I = 75$ mm/h; △, $I = 100$ mm/h.

relative effect of the vegetation on splash (equivalent to a C-factor for splash) can be calculated as:

$$C_M = \frac{(1-CC)M_R + C \cdot I \cdot (1-(SF+IC_{store})) \cdot DH}{M_R}.$$

(2.12)

Figures 2.5, 2.6 and 2.7 illustrate the calculated effect of different drop sizes, fall heights, stemflow percentages and rainfall intensities on the value of C_M. Figure 2.8 shows how important the drop-size distribution of the rain can be when interpreting the effects of vegetation. As splash erosion is proportional to the drop diameter raised to the sixth power, leaf drainage

Figure 2.6 Changes in C_M with changes in rainfall intensity (I) for different canopy heights (H) (after Styczen and Høgh-Schmidt, 1988). Curves are calculated for 100% canopy cover and a drop size for leaf drainage (δ) of 5.0 mm.

may result in serious soil breakdown. Contrary to ordinary opinion, a plant canopy situated more than about 1 m above the ground cannot be expected to decrease splash erosion by itself; indeed, it is more likely to enhance it. Figure 2.9 shows this effect measured under maize (Morgan, 1985) and calculated according to equations 2.10 and 2.12.

Similar conclusions were reached by Moss and Green (1987) who, on the basis of empirical data, divided vegetation layers into the following categories:

1. Layer 1, <0.3 m: where, owing to the often high density of plant–ground contacts, leaf drainage volumes are usually small and the impact velocities too low to allow significant damage.

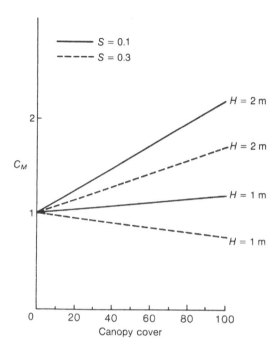

Figure 2.7 Changes in C_M with changing percentage permanent interception (S) for leaf drainage with a diameter (δ) of 5.0 mm and two heights of canopy (H) (after Styczen and Høgh-Schmidt, 1988). Rainfall intensity equals 50 mm/h.

2. Layer 2, 0.3–1.0 m: where there is a transition from small to significant leaf drip and soil damage.
3. Layer 3, 1.0–2.5 m: from which leaf drips reach high erosivity and achieve a marked ability to cause soil damage.
4. Layer 4, 2.5–6.0 m: in which the ability of leaf drips to cause erosion and soil damage continues to increase but more slowly than in layer 3.
5. Layer 5, >6 m: where the free fall height is sufficient for leaf drips to attain 90% or more of their terminal velocity; hence, above this height there is little further increase in either their ability to cause soil damage or their erosivity.

If, in contrast to the above, it is assumed that splash erosion on sand is proportional to the incoming kinetic energy instead of the sum of the squared momenta, the apparent effects of vegetation become less drastic. For very tall vegetation, the energy impact approximately doubles compared to that on bare soil, but for most agricultural crops, the impact is reduced. The relative change in energy impact is shown in Figure 2.10 for four rainfall intensities, five fall heights, 10% stemflow and different cover percentages. From such calculations, a soil under a 0.5 m tall soya bean crop (80% cover, 10% stemflow) receives only 50% of the energy received by a bare soil. In the case of 1.5 m tall maize (also 80% cover and 10% stemflow), the soil receives 85–90% of the energy. For 6 m tall trees without any ground cover, the energy received by the soil reaches 200%.

It may seem strange that the amount of energy reaching the ground under trees is more than 100%. This is due to the difference in frictional resistance on small and large drops. Leaf drips are not only larger and heavier, they also gain a higher velocity so that the final impact energy is increased.

Equation 2.10 contains two soil factors that may be influenced by vegetation. These are $\hat{e}$, the average energy required to break the bonds between two micro-aggregates, and Pr, the probability that the kinetic energy received by the micro-aggregate is large enough to make it measurable as splash. The term, k, in equation 2.9, which expresses the detachability of the soil, also encompasses these factors which are discussed in more detail in section 2.4.1.

2.2.3 INFILTRATION

For serious erosion to take place, some amount of runoff must occur. The amount of runoff generated is closely related to the infiltration rates (unsaturated and saturated hydraulic conductivity) of the soil, the antecedent moisture content and, indirectly, to the direction of water flow within the soil.

When rain water reaches the ground underneath vegetation, it may stand a better chance of infiltrating than on unvegetated soil. Organic

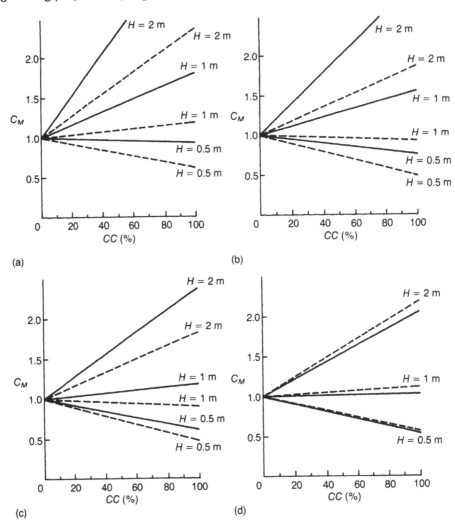

Figure 2.8 C_M as a function of percentage vegetation cover (*CC*) and canopy height (*H*), calculated for two drop-size-distributions (——, Marshall and Palmer, 1948; ----, Carter *et al.*, 1974) for leaf drainage of 5.0 mm diameter (δ), and percentage permanent interception storage and stemflow equal to 10% of the rainfall (after Styczen and Høgh-Schmidt, 1988). Rainfall intensities are (a) 35 mm/h; (b) 50 mm/h; (c) 75 mm/h; (d) 100 mm/h.

matter, root growth, decaying roots, earthworms, termites and a high level of biological activity in the soil help to maintain a continuous pore system and thereby a higher hydraulic conductivity. Through an increase in the infiltration rate, and perhaps also in the moisture storage capacity of the soil, vegetation may decrease the amount of runoff generated during

a storm; it will probably also increase the time taken for runoff to occur. A bare soil may be compared to a bucket with few or small holes in the bottom, while the vegetated soil is rather like a slightly larger bucket with more and bigger holes. It is necessary to apply more water at a greater rate to make the second bucket overflow. Thus, a higher infiltration may decrease

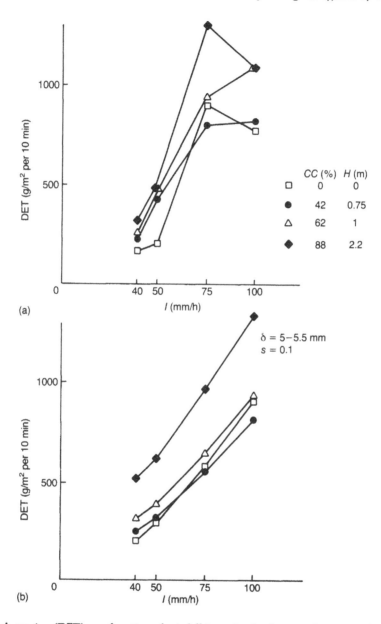

Figure 2.9 Splash erosion (DET) as a function of rainfall intensity for four combinations of percentage maize cover (*CC*%) and height (*H*) (after Styczen and Høgh-Schmidt, 1988). (a) Observed data (from Morgan, 1985); (b) calculated data. Based on equations 2.10 and 2.12 with a drop diameter (δ) of leaf drainage of 5.0–5.5 mm and percentage permanent interception storage and stemflow equal to 10% of the rainfall.

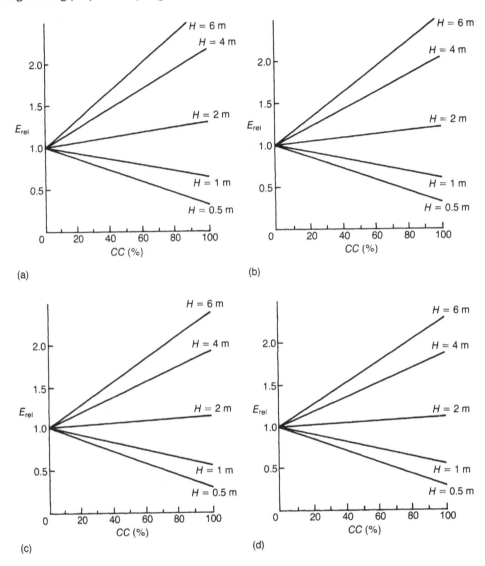

Figure 2.10 Relative change in kinetic energy of the rain (E_{rel}) reaching the soil as a function of percentage vegetation cover (CC%) and canopy height (H), calculated for the Marshall–Palmer drop-size distribution, drop diameter (δ) of leaf drainage of 5.0 mm and percentage permanent interception and stemflow of 10% of the rainfall. Rainfall intensities are (a) 35 mm/h; (b) 50 mm/h; (c) 75 mm/h; (d) 100 mm/h.

the number of erosive events per year because a greater storm is needed to produce the critical amount of runoff.

The saturated hydraulic conductivity of a soil (k_{sat}) depends on its texture and structure, the presence of cracks and the number of biopores it contains. McKeague, Wang and Coen (1986)

present some guidelines for estimating k_{sat} from soil morphology. These are of interest here because the descriptions help in visualizing the changes occurring in a soil as a result of biological interference. Rawls, Brakensiek and Soni (1983) and Brakensiek and Rawls (1983) estimate k_{sat} for soils with different pore-size distri-

Table 2.6 Morphological descriptions and corresponding values of saturated hydraulic conductivity (k_{sat}) for soils with a loamy texture but different structural properties and pore content (from McKeague, Wang and Coen, 1986)

$k_{sat}(mm/h)$	Soil description
1.5–5.0	Massive to weak coarse blocky or prismatic non-compact loamy or clayey material with tightly-accommodated peds (if any), < 0.02% channels > 0.5 mm, and few very fine voids visible with a hand lens
5.0–15	Structureless loamy material, friable, bulk density $1.5 \pm 0.1 \, Mg/m^3$, not compact, with < 0.02% channels
15–50	*Either* moderately packed loamy to clayey material with weakly developed pedality (adherent partly-formed peds); 0.02–0.1% channels $\geqslant$ 0.5 mm, some of which traverse the horizon
	or moderate medium to coarse blocky loamy or clayey material with firm dense peds, < 0.02% channels (= to 15 biopores/m^2 of 4 mm diameter)
50–150	*Either* approximately 0.1–0.2% channels > 0.5 mm, at least half of which extend through the horizon, < 0.02% large ($\geqslant$ 5 mm) channels, structureless or weak structure; texture finer than fine sandy loam, if not compact
	or moderate fine or medium blocky with weakly adherent peds or moderate to strong medium to coarse blocky; < 0.1% channels extend through the horizon; texture finer than fine sandy loam, if not compact

butions, textures and organic matter contents, with and without tillage, and with crusting.

In Table 2.6, six different morphological descriptions of loamy soil are given, together with corresponding values for saturated hydraulic conductivity (McKeague, Wang and Coen, 1986). The soils have the same texture but differ with respect to the number of biopores or the level of aggregation and structural development. According to this description, the saturated hydraulic conductivity of a loamy soil may range from 1.5 to 150 mm/h, depending on these soil characteristics.

It is evident that such large differences in hydraulic conductivity will result in large differences in the amount of runoff generated during a particular storm. Assuming that the soil is already wet and the rate of infiltration equals the saturated hydraulic conductivity of the soil,

a very simple way to estimate the amount of runoff is to compare rainfall intensity and infiltration ($I - k_{sat}$). Runoff amounts for different intensities and four values of k_{sat} are given in Table 2.7. Similar results can be calculated for soils of other textures.

For dry soils, the differences in runoff generation may be even larger, as the time taken to wet the soil to saturation varies. This also implies that the delay in time before runoff occurs is longer for soils with high hydraulic conductivities.

At present it is possible to account for the effects of vegetation on infiltration only very crudely, for example by an empirical adjustment to the value of k_{sat} for an unvegetated soil, so that (Holtan, 1961):

$$k_{satveg} = k_{sat} + a^{1.4}, \qquad (2.13)$$

Table 2.7 Runoff amounts (m^3/ha h) calculated for eight rainfall intensities and four infiltration rates

Rainfall intensity (mm/h)	Infiltration rate (mm/h)			
	1.5	5.0	15	50
10	85	50	0	0
20	185	150	50	0
30	285	250	150	0
40	385	350	250	0
50	485	450	350	0
75	735	700	600	250
100	985	950	850	500
150	1485	1450	1350	1000

Table 2.8 Basal areas for different vegetation types (after Holtan, 1961)

Land use or cover	Hydrological condition	Percentage basal area rating
Fallow (after row crop)	–	0.10
Fallow (after sod)	–	0.30
Row crops	poor	0.10
	good	0.20
Small grain	poor	0.20
	good	0.30
Hay (legumes)	poor	0.20
	good	0.40
Hay (sod)	poor	0.40
	good	0.60
	excellent	0.80
Pasture or range	poor	0.20
(bunch grass)	fair	0.30
	good	0.40
Temporary pasture (sod)	poor	0.40
	fair	0.50
	good	0.60
Permanent pasture or meadow (sod)	poor	0.80
	good	1.00
Woods and forest	–	1.00

where k_{satveg} = the saturated hydraulic conductivity of the soil with a vegetation cover; and a = percentage basal area of the vegetation.

Typical values of a are given in Table 2.8. This adjustment alters k_{sat} largely as a function of vegetation density and, broadly, lumps together all the effects of vegetation mentioned above that lead to higher infiltration rates and, probably, though it is by no means clear, to a reduction in the amount of rain reaching the ground after interception by the vegetation canopy. It does not take account of the effect of the vegetation cover on the spatial distribution of the rainfall at the ground surface which, through concentrations of water in leaf drips and stemflow, can lead, as seen above, to localized intensities which may exceed infiltration capacity and result in runoff generation. Alternatively, k_{satveg} could be measured in the field but it is often difficult to place infiltrometers over representative areas of the vegetation cover, particularly with shrubs and scrub. Also, the variability in k_{satveg} is invariably great, with coefficients of variation in excess of 200%, so that large numbers of replicates are required to obtain meaningful values.

A further way in which vegetation influences infiltration is through the difference in antecedent moisture content that may occur because more water is removed by evapotranspiration from a soil covered by vegetation than from a bare soil surface. Thus, the capillary pressure, ψ, within the soil at the onset of rain may be lower (ψ numerically higher) and the time before saturation is reached longer for a vegetated soil.

2.2.4 SURFACE CRUSTING

On silty soils, soils containing high proportions of fine sand, soils low in organic matter and soils which for some other reason have an unstable or poor structure, surface crusting or sealing may take place, as the finer particles detached by raindrop impact clog up the pores and cracks and reduce the infiltration rate. The speed of crusting seems to be related to either the incoming shear forces of the raindrops (Farres, 1978; Al-Durrah and Bradford, 1982) or the rainfall energy (Boiffin, 1985; Govers and Poesen, 1985).

The effect of raindrop impact on infiltration can be particularly spectacular on soils susceptible to crusting. Brakensiek and Rawls (1983) present data on soils with and without crusts, where the crusted soils sustain infiltration rates 15–20 mm/h lower than the uncrusted soils. Measurements on sandy soils in Israel show that crusting reduces the infiltration capacity from 100 to 8 mm/h, and on a loess soil from 45 to 5 mm/h (Morin, Benyamini and Michaeli, 1981). The infiltration capacity of sandy soils in Mali ranges from 100 to 200 mm/h but, when a crust has developed, it is reduced to 10 mm/h. Only a few storms are needed to bring about this change. A 50% reduction in infiltra-

tion can occur in one storm (Hoogmoed and Stroosnijder, 1984). Studies on the behaviour of loamy soils in northern France (Boiffin, 1985) show that surface crusting can reduce infiltration capacities from 20–50 mm/h to about 1 mm/h at a rate which is dependent upon the cumulative rainfall received since tillage.

Considering the effect of vegetation on the rainfall energy (Figure 2.10), it may be expected that the speed at which the infiltration rate is reduced is altered by a plant cover. Under relatively low vegetation, the infiltration may remain higher for a longer time than on bare soil, resulting in runoff occurring later and in smaller quantities. Under tall vegetation without undergrowth, the opposite situation may prevail, because sealing or crusting will take place shortly after the onset of rain.

2.3 HYDRAULIC EFFECTS

The passage of water across a bare soil surface may entrain and transport soil particles already detached and, particularly if the flow is concentrated in channels, may also detach additional particles. Erosion is said to be either detachment-limited or transport-limited, depending on how much detached material is available for transport at a given moment. If too much material exists, all of it will not be removed, the erosion rate will be controlled by the transport capacity and the erosion will be transport-limited. If the transport capacity exceeds the detachment rate, all the detached particles will be removed, the erosion rate will be controlled by the supply of detached material and will be detachment-limited.

Flow transport capacity and detachment by flow are often easily confused. Both are related to energy-spending processes within the flow at the interface between the flowing water and the bed. There are differences, however, in the way the energy is expended. Transport capacity is defined as the capacity of flow to carry material of a given noncohesive type (primary particles and individual soil aggregates) with the energy of the flow spent on lifting and carrying the sediment particles. Soil detachment by flow relates to the situation where the energy of the flow is spent on detachment as well as entrainment and transport of sediment particles. Vegetation can limit the capacity of flowing water to detach and transport sediment. The most obvious effect is through the reduction in flow velocity brought about by contact between the flow and the vegetation. The stems and leaves of the vegetation impart roughness to the flow.

2.3.1 SURFACE ROUGHNESS AND FLOW VELOCITY

Surface roughness is an important parameter controlling the speed of the generated runoff. It may be described by a coefficient of friction. The coefficient of friction is usually an 'effective' roughness coefficient that includes the effects of raindrop impact, concentration of the flow, obstacles such as litter, ridges, rocks and roughness from tillage, the frictional drag over the surface, and the erosion and transport of sediment (Engman, 1986). It is more a function of vegetation (plant arrangement, plant population, litter, mulch) and, on agricultural land, tillage methods, than it is a soil–vegetation interaction parameter. The roughness coefficient is normally considered as a summation of the roughness imparted by the soil particles, surface micro-topography (form roughness) and vegetation, acting independently of each other.

Surface roughness is inversely related to both the velocity and quantity of runoff as expressed by the following equations:

$$u = R^{0.667}S^{0.5}/n \qquad (2.14)$$

and

$$Q = R^{1.667}S^{0.5}/n \qquad (2.15)$$

where u = the velocity of the flow (m/s); R = the hydraulic radius (m), often taken as equal to flow depth in shallow flows; S = slope of the energy line (m/m); n = Manning's roughness coefficient ($m^{1/6}$); and Q = the quantity of runoff ($m^3/m\,s$).

Rewriting the equations shows that velocity is dependent on roughness to the power of -0.6 ($u \propto n^{-0.6}$).

For a given amount of runoff it may be calculated that doubling the roughness increases the water depth by 50% and decreases velocity by 34%.

Alternative friction factors to Manning's n are the dimensionless Darcy–Weisbach (f) and Chezy's (C). These are related to Manning's n and to each other as follows:

$$f = 8gn^2R^{-0.33} \qquad (2.16)$$

$$C = R^{0.167}/n \qquad (2.17)$$

and

$$C = (8g/f)^{0.5}. \qquad (2.18)$$

Engman (1986) has suggested a number of values for Manning's roughness coefficient. These are listed in Table 2.9. The possible range of n is large: for bare smooth soil, n is in the order of 0.01; for 5–10 t/ha of straw mulch, $n = 0.07$; and for grass, n ranges from 0.2 to 0.4. Thus, for a constant amount of runoff, surface roughness reduces flow velocity on a mulched field to approximately one-third and on a grass field to one-eighth of what it would be on bare smooth soil.

The level of roughness depends upon the morphology of the plant and its density of growth. Manning's n values can be related to a vegetation retardance index (CI) which is a function of the density and height of the plant stems (Temple, 1982; Table 2.10). The range in n values for each retardance class reflects the variation in roughness which occurs with flow depth (Figures 2.11 and 2.12). With shallow flows, the vegetation stands relatively rigid and roughness values are about 0.25–0.3, associated with distortion of the flow around the individual plant stems. As flow depth increases, the stems begin to oscillate, further disturbing the flow, and roughness values rise to around 0.4. When flow depth begins to submerge the vegetation, roughness values decline rapidly, often by an order of magnitude, because the plants tend to lay down in the flow and roughness is mainly

Table 2.9 Recommended values for Manning's *n* (after Engman, 1986)

Cover or treatment	Residue rate (t/ha)	Value recommended	Range
Concrete or asphalt		0.011	0.01–0.013
Bare sand		0.01	0.010–0.016
Gravelled surface		0.02	0.012–0.03
Bare clay-loam (eroded)		0.02	0.012–0.033
Fallow – no residue		0.05	0.006–0.16
Chisel plough	<0.6	0.07	0.006–0.17
	0.6–2.5	0.18	0.07–0.34
	2.5–7.5	0.30	0.19–0.47
	>7.5	0.40	0.34–0.46
Disc/harrow	<0.6	0.08	0.008–0.41
	0.6–2.5	0.16	0.10–0.25
	2.5–7.5	0.25	0.14–0.53
	>7.5	0.30	–
No tillage	<0.6	0.04	0.03–0.07
	0.6–2.5	0.07	0.01–0.13
	2.5–7.5	0.30	0.16–0.47
Mouldboard plough		0.06	0.02–0.10
Coulter		0.10	0.05–0.13
Range (natural)		0.13	0.01–0.32
Range (clipped)		0.10	0.02–0.24
Grass (blue grass sod)		0.45	0.39–0.63
Short grass prairie		0.15	0.10–0.20
Dense grass		0.24	0.17–0.30
Bermuda grass		0.41	0.30–0.48

due to skin resistance; as a result, velocities increase.

Greatest reductions in flow velocity occur with dense, spatially uniform vegetation covers. Very open, clumpy and tussocky vegetation is less effective and may even lead to localized increases in velocity and erosion as flow becomes concentrated between the clumps. Figure 2.13 shows typical flow paths around a clump of vegetation. As flow separates around the clump, pressure (normal stress) is higher on the upstream than on the downstream face, and eddying and turbulence are set up downstream; in addition, a zone of backflow is established just upstream of the clump (Babaji, 1987). Vortex erosion can occur both upstream and downstream. The combined effect of these pro-

cesses means that erosion rates under tussocky vegetation may remain high and match those prevailing on bare soil (De Ploey, Savat and Moeyersons, 1976).

2.3.2 SEDIMENT TRANSPORT CAPACITY

A number of equations have been proposed for calculating the transport capacity of flow. Traditionally, transport capacity has been defined with respect to sandy beds (river beds) and some confusion arises when dealing with cohesive beds, partly because energy is also spent in detaching the material and partly because the mechanisms of detachment are less well understood than is the case with sand particles.

Table 2.10 Values of Manning's n for different vegetation retardance classes (after Temple, 1982)

Vegetation type	CI^a	n
Very tall dense grass (>600 mm)	10.0	0.06–0.20
Tall grass (250–600 mm)	7.6	0.04–0.15
Medium grass (150–250 mm)	5.6	0.03–0.08
Short grass (50–150 mm)	4.4	0.03–0.06
Very short grass (< 50 mm)	2.9	0.02–0.04

$^a CI$ = index of vegetation retardance defined by Temple, (1982):

$$CI = 2.5\left(h\sqrt{M}\right)^{1/3}$$

where h is the height of the plant stems (m) and M is the density of stems (stems/m^2).

Reference stem densities for good uniform stands are:

Bermuda grass	5380
Buffalo grass	4300
Kentucky bluegrass	3770
Weeping love grass	3770
Alfalfa	5380
Common lespedeza	1610
Sudan grass	538

For legumes and large-stemmed or woody species, the reference stem density is about five times the actual count of stems very close to the bed.

The situation is complicated further because at least three main modes of transport may occur in the flow. Material may be transported as bed load, suspended load or wash load. Bed load is transported in a rolling manner along the soil surface. This mode of transport dominates when the ratio between the lifting forces and the stabilizing forces on the particle is below 0.2; in other words, when the particles are large and heavy compared to the forces within the flow. When this ratio is greater than 1.0, particles are transported as suspended load (Engelund and Hansen, 1967). For very small particles (less than approximately 2 μm), Brownian movements dominate over sedimentation following Stokes' Law, and this material is referred to as wash load.

A very large number of sediment transport equations have been published. Some were derived for sandy river beds, others for soils, but none covers all three modes of transport. The following are typical of those in which sediment transport is related to runoff.

Schoklitsch (1950): for bed load in rivers,

$$Qs = CsS^{1.4}\gamma_w^2 Q^{0.6}\left(Q^{0.6} - Qc^{0.6}\right)n^{-0.5}, \tag{2.19}$$

where Cs = a coefficient expressing soil characteristics; γ_w = unit weight of water; and Q_c is

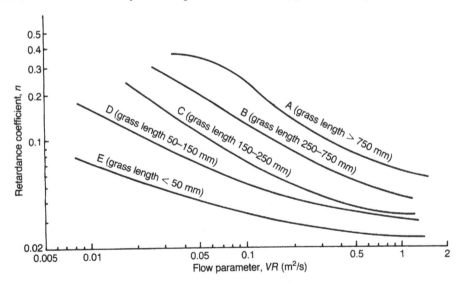

Figure 2.11 Variation in the frictional resistance of grass swards expressed by Manning's n for different retardance categories (after US Soil Conservation Service, 1954).

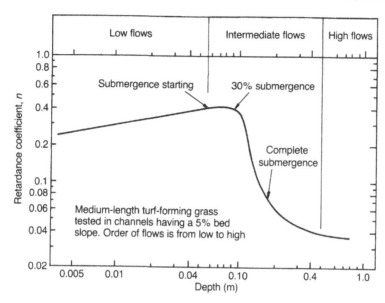

Figure 2.12 Relationship between Manning's *n* and depth of water flow for a medium length grass (after Ree, 1949).

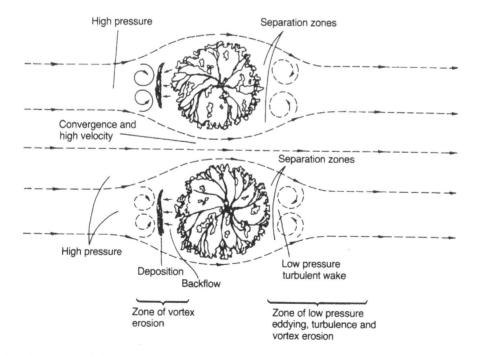

Figure 2.13 Plan view of the pattern of water flow around tussocky vegetation (after Babaji, 1987).

Table 2.11 Relative transport capacities $(m^3/m\,s)$ for runoff generated on loamy soil with different hydraulic properties. The runoff amounts used are given in Table 2.7

Precipitation (mm/h)	Saturated hydraulic conductivity (mm/h)			
	1.5	5.0	15	50
10	0.055	0.023	0	0
20	0.20	0.14	0.023	0
30	0.41	0.33	0.14	0
40	0.68	0.58	0.33	0
50	1.0	0.88	0.58	0
75	2.0	1.8	1.4	0.33
100	3.3	3.1	2.5	1.1
150	6.5	6.2	5.5	3.3

the critical discharge at which sediment transport takes place.

Meyer and Wischmeier (1969): for shallow flows on hillslopes,

$$Qs = kS^{5/3}Q^{5/3}, \qquad (2.20)$$

where k = an experimental coefficient largely related to soil characteristics.

Engelund and Hansen (1967): for bed load transport in rivers but applied by Nielsen and Styczen (1986) to runoff on hillslopes,

$$Qs = 0.04(2g/0.25f)^{1/6}(QS)^{5/3}/$$
$$(J-1)^2 g^{1/2}d_{50}, \qquad (2.21)$$

where f = Darcy-Weisbach roughness coefficient; J = relative sediment density, defined as the ratio γ_s/γ_w, where γ_s is the volume weight of transported sediment; and d_{50} = the diameter at which 50% of the soil particles are finer.

De Ploey (1984): for overland flow and rill flow on hillslopes,

$$Qs/Q \propto S^{1.25}Q^{0.625}. \qquad (2.22)$$

Govers and Rauws (1986): for overland flow,

$$Qs/Q = ASu + B, \qquad (2.23)$$

where Su = unit stream power (product of slope and mean flow velocity); and A, B = empirical coefficients which vary in value with sediment particle size.

Carson and Kirkby (1972): for overland flow,

$$Qs = 0.0085Q^{1.75}S^{1.625}d_{84}^{-1.11} \qquad (2.24)$$

where d_{84} = the diameter at which 84% of the soil particles are finer.

Morgan (1980): for overland flow,

$$Qs = 0.0061Q^{1.8}S^{1.13}n^{-0.15}d_{35}^{-1}, \qquad (2.25)$$

where d_{35} = the diameter at which 35% of the soil particles are finer.

In these equations, Qs = sediment transport capacity $(m^3/m\,s)$ and Q = volume of runoff $(m^3/m\,s)$, so that Qs/Q = sediment concentration.

Despite their differences and shortcomings, the equations consider transport capacity to be proportional to the volume of runoff raised to the power of between 1.6 and 1.8, or to $Q(Su - B)$. Transport capacity is also inversely related to both roughness, n, raised by powers of between 0.15 and 0.5, and particle size, raised by the power of approximately 1. As indicated above, vegetation will affect the transport capacity of runoff by controlling its volume and, through the effect on surface roughness, its velocity.

As an example of the type of calculations that can be made using the above equations, Table 2.11 lists some relative transport capacities calculated using the Engelund–Hansen equation for the runoff amounts given in Table 2.7. The values are approximate and are presented only to show the differences in transport capacity that may arise from differences in infiltration rates. The effects of soil properties on transport capacity are discussed in section 2.4.

In reality, the calculation of transport capacity is much more complicated because the sediment consists of a mixture of primary particles and soil aggregates of different sizes. For soils with a wide aggregate-size distribution, the single particle-size parameter included in the above equations is inadequate, and the transport capacity may have to be calculated separately for different particle or aggregate size classes. For a small amount of runoff, the area of 'attack' is limited to the area covered by small particles. When the runoff increases, more particle sizes and thus a larger surface area become accessible. When an increasing amount of fine material is removed from the surface and only the larger particles are left behind, hardly any material can be removed because the larger particles protect the soil surface. This effect is called armouring.

2.3.3 SOIL DETACHMENT BY FLOW

Soil detachment by flow is often considered as a function of the shear stress of the flow ($\tau = \rho_w g S Q$), raised to the third or fifth power, or as a linear function of the shear stress above a critical value (τ_c) which is related to the shear strength of the soil (τ_s). One example of an equation based on this view is:

Rose *et al.* (1983),

$$DF = 0.276\eta(\rho_w g S Q - \tau_s), \quad (2.26)$$

where DF = the rate of soil particle detachment by flow, and η = the efficiency of bedload transport.

Alternatively, detachment is viewed as a function of the grain shear velocity (u^*g) above a critical value (u^*g_{crit}) which is dependent upon the cohesion of the soil. Total shear velocity (u^*) of the flow is defined by:

$$u^* = (gRS)^{0.5} \quad (2.27)$$

and the grain shear velocity represents that portion which is associated with the roughness of the soil grains. The remaining shear velocity relates to microtopographic (form) roughness and the vegetation. The following equation is an example of this approach.

Rauws and Govers (1988):

$$DF = A(u^*g - u^*g_{crit}), \quad (2.28)$$

where A = a coefficient expressing soil, including grain size characteristics, and

$$u^*g_{crit}(cm/s) = 0.89 + 0.56c, \quad (2.29)$$

where c = cohesion of the soil (kPa) at saturation as measured with a torvane.

Several authors (Meyer and Wischmeier, 1969; Foster and Meyer, 1975; Beasley, Huggins and Monke, 1980; Park, Mitchell and Scarborough, 1982) consider that the detached material fills up the transport capacity of the flow and that sedimentation occurs only when the flow is overloaded. Alternatively, it might be assumed that the detached material settles continuously and that the amount present in the flow represents a balance between detachment and sedimentation (Rose *et al.* 1983; Nielsen and Styczen, 1986). The net effect when the situation is viewed as a balance of processes is described by the following equation:

$$DF = \eta Qs_{EH}, \quad (2.30)$$

where = η is the ratio of energy spent by the flow on lifting particles to the total amount of energy spent on lifting plus detachment; and Qs_{EH} = flow transport capacity calculated according to Engelund and Hansen (equation 2.21). Equation 2.26 (Rose *et al.* 1983) follows the same line of reasoning but is derived in a different way.

On the basis of laboratory tests with sandy, clay loam and clay soils and with sand, Quansah (1985) obtained the empirical relationship:

$$DF = e^{16.37}Q^{1.5}S^{1.44}d_{50}^{-1.54}, \quad (2.31)$$

where DF is in kg/m^2, Q in m^3/m s and d_{50} in mm.

These equations show that whether detachment by flow is described by itself or sedimentation is included, the quantity of runoff generated plays an important role. The effect of runoff, as measured by the value of the power exponent, is greatest when sedimentation is considered. The empirically derived value of 1.5 (equation 2.31) is closer to the value of 1.67 (equation 2.30) suggested when sedimentation is taken into account than to the value of 1.0 (equation 2.26) when it is not. Inclusion of sedimentation may therefore be the most promising approach.

As described earlier, differences in interception and infiltration due to vegetation will have an important influence on runoff generation and, therefore, on detachment by flow. Roughness is accounted for only in equation 2.30. According to this equation, doubling the roughness will decrease the amount of detachment by 20% for a given quantity of runoff.

2.3.4 TRANSPORT OF SPLASHED MATERIAL

Splash erosion may take place in the absence of runoff. However, the amount of material removed from a hillslope or a catchment in this manner is very small, even though it may be the dominant erosion process where runoff amounts are negligible.

When runoff occurs, the transport of the splashed particles depends upon the quantity of the splashed material which goes into the flow and on the flow velocity. Viewed in a very simple manner, splashed material, except for very small particles, behaves as suspended load and moves with the water until it settles. The average fall height will be a function of water depth, D (e.g. $D/2$, because the splashed material could be rather evenly distributed due to turbulence). The time (t) taken to settle becomes $D/2w$, where w is the average fall velocity of a particle of a given size and density. The distance moved by this particle becomes $t \cdot u = uD/(2w) = Q/(2w)$ (Styczen and Nielsen, 1989).

The distance upstream from which splashed material is transported over a given point thus varies with runoff quantity, particle size and particle density. The greater the velocity of runoff generated, the more splashed material is likely to be moved out of an area. The larger the amount of detachment, the greater will be the speed of sedimentation and the shorter the distance moved by the detached particle. Thus infiltration and surface roughness, through their effects on runoff generation, are important controls over the splash process.

However, if the soil surface is totally covered by water of some depth, splash does not take place at all. Relations suggested by Park, Mitchell and Scarborough (1982) and Rose *et al.* (1983) indicate that with a water depth of 6–8 mm and raindrops of 2.0–2.5 mm, splash is reduced by 80–90% compared with that for zero water depth. Equation 2.9 proposes that splash decays exponentially with increasing water depth. Vegetation in contact with the soil surface can increase water depths by reducing flow velocity.

2.3.5 SEDIMENTATION

Vegetation not only retards flow but acts as a filter to sediment being carried in the flow. The denser the vegetation, the more sediment can be trapped and removed from the flow. The effectiveness of close-growing vegetation in causing sedimentation has been modelled for grass barriers using laboratory experiments (Tollner, Barfield and Hayes, 1982; Hayes, Barfield and Tollner, 1984). The sediment wedge created by the barrier consists of three zones (Figure 2.14): (A) the surface slope, (B) the foreslope, and (C) the bottom slope. As the sediment is trapped and the sediment wedge builds up, these zones migrate downslope or downstream.

The model is most easily understood by starting with zone C and calculating its trapping efficiency (fC) over time (t):

$$fC(t) = \exp - 1.05 \times 10^{-3} (u_m \cdot R_s)/v)^{0.82}$$
$$(w \cdot L(t)/u_m \cdot H)^{-0.91}, \qquad (2.32)$$

where $f(C) = (Q_{sd} - Q_{so})/Q_{sd}$; Q_{so} = sediment outflow from the filter; Q_{sd} = sediment trans-

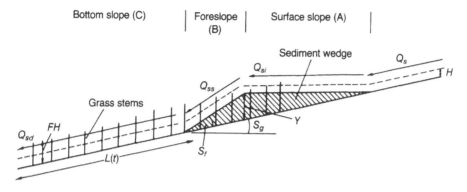

Figure 2.14 Schematic representation of sedimentation in a grass filter strip. Q = runoff; H = flow depth; Y = height of sediment wedge; FH = height of the grass in the filter; S_g = ground slope; S_f = foreslope of sediment wedge. Other notation described in the text (after Hayes, Barfield and Tollner, 1984).

port over zone C; $u_m = 1/n R_s^{2/3} S_g^{1/2}$; $R_s = (SS \cdot H)/(2H + SS)$; ν = kinematic viscosity of the water; w = settling velocity of the sediment; $L(t)$ = length of zone C along the slope; S_g = slope of the ground surface; SS = average spacing of the grass stems; H = depth of flow; and n = a modified Manning's n (S/cm$^{1/3}$) approximated as 0.012 for grass stems.

This relationship is based on an application of the Manning equation of flow velocity using an analogy between flow in a rectangular channel and flow between grass stems to give a channel with a width equal to the average spacing of the stems. The critical factors determining the efficiency of the filter are the density, shape and resilience of the grass stems as these affect the surface roughness. The hydraulic radius (R_s) is defined here as the spacing hydraulic radius.

The term Q_{sd} is estimated using an Einstein-type sediment transport equation:

$$\Delta (d/S_g R_s) = 1.08 (Q_{sd}/(\gamma_s \Delta g \, d^3))^{-0.28}, \tag{2.33}$$

where $\Delta = (\gamma_s - \gamma_w)/\gamma_s$.

Knowing Q_{sd} and the right-hand side of equation 2.32, the sediment outflow (Q_{so}) and the trap efficiency (fC) are determined.

Sediment transport over zone B (Q_{ss}) is represented by:

$$Q_{ss} = (Q_{si} - Q_{sd})/2, \tag{2.34}$$

and the trapping efficiency of zone B (fB) is calculated from:

$$fB = (Q_{si} - Q_{sd})/Q_{si}. \tag{2.35}$$

The term Q_{ss} is estimated using equation 2.33 but substituting Q_{ss} for Q_{sd} and S_f (slope of the foreslope of the sediment wedge) for S_g.

Sediment transport over the surface slope of zone A is represented by Q_{si} and the mass deposition rate for the wedge is Q_{su}. Then

$$Q_{si} = Q_s - Q_{su}, \tag{2.36}$$

$$Q_{su} = (\gamma_{sdep}(Y_f^2 - Y_i^2))/(2(t_f - t_i)S_g), \tag{2.37}$$

$$Y_f(t) = FH \qquad \text{for } Y_f(t) \geqslant FH, \tag{2.38}$$

$$Y_f(t) = (2/\gamma_{sdep} (fB \cdot Q_{si} \cdot S_e) (t_f - t_i) + Y_i(t)^2)^{1/2} \qquad \text{for } Y_f(t) < FH, \tag{2.39}$$

where Q_s = sediment transport upslope of the barrier; Y = height of the sediment wedge; $(t_f - t_i)$ = difference in time between time periods f and i; γ_{sdep} = unit weight of deposited sediment; S_e = angle between foreslope of the sediment wedge and the ground slope; and FH = height of the grass in the filter.

The extent of the sediment wedge upslope (Z_f) is calculated from:

$$Z_f = Y_f/S_g. \tag{2.40}$$

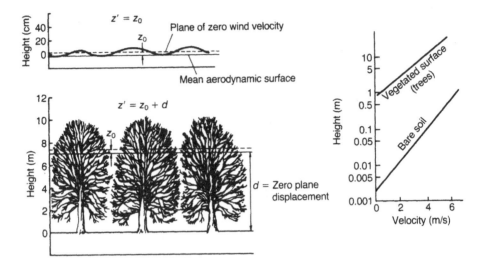

Figure 2.15 Wind velocity as a function of height above a bare and a vegetated soil surface.

The total amount of sediment trapped in the filter (T) is determined from:

$$T = Q_{su} + (Q_{sd} - Q_{so}) + (Q_{si} - Q_{sd}). \quad (2.41)$$

The trapping efficiency of the filter is:

$$f = (Q_s - Q_{so})/Q_s. \quad (2.42)$$

In addition to the properties of the grass stems mentioned above, the effectiveness of the filter will depend upon the height of the grass as this will influence the volume of sediment that can be trapped.

Sedimentation in the filter causes slope steepness to decline as the ground slope is replaced by the slope of zone A in the sediment wedge. Flow velocities are thus decreased and the erosive capacity of the flow reduced. However, the foreslope of the sediment wedge is steeper than the ground slope. Whilst the foreslope remains within the barrier, the potential increase in velocity over this steeper slope is largely offset by the roughness imparted by the grass stems. When the foreslope has migrated downslope to the edge of the barrier, however, flow leaving the barrier will have its velocity increased. Since, as shown in laboratory experiments by Emama (1988), most of the sediment originally carried in the flow is deposited within and upslope of the barrier, the flow is now largely sediment free and is therefore able to effect considerable erosion.

2.3.6 MODIFICATION OF AIR FLOW

Shear velocity in open ground

In the absence of convective eddies generated by vertical temperature gradients, wind speed over uniform level open ground increases logarithmically with height from a height (z_0) which is defined as the height above the mean aerodynamic surface at which wind velocity is zero (Figure 2.15). According to Bagnold (1941), the open field wind velocity profile is described by the equation:

$$u(z) = 2.3/k \cdot u^* \cdot \log(z/z_0), \quad (2.43)$$

where u = mean wind velocity at height z; k = von Karman universal constant for turbulent flow (= 0.4 for clear fluids); and u^* = drag or shear velocity.

The term z_0 is known as the roughness length and is a measure of the ground surface roughness. Bagnold (1941) found that z_0 was equal to

Table 2.12 Drag coefficients for resistance of vegetation in moving air (after Wright and Brown, 1967; Randall, 1969; Voetberg, 1970; Morgan and Finney, 1987)

Cover	z_0 (cm)	CD	Cd
Grass	0.2–0.3	0.005–0.009	
Sugar beet	0.4–1.6		0.003–0.33
Wheat	1.2–3.0		0.001–0.08
Barley			0.001
Planted straw strips	2.1		0.001–0.08
Onions	0.8		0.006–0.50
Peas	0.4		0.001–0.23
Potatoes	5.4		0.001–0.07
Broad beans			0.01–0.05
Apple orchard (winter)		0.02–0.03	
Apple orchard (summer)		0.06–0.07	
Maize		0.01–0.10	0.02–0.15
Rice		0.01–0.10	
Coniferous forest	1.0	0.03–0.10	
Deciduous forest	1.8	0.01–0.03	

z_0, CD and Cd are defined in the text.
  Values for Cd, except for maize, are for crop biomass in the lower 5 cm of the atmosphere and for a wind velocity at 5 cm height of 1 m/s.
  The values of z_0, CD and Cd shown here are typical for the cover types specified. In practice, values of z_0 decrease with increasing wind velocity whereas those of CD and Cd can both increase and decrease with increasing wind velocity. Values of all three coefficients increase with increasing plant growth.

about 1/30th of the height of the sand particles or stones that caused the roughness. Other workers, however, indicate that z_0 approximates 1/10th of the height of the roughness elements (Monteith, 1973; Bache and MacAskill, 1984).

The shear velocity can be calculated by rearranging equation 2.43. It is thus directly proportional to the rate of increase in wind velocity with the logarithm of height. Since it is equivalent to the slope of the line in Figure 2.15, it can be determined by measuring the wind speed at two different heights, plotting the results on a graph of velocity versus the logarithm of height, joining the points with a straight line and calculating the slope of the line expressing the change in velocity for a unit change in log height. It should be noted that the convention of plotting the dependent variable on the y-axis gives way to the convention of plotting height vertically. Thus, higher shear velocities appear as gentler-sloping lines on the graph since they represent a high value of change in the x-axis for a unit change in the y-axis.

Shear velocity is not an actual velocity but has the same units as velocity. It is defined by

$$u^* = \tau/\rho_a, \tag{2.44}$$

where τ = surface shear stress exerted by the air flow; and ρ_a = density of the air (= 0.00123 Mg/m^3 as an average value at sea-level).

Shear velocity with vegetation

Vegetation reduces the shear velocity of the wind by exerting a drag on the air flow. This is compensated for by a transfer of momentum from the air to the vegetation which, for an incompressible fluid, implies a reduction in velocity. Vegetation thus acts as a momentum sink. Vegetation increases the roughness length, z_0, which can be approximated in value as 1/10th of the height of the plant canopy. Typical values of z_0 for a range of surfaces are given in Table 2.12. A vegetation cover also displaces the height of the mean aerodynamic surface above the ground by a distance, d, known as the zero plane displacement (Figure 2.15). The value of d is usually approximated as 0.7 times the height of the plant canopy. Wossenu Abtew, Gregory and Borrelli (1989) have shown that a more accurate assessment can be obtained from $d = H \times F$, where H is the average height of the individual roughness elements and F is the fraction of the total surface covered by those elements. They also show that the roughness length can be estimated from $z_0 = 0.13\,(H - d)$. The term z_0 may be viewed as a measure of the bulk effectiveness of the vegetation cover in absorbing momentum, and the term d is a measure of the mean height at which the absorption takes place (Thom, 1975).

Drag coefficients

The frictional drag exerted on the atmosphere by a vegetation canopy in bulk can be expressed by an equation derived from a simplification of the Navier–Stokes equation for the conservation of momentum in incompressible, steady, two-dimensional air flow (Seginer, 1972; Skidmore and Hagen, 1977; Hagen *et al.*, 1981):

$$\tau = 1/2\rho_a u(z)^2 CD, \qquad (2.45)$$

where τ = the drag force per unit horizontal area of vegetation; and CD = a bulk drag coefficient.

Equating the expressions of τ in equations 2.44 and 2.45 gives:

$$CD = 2u^{*2}/u(z)^2. \qquad (2.46)$$

Considering this equation alongside equation 2.43, it is clear that a relationship exists between the bulk drag coefficient and the aerodynamic properties of the crop canopy as expressed by z_0 and d. In general terms, the rougher the surface and the higher the zero plane displacement, the greater is the drag coefficient. Since both z_0 and d vary with vegetation type and its stage of growth, the drag coefficient, CD, will also vary. All three terms are also dependent upon wind speed. As wind velocity increases, CD and z_0

should fall as a result of streamlining of the foliage elements downwind and d will fall because of the greater penetration of wind into the canopy. The drag coefficient is therefore dynamic and cannot be represented by a single value as is normally the case with a rigid body. Typical values are given in Table 2.12.

Contrary to the above, the bulk drag coefficient has been found by Randall (1969) in apple orchards and Bache (1986) with cotton canopies to increase with increasing wind speed. This may be explained by the dependence of the above on the assumption that the whole leaf area contributes to the momentum transfer whereas, in reality, the effective foliage area for momentum absorption changes in a complex manner through streamlining, leaf flutter and plant vibration as the wind speed alters. In such cases, the terms CD, z_0 and d, which express the effect of the vegetation in bulk, are only broad and not necessarily truthful indicators of what is happening. Also, their emphasis on conditions at the interface between the plant canopy and the atmosphere above, rather than on the soil surface beneath the vegetation, limits their value for determining the likelihood of wind erosion occurring.

An alternative and arguably more meaningful approach is to derive a drag coefficient (Cd) to describe the effect of the vegetation on the air within the plant layer. This is achieved by balancing the drag force of the wind profile exerted on the vegetation at height (h) with the extraction of momentum due to the frictional surface area of the individual foliage elements. This gives (Wright and Brown, 1967):

$$\tau(h) = 0.5 \int_0^h Cd\,A(z)\,u(z)^2 dz, \qquad (2.47)$$

where $A(z)$ is the leaf area density (i.e. leaf area per unit volume).

Substituting equation 2.44 and rearranging yields:

$$Cd = 2u^{*2} \Big/ \int_0^h u^2 A(z) dz. \qquad (2.48)$$

Wright and Brown (1967) found that values of Cd for maize leaves varied with height within

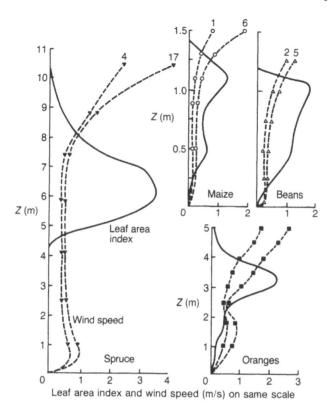

Figure 2.16 Wind velocity as a function of height and leaf area index for four crop types (after Landsberg and James, 1971). Two profiles are shown for each crop.

the canopy and increased with increasing wind velocity. These results emphasize the importance of variations in vegetation structure and effective foliage area in controlling the amount of drag and, thereby, the form of the wind profile.

Figure 2.16 shows how wind velocity changes with height within the vegetative layer for a range of vegetation or crop types as a function of leaf area density (Landsberg and James, 1971). Equation 2.43 is only valid as an expression of the velocity profile in the air above the zero plane displacement. Below this, the wind profile may be fitted by one of the following equations:

$$u(z) = u(h)(1 + m(1 - z/h))^{-2} \quad (2.49)$$

$$u(z) = u(h)\exp(-n(1 - z/h)), \quad (2.50)$$

where m is an experimental parameter which is characteristic of the vegetation type (Thom, 1971) and n is an attenuation coefficient which typically varies between 2 and 5 in value depending upon the foliage density and the type of vegetation (Inoue, 1963; Cionco, 1965).

Equations 2.49 and 2.50 are normally valid with tall vegetation such as trees. With shorter vegetation, for example, maize and rice, equation 2.50 describes the wind profile in the top half of the plant layer reasonably well but in the lower 20% of the profile the wind speeds decrease more slowly than the equation predicts (Denmead, 1976). Indeed, they may even increase close to the ground surface because of the sparser vegetation cover at the bottom of the plant layer. Wind tunnel studies (Morgan, Finney and Williams, 1986) on crops less than

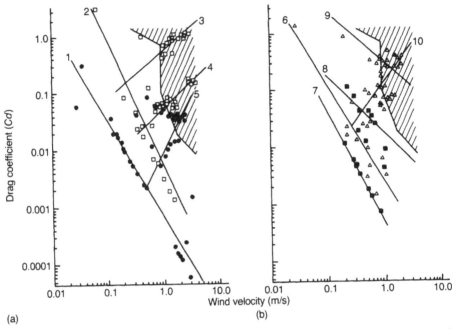

Figure 2.17 Relationships between drag coefficients (*Cd*) for the 5 cm layer of vegetation close to the ground surface and wind velocity measured at 5 cm height for (a) planted straw strips (●) and onions (□); and (b) live barley strips (■) and sugar beet (△). Shaded area represents risk of soil particle movement (from Morgan, 1989.)

0.15 m tall showed that the wind profile within the vegetation layer was better described by the equation:

$$u(z)/u(h) = 0.9 + 0.1\ln(z/h). \quad (2.51)$$

Shaw and Pereira (1982) and Hagen and Lyles (1988) also apply a log height–velocity relationship to the air in the lowest part of a vegetation cover close to the ground surface.

Morgan and Finney (1987) attempted to examine conditions in the lower part of the atmosphere by using equation 2.48 to calculate drag coefficients for the vegetation in the lowest 5 cm of single crop rows from field measurements of average 10 s wind velocities made with cup anemometers. They found that the drag coefficients both increased and decreased in value with wind speed (Figure 2.17), depending upon the consistency of the wind. If the latter is expressed by an index of turbulence (*TU*), defined as the ratio of the values of the standard deviation to the mean of a series of

consecutive wind velocity recordings, the drag coefficient decreases with increasing wind speed when $TU > 0.2$. For $TU \leqslant 0.2$, the drag coefficient increases with wind speed, presumably as a result of leaf flutter in the more continuous wind disturbing the atmosphere surrounding the foliage and setting up a 'wall' effect.

Although the correlation coefficient (*r*) between values of the drag coefficient and wind speed was always higher than -0.80 ($P < 0.02$) for the negative relationship, it was generally lower for the positive one. This implies that wind speed alone does not adequately explain the variability in drag coefficients. Further insight is provided by wind tunnel studies (Morgan, Finney and Williams, 1988) of the drag coefficient of individual leaves (*CL*) as a function of their morphological properties of size, shape, fragmentation, orientation and rigidity. These showed that the single-leaf drag coefficient increased as the projected area and the deflection angle decreased and the down wind alignment increased. The results imply that highest drag is associated with greatest contact length between the wind and the air flow in a downwind direction and not with the area of foliage facing the wind; a finding which points to the importance of flow separation around the leaf and skin friction from the leaf surface over form drag in contributing to wind resistance. Since bladed leaves were found to have lower deflection angles than round or ovate leaves, it follows that high drag is associated with bladed leaves aligned downwind not as a result of streamlining but because of their natural growth position.

Information on foliage properties can add substantially to the understanding of drag coefficient (Cd) values in the lowest 5 cm of the atmosphere. From the field data of Morgan and Finney (1987), two relationships are obtained:

$$\log Cd = -1.648 - 1.406 \log u - 378.4\,PA +$$

$$0.00466\,H + 0.01045\,V$$

$$\text{for } TU > 0.2$$

$$(R = 0.839;\ n = 159), \qquad (2.52)$$

$$\log Cd = -0.139 + 0.316 \log u - 369.1\,PA +$$

$$0.1167\,BM - 1.757\,TU$$

$$\text{for } TU \leqslant 0.2$$

$$(R = 0.727;\ n = 130), \qquad (2.53)$$

where u = wind velocity (m/s) at 5 cm height; PA = projected area of the foliage facing the wind (m^2); H = average angle of the leaves from the vertical in a downwind direction (degrees); V = average angle of the leaves from the vertical in a crosswind direction (degrees); BM = biomass (kg DM/m^3); and TU = turbulence index (described above).

The terms PA, H, V and BM are determined for a representative 10 cm length of a plant row in the lowest 5 cm of the atmosphere. Both equations show that the drag coefficient increases as the projected foliage area facing the wind decreases, again demonstrating the importance of contact length downwind between the leaf surfaces and the air. This is also implicit in the positive effects of downwind leaf alignment and increasing biomass, the latter being another indication of greater surface area of foliage. Typical values for Cd are given in Table 2.12.

This analysis shows that plant properties can have an important influence over values of the drag coefficient (Cd). This, in turn, as combining equations 2.48 and 2.43 shows, has an effect on shear velocity:

$$u^* = 0.71 \int_0^h \left(Cd A(z)\, u(z)\, \mathrm{d}z \right)^{0.5}. \qquad (2.54)$$

2.4 MECHANICAL EFFECTS

2.4.1 SOIL REINFORCEMENT

The roots and rhizomes of the vegetation interact with the soil to produce a composite material in which the roots are fibres of relatively high tensile strength and adhesion embedded in a matrix of lower tensile strength. The shear strength of the soil is therefore enhanced by the root matrix.

Field studies of forested slopes (O'Loughlin, 1984) indicate that it is the fine roots, 1–20 mm in diameter, that contribute most to soil reinforcement and that the larger roots play no significant role. Grasses, legumes and small shrubs can have a significant reinforcing effect down to depths of 0.75–1.5 m. Trees have deeper-seated effects and can enhance soil strength to depths of 3 m or more depending upon the root morphology of the species (Figure 2.18; Yen, 1972). Root systems lead to an increase in soil strength through an increase in cohesion brought about by their binding action in the fibre/soil composite and adhesion of the soil particles to the roots. It is generally held that roots have no effect on soil friction angle but Tengbeh (1989) found that grass roots increased the angle of internal friction of a sandy soil but had no such effect on a sandy clay loam.

The pattern of the relationship between soil cohesion and the roots is not known. Tengbeh (1989) found that Loretta grass (*Lolium perenne*)

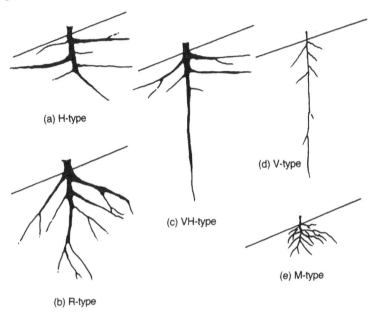

(a) H-type

(d) V-type

(c) VH-type

(b) R-type

(e) M-type

Figure 2.18 Patterns of root growth in trees (after Yen, 1987). (a) H-type: maximum root development occurs at moderate depth, with more than 80% of the root matrix found in the top 60 cm; most of the roots extend horizontally and their lateral extent is wide. (b) R-type: maximum root development is deep, with only 20% of the root matrix found in the top 60 cm; most of the main roots extend obliquely or at right angles to the slope and their lateral extent is wide. (c) VH-type: maximum root development is moderate to deep but 80% of the root matrix occurs within the top 60 cm; there is a strong tap root but the lateral roots grow horizontally and profusely, and their lateral extent is wide. (d) V-type: maximum root development is moderate to deep; there is a strong tap root but the lateral roots are sparse and narrow in extent. (e) M-type: maximum root development is deep but 80% of the root matrix occurs within the top 30 cm; the main roots grow profusely and massively under the stump and have a narrow lateral extent. H- and VH-types are considered beneficial for slope stabilization and wind resistance. H- and M-types are beneficial for soil reinforcement. The V-type is wind resistant.

increased the cohesion (c; kPa) of soil as a function of root density (RD; Mg/m^3) in an exponential relationship so that for a sandy clay loam soil:

$$c = 10.54 + 8.63 \log RD \qquad r = 0.99, n = 16 \tag{2.55}$$

and for a clay soil:

$$c = 11.14 + 9.9 \log RD \qquad r = 0.99, n = 11. \tag{2.56}$$

In contrast, linear relationships have been obtained by Waldron (1977) between the change in soil shear strength and the root area ratio of barley roots in a silty clay loam soil, and by Ziemer (1981) between shear strength and root biomass of *Pinus cordata* in a sand.

These studies show that root reinforcement can make significant contributions to soil strength, even at low root densities and low shear strengths. Equations 2.55 and 2.56 indicate that cohesion increases rapidly with increasing root density at low root densities but that increasing root density above 0.5 Mg/m^3 on the clay soil and 0.7 Mg/m^3 on the sandy clay loam soil has little additional effect. This implies that vegetation can have its greatest effect close to the soil surface where the root density is generally highest and the soil is otherwise weakest.

Since shear strength affects the resistance of the soil to detachment by raindrop impact (Cruse and Larson, 1977; Al-Durrah and Bradford, 1982), and the susceptibility of the soil to rill erosion (Laflen, 1987; Rauws and Govers, 1988) as well as the likelihood of mass soil failure, root systems can have a considerable influence on all these processes. The maximum effect on resistance to soil failure occurs when the tensile strength of the roots is fully mobilized and that, under strain, the behaviour of the roots and the soil are compatible. This requires roots of high stiffness or tensile modulus to mobilize sufficient strength and the 8–10% failure strains of most soils. The tensile effect is limited with shallow-rooted vegetation where the roots fail by pullout, i.e. slipping due to loss of bonding between the root and the soil, before peak tensile strength is reached (Waldron and Dakessian, 1981). The tensile effect is most marked with trees where the roots penetrate several metres into the soil and their tortuous paths around stones and other roots provide good anchorage. Root failure may still occur, however, by rupture, i.e. breaking of the roots when their tensile strength is exceeded. The strengthening effect of the roots will also be minimized in situations where the soil is held in compression instead of tension, e.g. at the bottom of hillslopes. Root failure here occurs by buckling.

2.4.2 ROLE OF ORGANIC MATTER

The return of vegetative material to the soil as organic matter plays a vital role in aggregation of the soil particles. Aggregate-stabilizing compounds are formed during the degradation of organic material, such as manure, plant roots, leaves and stems, and straw, by microbial and faunal activity within the soil. Thus, the level of biological activity or the speed of degradation of organic matter are probably better indicators of the relative stability of soil aggregates than the content of organic matter as such. A good vegetative cover is likely to increase the biological activity and the rate of aggregate forma-

tion, but no quantification of this effect can be given. Increased aggregate stability of a soil increases permeability and infiltration which, in turn, reduces surface runoff and enhances the available water content for plant growth. This promotes better vegetation growth with greater protection of the soil surface and a drier soil environment.

The size and stability of the soil aggregates affects their detachability by raindrop impact and their detachability and transportability by surface runoff. Several of the transport equations (section 2.3.2) contain factors of the type d^{-x}, where x varies between 1.0 and 1.54, showing that transport capacity becomes inversely related to the diameter of the aggregates. As the Engelund–Hansen sediment transport relationship (equation 2.21) indicates, transportability is also dependent on the density of the soil particles. Thus, large soil aggregates are moved before primary particles of the same diameter because their densities are approximately 1.8–2.0 Mg/m^3 compared with 2.6 Mg/m^3 for sand.

2.4.3 ROOT WEDGING

Root wedging is a potentially destabilizing process whereby fissures and joints in rocks are opened up by the advance and growth of roots. Trees create the biggest problem, though grass roots can also force open small cracks. Where vegetation gains a hold on steep slopes with steeply-inclined joint planes or fissures, the wedging action of plant roots can dislodge and topple blocks or sections of the rock. Earth (soil) slopes are less likely to be affected.

Root wedging may not cause instability during the lifetime of a tree as the rocks may be enveloped within the roots and trunk. It is on the death of the tree that dislodged blocks are likely to fall free.

2.4.4 ARCHING AND BUTTRESSING

The tap and sinker roots of many tree species extend through the soil layers and into the

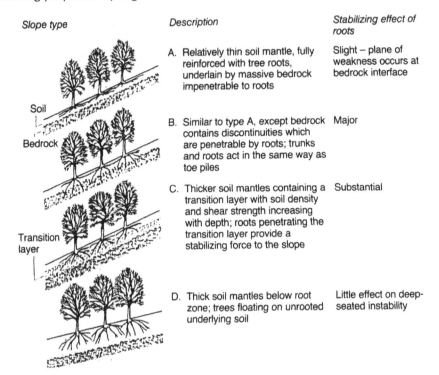

Slope type	Description	Stabilizing effect of roots
Soil / Bedrock	A. Relatively thin soil mantle, fully reinforced with tree roots, underlain by massive bedrock impenetrable to roots	Slight – plane of weakness occurs at bedrock interface
	B. Similar to type A, except bedrock contains discontinuities which are penetrable by roots; trunks and roots act in the same way as toe piles	Major
Transition layer	C. Thicker soil mantles containing a transition layer with soil density and shear strength increasing with depth; roots penetrating the transition layer provide a stabilizing force to the slope	Substantial
	D. Thick soil mantles below root zone; trees floating on unrooted underlying soil	Little effect on deep-seated instability

Figure 2.19 Classes of plant-root reinforced and anchored slopes (after Tsukamoto and Kusakabe, 1984).

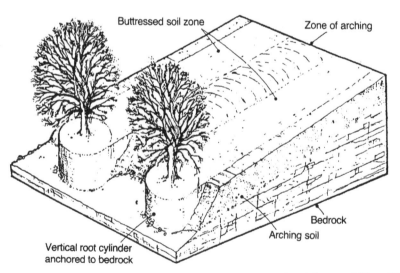

Figure 2.20 Schematic representation of soil buttressing and arching (after Wang and Yen, 1974).

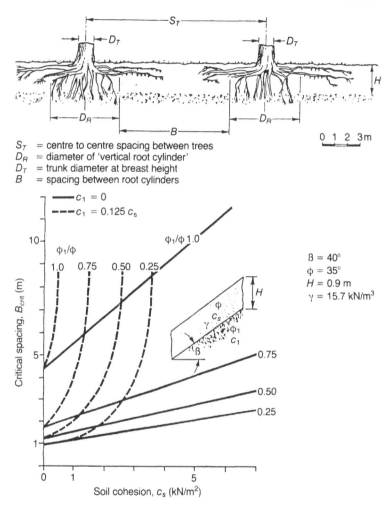

S_T = centre to centre spacing between trees
D_R = diameter of 'vertical root cylinder'
D_T = trunk diameter at breast height
B = spacing between root cylinders

Figure 2.21 Critical spacing of trees for arching based on the theory of Wang and Yen (1974) applied to a steep sandy slope (after Gray and Leiser, 1982).

underlying bedrock, anchoring them to the slope. The trunks and large roots then act in the same way as stabilizing piles and buttress the soil, restraining it from movement downslope. The extent to which buttressing can contribute to the stability of the soil mass on a slope depends upon the depth of the soil mantle and the groundwater as well as on the penetrability of the bedrock by roots (Figure 2.19; Tsukamoto and Kusakabe, 1984).

Where trees are sufficiently close together, the soil between the unbuttressed parts of the slope may gain strength by arching (Figure 2.20). Based on work by Wang and Yen (1974), Gray (1978) has produced a plot of the theoretical critical (minimum) spacing required for arching to occur on a 40° slope with a 0.9 m deep sandy soil mantle (Figure 2.21). This shows that the critical spacing depends upon the cohesiveness of the underside of the supported soil mass. If cohesion is zero and the residual friction along the underside is half the peak friction, the critical spacing is only 1.2 m. If a cohesion of 2.4 kN/m² is assumed, with a residual cohesion of 12.5% of this value, the critical spacing increases to 6.4 m. Tree spacings on such slopes in the field are often of the right order for arching to develop.

2.4.5 SURCHARGING

Surcharge arises from the additional weight of the vegetation cover on the soil. This effect is normally considered only for trees, since the weight of grasses and most herbs and shrubs is comparatively small. Surcharge increases the downslope forces on a slope, lowering the resistance of the soil mass to sliding, but it also increases the frictional resistance of the soil. Bishop and Stevens (1964) show that large trees can increase the normal stress on a slope by up to 5 kN/m^2 but that no more than half contributes to an increase in shear stress. Generally, the second effect outweighs the first, so that, overall, surcharge is beneficial. Nevertheless, surcharge at the top of a slope can reduce overall stability whereas, at the bottom of the slope, it will increase stability.

De Ploey (1981) invokes surcharge combined with lowering of the cohesion of the soil mass through increased infiltration and, therefore, increased soil water content, as contributing to landslides on the forested slopes of the Serra do Mar, east of Santos, in Brazil. The surcharge becomes critical when rainfall of several hundreds of millimetres occurs in a wet spell of a few days; for example, on 17 and 18 March 1967 when daily rainfalls totalled 260 and 420 mm respectively. In such events, interception and evapotranspiration are reduced virtually to zero and the soil is unable to either dry out or drain. The critical factors here may well be the low angle of internal friction of the soil material which, when close to waterlogging, is reduced to less than 20°, and the steepness of the slopes, which are over 20°. In contrast, Gray and Megahan (1981) state that surcharge is beneficial when cohesion is low and groundwater levels are high provided that the angle of internal friction of the soil is also high and the slope angles are low.

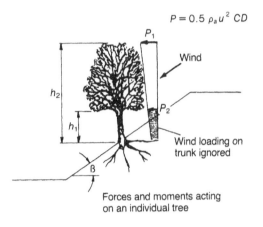

$$P = 0.5\,\rho_a u^2\,CD$$

Forces and moments acting on an individual tree

Figure 2.22 Effect of wind loading on a single tree (from Coppin and Richards, 1990).

2.4.6 WIND LOADING

The pressure (P) exerted on a vegetation cover by wind can be transmitted to the soil as an increased loading (D), reducing its resistance to failure. From the work of Hsi and Nath (1970) and Brown and Sheu (1975), for a single tree (Figure 2.22):

$$D = \sum_{i=h1}^{i=h2} (0.5\rho_a u^2 CD \cos^2 \beta b)\,i, \qquad (2.57)$$

where D = drag force (kg) transmitted into the slope; $h1$ = height of the bottom of the tree canopy above the ground (m); $h2$ = height of the top of the tree canopy above the ground; ρ_a = density of the air (kg m^3); u = wind velocity (m/s); CD = the bulk drag coefficient of the vegetation; β = slope angle (degrees); and b = transverse width of the crown (m) at each height increment, i.

Wind loading is only significant for trees and when the wind velocity exceeds 11 m/s.

Wind pressure on a tree can also produce a destabilizing moment which, if the tree is not well anchored, will cause it to topple over. Increased infiltration of water into the soil through the scar created by the uprooted tree can then lower the resistance of the whole soil mass to failure.

2.4.7 SURFACE PROTECTION

Vegetation protects the soil mechanically by absorbing directly the impact of walkers, livestock and vehicles. Most studies of this effect have concentrated on the resistance of vegetation to damage by walking. When an individual walks over the ground surface, the soil and vegetation are compacted in the early part of each step under the pressure of the heel; at the end of the step, they are sheared by the movement of the toe. The shearing action is the most damaging (Quinn, Morgan and Smith, 1980).

Broadly, grasses are reasonably resistant and can withstand between 1000 and 2000 passes by walkers before the density of cover falls below 50%. In contrast, alpine plant communities can withstand about 60 passes and arctic tundra communities only eight passes (Liddle, 1973).

The effect of walking on shrubs is greater than that on grasses because plants that produce buds and shoots at or below ground level have their growth points protected by the overlying foliage as the vegetation is flattened underfoot. They are therefore less easily damaged. For this reason, heath and bracken disappear more rapidly than grasses in upland areas under heavy recreational use.

2.5 VISUALIZATION

In order to visualize the combined effect when several vegetation-related parameters are varied simultaneously, some simple calculations are presented. The numerical values should not be taken too seriously, as the overall results only illustrate the theoretical principles previously discussed. Nevertheless, the influence of vegetation is demonstrated to about the right order of magnitude.

2.5.1 WATER EROSION

The effect of vegetation cover on water erosion is illustrated for seven conditions for which typical values of percentage cover and plant height are presented in Table 2.13. A value of saturated hydraulic conductivity has been chosen for a loamy soil and then varied, on the basis of McKeague, Wang and Coen (1986), taking account of the number of biopores and the level of soil aggregation expected under each condition. Values of Manning's *n* are selected from Engman (1986). The soil is assumed wet, so the infiltration rate is regarded as equal to the saturated hydraulic conductivity.

The accumulated runoff water is calculated for four rainfall intensities, three slope lengths or distances downslope, and two slope steepnesses. For each combination, the water depth, flow velocity and, using the Engelund and Hansen (1967) formula, transport capacity relative to that on bare soil are determined. Soil detachment by raindrop impact is calculated relative to that for bare soil using the procedure of Styczen and Høgh-Schmidt (1988). None of the calculations take into consideration an

Table 2.13 Parameters used for assessment of the effect of vegetation on erosion by water

Cover type	Saturated hydraulic conductivity (mm/h)	Manning's n	Percentage cover	Height (m)
Bare soil	10	0.01	0	
Grass	50	0.20	90	0
Soya beans	25	0.04	80	0.5
Maize	25	0.02	80	1.5
Agricultural crops and crusted soil	5	0.02	80	1.5
Eucalyptus and crusted soil	5	0.01	80	3–8
Eucalyptus with grass	50	0.20	80–90	3–8/0

aggregate size distribution of the soil. A crusting index (Table 2.14) is developed based on the

Table 2.14 Crusting index calculated as a function of rainfall energy. The index unit is the energy received relative to that on a bare soil when the rainfall intensity is 25 mm/h

Cover	Rainfall intensity (mm/h)			
	25	50	75	100
Bare soil	1	2.2	3.6	5.0
Grass, 90% cover	0.1	0.2	0.4	0.5
Soya beans	0.5	1.0	1.5	2.1
Maize	1.1	2.0	3.2	4.3
Eucalyptus, crusted soil	2.8	5.4	7.9	10.4
Eucalyptus with grass	0.3	0.5	0.8	1.0

energy of the rainfall received at the ground relative to that received by a rain of 25 mm/h intensity on a bare soil. A rilling index (Table 2.15) is calculated as the product of runoff volume, slope length and slope steepness.

In Figure 2.23, curves representing relative transport capacities and soil detachment by raindrop impact are presented for each condition. According to the value chosen for soil erodibility, the curves for soil detachment may be multiplied by a constant and the cross-over points of the two curves will change accordingly. Detachment by runoff has not been added as its relative importance depends upon erodibility. It is, however, probably proportional to the transport capacity. The curves for soil detachment differ from those presented in a similar

Table 2.15 Rilling index calculated as the product of runoff, slope length and slope steepness for a 5% slope and different rainfall intensities

Cover type	Rainfall intensity (mm/h)			
	25	50	75	100
Slope length 20 m				
Bare soil	0.15	0.4	0.65	0.9
Grass	0	0	0.25	0.5
Soya beans	0	0.25	0.5	0.5
Maize	0	0.25	0.5	0.75
Agricultural crops with crusted soil	0.2	0.45	0.7	0.95
Eucalyptus with crusted soil	0.2	0.45	0.7	0.95
Eucalyptus with grass	0	0	0.25	0.5
Slope length 50 m				
Bare soil	0.38	1.0	1.63	2.25
Grass	0	0	0.63	1.25
Soya beans	0	0.63	1.25	1.88
Maize	0	0.63	1.25	1.88
Agricultural crops with crusted soil	0.5	1.13	1.75	2.38
Eucalyptus with crusted soil	1.0	1.13	1.75	2.38
Eucalyptus with grass	0	0	0.63	1.25
Slope length 100 m				
Bare soil	0.75	2.0	3.25	4.5
Grass	0	0	1.25	2.5
Soya beans	0	1.25	2.5	3.75
Maize	0	1.25	2.5	3.75
Agricultural crops with crusted soil	1.0	2.25	3.5	4.75
Eucalyptus with crusted soil	1.0	2.25	3.5	4.75
Eucalyptus with grass	0	0	1.25	2.5

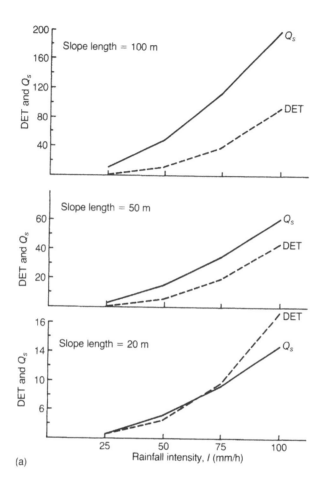

Figure 2.23 Relative transport capacities (Q_s) and soil detachment (DET) by splash as a function of rainfall intensity on a 5% slope for slope lengths of 100 m, 50 m and 20 m on (a) bare soil; (b) grass with and without a cover of Eucalyptus; (c) soya beans; (d) maize; (e) soya beans and maize on a crusted soil; and (f) bare soil with and without a crust under Eucalyptus. Note that the scales of the y-axis differ for the three slope lengths.

exercise by Foster and Meyer (1975) because the procedure used here relates the splashed material transported away to the quantity of the runoff instead of assuming that all the splashed particles are available for transport.

At the top of the slope (represented by the 20 m slope length), where only small amounts of runoff occur, the erosion is transport-limited except under grass. Further downslope (50 m slope length), where more runoff accumulates, the transport capacity exceeds soil detachment and the erosion becomes detachment-limited

for conditions with vegetation close to the surface and for bare soil. At the foot of the slope, after 100 m slope length, erosion is detachment-limited except for trees without undergrowth and litter. However, even though erosion remains transport-limited under trees, the large amounts of runoff water accumulated may cause rilling to occur. In this case, detachment in the rills may dominate the erosion completely. As the transport capacity and the erosive capacity of rill flow are much larger than for sheet flow, much more material can be expected to be

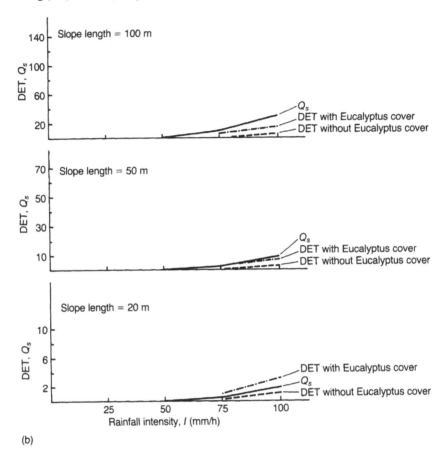

(b)

Figure 2.23 (cont.)

removed from an area when rills form. Passing the threshold for rill formation means a serious acceleration in water erosion.

A major difference between the curves is the starting point at which runoff commences. This hydrological effect may be just as important in determining the magnitude of erosion as the effects of vegetation on runoff volume.

The difference in the transport capacity for maize and soya bean is due to differences in surface roughness, while the difference in soil detachment by raindrop impact is due to the different plant heights. The curves for grass show a very low amount of runoff generated, a high degree of roughness and almost no detachment.

The most serious condition is for the trees

without cover. This is because of the large amounts of runoff which are generated very quickly after the onset of the storm, due to high values of the crusting index, and the high flow velocities. Despite the large amounts of runoff, soil loss is transport-limited, indicating extremely high detachment rates which arise from the high fall heights of the leaf drainage. The rilling index is larger here than in any of the other cases.

Two studies which support the validity of the last case have been carried out in Java, Indonesia, by Coster (1938) and Wiersum (1985). Both authors studied the effects of various vegetation layers in an *Acacia auriculi-formis* forest on surface erosion by removing

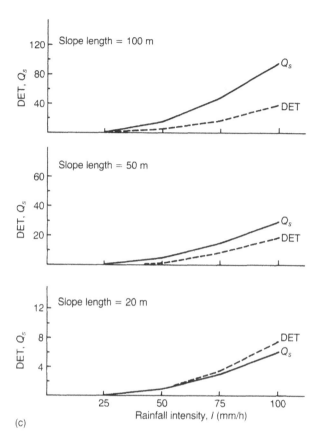

Figure 2.23 (cont.)

one vegetation layer at a time. Their findings clearly illustrate the importance of the height of the canopy above the soil surface and the importance of a ground vegetation or litter cover.

Coster (1938; Table 2.16) found that far more erosion occurred under trees without undergrowth and litter than on bare soil. When the trees were removed but litter and undergrowth kept intact, the soil loss was only 1/15th of that from bare soil. In the undisturbed forest, the soil loss was very low. Wiersum (1985) obtained less drastic differences but at least 20 times as much soil was lost when litter was removed compared to when litter was present. When both litter and undergrowth are intact, hardly any erosion occurs.

In both cases, it is the vegetation close to the soil surface and the litter that play the important role in controlling the erosion. Although the vegetation layers in the canopy catch rainwater and divert some to stemflow, these effects are more than offset by the increase in drop size of the rain which reaches the ground surface as leaf drainage.

2.5.2   SLOPE STABILITY

The stability of a slope against failure is evaluated by the factor of safety (F), which is defined as the ratio of the resistance of the soil mass to shear along a potential slip plane to the shear force acting on that plane. Soil failure occurs when the ratio falls to unity. The simple case of a translational failure along a sliding surface

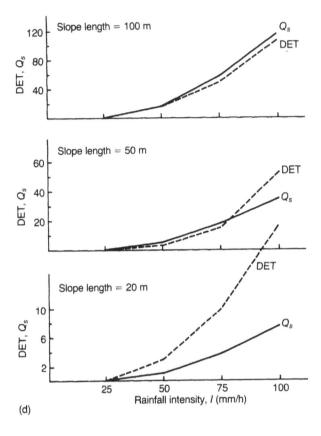

(d)

Figure 2.23 (cont.)

parallel to the ground over a relatively long uniform slope can be analysed by infinite slope analysis. In this case, a single element or segment (Figure 2.24) of the slope can be considered as representative of the whole, and the head and toe portions of the slope are ignored as being negligible in extent.

Using effective stress analysis, the factor of safety without vegetation can be defined by:

$$F = \frac{c' + (\gamma z - \gamma_w h_w) \cos^2 \beta \tan \phi'}{\gamma z \sin \beta \cos \beta},$$

$$(2.58)$$

where c' = effective soil cohesion (kN/m^3); γ = unit weight of soil (kN/m^3); z = vertical height of soil above the slip plane (m); β = slope angle (degrees); γ_w = unit weight of water ($= 9.8\,kN/m^3$); h_w = vertical height of groundwater table above the slip plane (m); and ϕ' = effective angle of internal friction of the soil material (degrees).

Figure 2.25, based on Coppin and Richards (1990), shows the main influences of vegetation on the stability of the slope segment. They can be included in the calculations of the factor of safety as follows:

$$F = \frac{(c' + c_R') + \{[(\gamma z - \gamma_w h_v) + W] \cos^2 \beta + T \sin \theta\} \tan \phi' + T \cos \theta}{[(\gamma z + W) \sin \beta + D] \cos \beta},$$

$$(2.59)$$

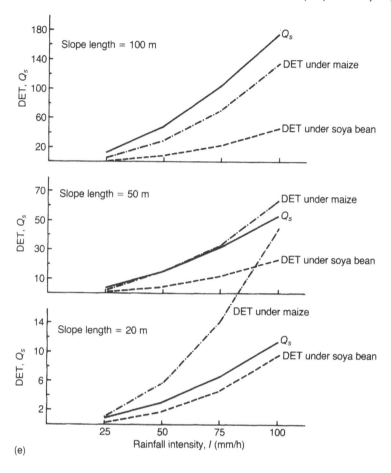

(e)

Figure 2.23 (cont.)

where c_R' = enhanced effective soil cohesion due to soil reinforcement by roots (kN/m³); W = surcharge due to weight of the vegetation (kN/m²); h_v = vertical height of groundwater table above the slip plane with the vegetation (m); T = tensile root force acting at the base of the slip plane (kN/m); θ = angle between roots and slip plane (degrees); and D = wind loading force parallel to the slope (kN/m).

Appendix 2.A gives the calculations for the factor of safety for a sample slope segment with and without vegetation. The calculations are purely illustrative but they show that the vegetation increases the factor of safety by 55%, assuming that the tensile strength of the roots is fully mobilized, and by 17%, if this effect

(T acting over angle θ) is ignored. The greatest effects are due to the increase in cohesion through root reinforcement of the soil and to the tensile strength of the roots themselves across the potential slip surface. Although field studies of the effect of vegetation on slope stability are rare, Greenway (1987) found that the additional cohesion brought about by tree roots increased the factor of safety on wooded slopes in Hong Kong by 29%.

2.6 SALIENT PROPERTIES OF VEGETATION

The calculations presented in the previous section demonstrate that the overall effect of vegetation is the result of a balance between

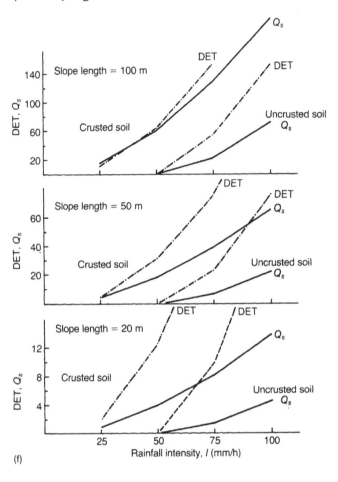

(f)

Figure 2.23 (cont.)

Table 2.16 Erosion (kg/m² y) in a montane forest, Java, Indonesia (after Wiersum, 1985 from table in Coster, 1938)

Cover	Number of observations (plot years)	Measured erosion	Erosion adjusted for equivalent slope and rainfall
Undisturbed forest	2	0.03	0.01
Trees removed	5	0.04	0.03
Undergrowth removed	4	0.06	0.05
Trees and undergrowth removed	1	0.08	0.02
Undergrowth and litter removed	10	4.32	2.61
Trees, undergrowth and litter removed	2	1.59	0.44
Shrub vegetation	4	0	0
Shrubs removed	3	0.20	0.23

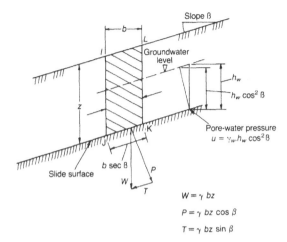

Figure 2.24 Factors involved in the infinite slope method for analysing slope stability (Notation in text.)

beneficial and adverse influences. These have been summarized by Coppin and Richards (1990), as shown in Table 2.17. The nature of the balance and, therefore, the engineering function which individual plants perform will depend upon their structure or architecture. Plants with strong tap and sinker roots will help stabilize a slope through arching and buttressing, whereas plants with a dense lateral rooting system will increase the strength of the top layer of the soil by adding to cohesion. In contrast, surface erosion processes are more strongly influenced by the above-ground growth of the vegetation. The properties of the vegetation which influence its engineering function are listed in Table 2.18, also taken from Coppin and Richards (1990). Many of these properties vary with the stage of vegetation growth and therefore alter both seasonally and, through ecological succession, over a longer time. Increasing cohesion of the soil through vegetation growth can offset long-term decreases in soil strength brought about by weathering and the fissuring and progressive softening of overconsolidated clays.

There is a close relationship between vegetation and erosion. On the one hand, vegetation, through its engineering functions, can control the amount of erosion which takes place. On the other hand, erosion can produce such a hostile and unstable environment that vegetation will not grow. The balance and competitiveness of the erosion–vegetation system has been analysed by Thornes (1988a,b, 1990) with respect to southeast Spain. He assumes that erosion limits plant growth through water and nutrient stress but that vegetation also limits erosion. Such an ecosystem may be in balance, or it may be self-reinforcing in one direction or another. For example, erosion will result in less vegetation which will produce a deteriorating water balance with less water available for plant growth and more water contributing to runoff and erosion. Alternatively, an increase in vegetation growth will lead to less erosion and a more favourable water balance for further vegetation. If grazing pressure is added to the ecosystem, a higher biomass production is necessary to keep the system in balance. Otherwise the situation may change from equilibrium to deterioration. The details of the system in southeast Spain are very complex because litter fall occurs at different times of year for different plant species and because a certain amount of the litter is removed by the runoff. One of the important effects of the ground vegetation is to keep the litter in place.

Most vegetation is self-regenerating but human and animal interference may destroy the natural cycles of plant growth. A good understanding of both natural and man-influenced ecosystems is therefore essential for analysing and predicting the engineering role of vegetation. Continuous hard grazing generally leads to loss of the vegetation cover; the plants lack sufficient leaves for photosynthesis and die, whilst damage to the growth points prevents their regeneration. These relationships are analysed theoretically in THEPROM, an erosion–productivity model being developed for rangelands in Botswana (Biot, 1990). Under special circumstances, however, the interactions may react another way. Valentin (1985) showed that, under very special ecological conditions, grassland vegetation may be improved by cattle grazing.

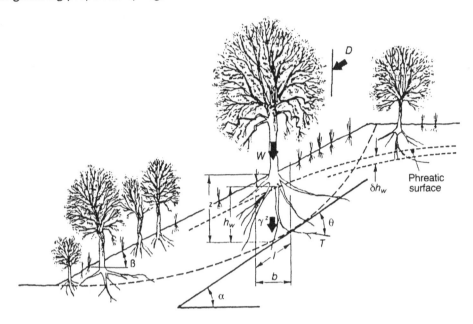

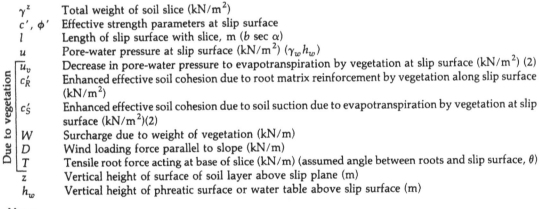

Figure 2.25 Main influences of vegetation on slope stability (after Coppin and Richards). Parameters applied in slope stability analysis:

γ^z Total weight of soil slice (kN/m^2)

c', ϕ' Effective strength parameters at slip surface

l Length of slip surface with slice, m (b sec α)

u Pore-water pressure at slip surface (kN/m^2) ($\gamma_w h_w$)

Due to vegetation

u_v Decrease in pore-water pressure to evapotranspiration by vegetation at slip surface (kN/m^2) (2)

c_R' Enhanced effective soil cohesion due to root matrix reinforcement by vegetation along slip surface (kN/m^2)

c_S' Enhanced effective soil cohesion due to soil suction due to evapotranspiration by vegetation at slip surface (kN/m^2)(2)

W Surcharge due to weight of vegetation (kN/m)

D Wind loading force parallel to slope (kN/m)

T Tensile root force acting at base of slice (kN/m) (assumed angle between roots and slip surface, θ)

z Vertical height of surface of soil layer above slip plane (m)

h_w Vertical height of phreatic surface or water table above slip surface (m)

Notes
1. The value of many of these parameters varies with depth and soil type.
2. In certain slope stability analyses the decrease in pore-water pressure due to vegetation (i.e. increased soil suction from evapotranspiration) is expressed as an enhanced effective soil cohesion, as distinct from a pore pressure reduction.

Table 2.17 Beneficial and adverse effects of vegetation (from Coppin and Richards, 1990)

Hydrological effects		*Mechanical effects*	
Foliage intercepts rainfall causing:		Roots bind soil particles and permeate the soil, resulting in:	
1. absorptive and evaporative losses, reducing rainfall available for infiltration	B	1. restraint of soil movement reducing erodibility	B
2. reduction in kinetic energy of raindrops and thus erosivity	B	2. increase in shear strength through a matrix of tensile fibres	B
3. increase in drop size through leaf drip, thus increasing localized rainfall intensity	A	3. network of surface fibres creates a tensile mat effect, restraining underlying strata	B
Stems and leaves interact with flow at the ground surface, resulting in:		Roots penetrate deep strata, giving:	
1. higher depression storage and higher volume of water for infiltration	A/B	1. anchorage into firm strata, bonding soil mantle to stable subsoil or bedrock	B
2. greater roughness on the flow of air and water, reducing its velocity, but	B	2. support to up-slope soil mantle through buttressing and arching	B
3. tussocky vegetation may give high localized drag, concentrating flow and increasing velocity	A	Tall growth of trees, so that:	
Roots permeate the soil, leading to:		1. weight may surcharge the slope, increasing normal and down-slope force components	A/B
1. opening up of the surface and increasing infiltration	A	2. when exposed to wind, dynamic forces are transmitted into the ground	A
2. extraction of moisture which is lost to the atmosphere in transpiration, lowering pore-water pressure and increasing soil suction, both increasing soil strength	B	Stems and leaves cover the ground surface, so that:	
		1. impact of traffic is absorbed, protecting soil surface from damage	B
3. accentuation of dessication cracks, resulting in higher infiltration	A	2. foliage is flattened in high velocity flows, covering the soil surface and providing protection against erosive flows	B

A = adverse effect
B = beneficial effect

Table 2.18 Salient properties of vegetation and their engineering significance (from Coppin and Richards, 1990)

Effect on	Influence	Ground cover (%)	Height	Leaf shape and length	Stem/leaf density	Stem/leaf robustness	Stem/leaf flexibility	Root depth	Root density	Root strength	Annual growth cycle	Weight
Surface competence												
	Soil detachment	●	●	●	●						●	
	Mechanical strength	●	●		●	●			●	●	●	
	Insulation	●			●						●	
	Retarding/arresting		●		●	●	●					
	Erosion	●			●						●	
Surface water regime												
	Rainfall interception	●		●	●							
	Overland flow/runoff	●			●							
	Infiltration				●			●	●			
	Subsurface drainage				●			●	●			
	Surface drag	●	●	●	●				●		●	
Soil water												
	Evapotranspiration		●		●				●		●	
	Soil moisture depletion leading to increased soil suction, reduced pore-water and soil weight							●			●	
Properties of soil mass												
	Root reinforcement							●	●	●	●	
	Anchorage/restraint							●	●	●		
	Arching/buttressing							●		●		
	Surface mat/net									●	●	●
	Surcharge		●									●
	Windthrow		●		●	●		●			●	●
	Root wedging							●	●			
Air flow												
	Surface drag		●	●	●		●				●	
	Flow deflection		●		●	●	●				●	
	Noise attenuation		●	●	●						●	
	Suspended particulates		●		●						●	

APPENDIX 2.A:

Estimation of the effect of vegetation on the factor of safety on a slope using the infinite slope method. Notation is given in the text.

FACTOR OF SAFETY WITHOUT VEGETATION:

$$F = \frac{c' + (\gamma z - \gamma_w h_w) \cos^2 \beta \tan \phi'}{\gamma z \sin \beta \cos \beta}$$

$c' = 10 \, \text{kN/m}^3$

$\gamma = 18 \, \text{kN/m}^3$

$z = 1.0 \, \text{m}$

$\beta = 35°$

$\phi' = 35°$

$\gamma_w = 9.8 \, \text{kN/m}^3$

$h_w = 0.5 \, \text{m}$

$$F = \frac{10 + \{(18 \times 1) - (9.8 \times 0.5)\} \times 0.6710 \times 0.7002}{18 \times 0.5736 \times 0.8192}$$

$$F = \frac{10 + \{18 - 4.9\} \times 0.6710 \times 0.7002}{8.4581}$$

$$F = \frac{16.1548}{8.4581}$$

$$F = 1.91$$

FACTOR OF SAFETY WITH VEGETATION

$$F = \frac{(c' + c_R') + [\{(\gamma z - \gamma_w h_v) + W\} \cos^2 \beta + T \sin \theta] \tan \phi' + T \cos \theta}{\{(\gamma z + W) \sin \beta + D\} \cos \beta}$$

$c_R' = 5 \, \text{kN/m}^3$

$W = 2.5 \, \text{kN/m}^3$

$D = 0.1 \, \text{kN/m}^2$

$h_v = 0.4 \, \text{m}$

$T = 5 \, \text{kN/m}$

$\theta = 45°$

$$F = \frac{(10 + 5) + [\{(18 - (9.8 \times 0.4)) + 2.5\} \times 0.6710 + (5 \times 0.7071)] \times 0.7002 + (5 \times 0.7071)}{[\{(18 + 2.5) \times 0.5736\} + 0.1] \times 0.8192}$$

$$F = \frac{15 + [\{((18 - 3.92) + 2.5) \times 0.6710\} + 3.5355] \times 0.7002 + 3.5355}{9.7147}$$

$$F = \frac{15 + 10.2654 + 3.5355}{9.7147}$$

$$F = \frac{28.8009}{9.7147}$$

$$F = 2.96$$

SENSITIVITY ANALYSIS

Increase in c_R' of $5\,kN/m^3$ increases F by 0.59
Increase in W of $2.5\,kN/m^3$ decreases F by 0.04
Increase in D of $0.1\,kN/m^2$ decreases F by 0.02
Increase in T of $5\,kN/m$ increases F by 0.71
Increase in h_w of $0.1\,m$ increases F by 0.08

REFERENCES

Al-Durrah, M. M. and Bradford, J. M. (1982) Parameters for describing soil detachment due to single waterdrop impact. *Soil Sci. Soc. Am. J.*, **46**, 836–40.

Appelmans, F., van Hove, J. and De Leenheer, L. (1980) Rain interception by wheat and sugar beet crops, in *Assessment of Erosion* (eds M. De Boodt and D. Gabriels). Wiley, Chichester, pp. 227–35.

Armstrong, C. L. and Mitchell, J. K. (1987) Transformations of rainfall by plant canopy. *Trans. Am. Soc. Agric. Engnrs.*, **30**, 688–96.

Babaji, G. A. (1987) Some plant stem properties and overland flow hydraulics. A laboratory simulation. PhD Thesis, Silsoe College, Cranfield Institute of Technology.

Bache, D. H. (1986) Momentum transfer to plant canopies: influence of structure and variable drag. *Atmospheric Environment*, **13**, 1681–7.

Bache, D. H. and MacAskill, I. A. (1984) *Vegetation in Civil and Landscape Engineering*. Granada Publishing, London.

Bagnold, R. A. (1941) *The Physics of Blown Sand and Desert Dunes*. Chapman and Hall, London.

Beasley, D. B., Huggins, L. F. and Monke, E. J. (1980) ANSWERS: a model for watershed planning. *Trans. Am. Soc. Agric. Engnrs.*, **23**, 938–44.

Biot, Y. (1990) THEPROM – An erosion productivity model, in *Soil Erosion on Agricultural Land* (eds J. Boardman, I. D. L. Foster and J. A. Dearing). Wiley, Chichester, pp. 465–79.

Bishop, D. M. and Stevens, M. E. (1964) Landslips on logged areas in southeast Alaska. *US Forest Service Research Paper NOR-1*, Northern Forest Experiment Station, Juneau, Alaska.

Boiffin, J. (1985) Stages and time-dependency of soil crusting in situ, in *Assessment of Soil Surface Sealing and Crusting* (eds F. Callebaut, D. Gabriels and M. De Boodt). Flanders Research Centre for Soil Erosion and Soil Conservation, Gent, pp. 91–8.

Brakensiek, D. L. and Rawls, W. J. (1983) Agricultural management effects on soil water processes. Part II. Green and Ampt parameters for crusting soils. *Trans. Am. Soc. Agric. Engnrs.*, **26**, 1753–7.

Brandt, C. J. (1989) The size distribution of throughfall drops under vegetation canopies. *Catena*, **16**, 507–24.

Brandt, C. J. (1990) Simulation of the size distribution and erosivity of raindrops and throughfall drops. *Earth Surf. Proc. Landf.*, **15**, 687–98.

Brown, C. B. and Sheu, M. S. (1975) Effects of deforestation on slopes. *J. Geotechnical Engineering Division, ASCE*, **101**, 147–65.

Bui, E. N. and Box, J. E. (1992) Stemflow, rain throughfall and erosion under canopies of corn and sorghum. *Soil Sci. Soc. Am. J.*, **56**, 242–7.

Carson, M. A. and Kirkby, M. J. (1972) *Hillslope Form and Process*. Cambridge University Press, Cambridge.

Carter, C. E., Greer, J. D., Braud, H. J. and Floyd, J. M. (1974) Raindrop characteristics in south central United States. *Trans. Am. Soc. Agric. Engnrs.*, **17**, 1033-7.

Chapman, G. (1948) Size of raindrops and their striking force at the soil surface in a red pine plantation. *Trans. Am. Geophys. Un.*, **29**, 664-70.

Cionco, R. M. (1965) A mathematical model for air flow in a vegetative canopy. *J. Appl. Met.*, **6**, 185-93.

Coppin, N. J. and Richards, I. G. (eds) (1990) *Use of Vegetation in Civil Engineering*. Construction Industry Research and Information Association/ Butterworths, London.

Coster, C. (1938) Surficial runoff and erosion on Java. *Tectona*, **31**, 613-728.

Cruse, R. M. and Larson, W. E. (1977) Effect of soil shear strength on soil detachment due to raindrop impact. *Soil Sci. Soc. Am. J.*, **41**, 777-81.

Denmead, O. T. (1976) Temperate cereals, in *Vegetation and the Atmosphere*, Vol. 2 (ed. J. L. Monteith). Academic Press, London, pp. 1-31.

De Ploey, J. (1981) The ambivalent effects of some factors of erosion. *Mem. Inst. Geol. Univ. Louvain*, **31**, 171-81.

De Ploey, J. (1982) A stemflow equation for grasses and similar vegetation. *Catena*, **9**, 139-52.

De Ploey, J. (1984) Hydraulics of runoff and loess loam deposition. *Earth Surf. Proc. Landf.*, **9**, 533-9.

De Ploey, J., Savat, J. and Moeyersons, J. (1976) The differential impact of some soil factors on flow, runoff creep and rainwash. *Earth Surf. Proc.*, **1**, 151-61.

Doorenbos, J. and Pruitt, W. O. (1977) Guidelines for predicting crop water requirements. *FAO Irrigation and Drainage Paper No. 24*.

Emama, E. L. (1988) Use of grass for control of hillside erosion. MSc Thesis, Silsoe College, Cranfield Institute of Technology.

Engelund, F. and Hansen, E. (1967) *A Monograph on Sediment Transport in Alluvial Streams*. Teknisk Forlag, Kφbenhavn.

Engman, E. T. (1986) Roughness coefficients for routing surface runoff. *J. Irrigation and Drainage Division, ASCE.*, **112**, 39-53.

Farres, P. (1978) The role of time and aggregate size in the crusting process. *Earth Surf. Proc.*, **3**, 243-54.

Finney, H. J. (1984) The effect of crop covers on rainfall characteristics and splash detachment. *J. Agric. Engng. Res.*, **29**, 337-43.

Foster, G. R. and Meyer, L. D. (1975) Mathematical simulation of upland erosion by fundamental erosion mechanics, in *Present and Prospective Technology for Predicting Sediment Yields and Sources*. USDA Agr. Res. Serv. Pub., ARS-S-40, pp. 190-207.

Free, G. R. (1960) Erosion characteristics of rainfall. *Agric. Engng.*, **41**, 447-9, 455.

Govers, G. and Poesen, J. (1985) A field scale study of surface sealing on loam and sandy loam soils. Part I. Spatial variability of soil surface sealing and crusting, in *Assessment of Soil Surface Sealing and Crusting* (eds F. Callebaut, D. Gabriels and M. De Boodt). Flanders Research Centre for Soil Erosion and Conservation, Gent, pp. 171-82.

Govers, G. and Rauws, G. (1986) Transporting capacity of overland flow on a plane bed and on irregular beds. *Earth Surf. Proc. Landf.*, **11**, 515-24.

Gray, D. H. (1978) Role of woody vegetation in reinforcing soils and stabilising slopes. *Proceedings, Symposium on Soil Reinforcing and Stabilising Techniques in Engineering Practice*. NSW Institute of Technology, Sydney, Australia, pp. 253-306.

Gray, D. H. and Leiser, A. J. (1982) *Biotechnical Slope Protection and Erosion Control*. Van Nostrand Reinhold, New York.

Gray, D. H. and Megahan, W. F. (1981) Forest vegetation removal and slope stability in the Idaho batholith. *US Forest Research Paper INT-271*. Intermountain Forest and Range Experiment Station, Ogden, UT.

Greenway, D. R. (1987) Vegetation and slope stability, in *Slope Stability* (eds M. G. Anderson and K. S. Richards). Wiley, Chichester, pp. 187-230.

Grindley, J. (1969) The calculation of actual evaporation and soil moisture deficit over specified catchment areas. *Hydrological Memorandum No. 38*, Meteorological Office, Bracknell.

Hagen, L. J. and Lyles, L. (1988) Estimating small grain equivalents of shrub-dominated rangelands for wind erosion control. *Trans. Am. Soc. Agric. Engnrs.*, **31**, 769-75.

Hagen, L. J., Skidmore, E. L., Miller, P. L. and Kipp, J. E. (1981) Simulation of effect of wind barriers on airflow. *Trans. Am. Soc. Agric. Engnrs.*, **24**, 1002-8.

Hayes, J. C., Barfield, B. J. and Tollner, E. W. (1984) Performance of grass filters under laboratory and field conditions. *Trans. Am. Soc. Agric. Engnrs.*, **27**, 1321-31.

Herwitz, S. R. (1985) Interception storage capacities of tropical rainforest canopy trees. *J. Hydrology*, **77**, 237-52.

Herwitz, S. R. (1986) Infiltration-excess caused by stemflow in a cyclone-prone tropical rainforest. *Earth Surf. Proc. Landf.*, **11**, 401-12.

Herwitz, S. R. (1987) Raindrop impact and water flow on the vegetative surfaces of trees and the effects of stemflow and throughfall generation. *Earth Surf. Proc. Landf.*, **12**, 425–432.

Holtan, H. N. (1961) A concept for infiltration estimates in watershed engineering. *USDA Agric. Res. Serv. Pub.*, ARS-41-51.

Hoogmoed, W. B. and Stroosnijder, L. (1984) Crust formation on sandy soils in the Sahel. I. Rainfall and infiltration. *Soil and Tillage Research*, **4**, 5–24.

Horton, R. E. (1919) Rainfall interception. *Monthly Weather Review*, **47**, 603–23.

Hsi, G. and Nath, J. H. (1970) Wind drag within a simulated forest. *J. Appl. Meteorology*, **9**, 592–602.

Hudson, N. W. (1963) Raindrop size distribution of high intensity storms. *Rhod. J. Agric. Res.*, **1**, 6–11.

Inoue, E. (1963) On the turbulent structure of air flow within crop canopies. *J. Met. Soc. Japan, Series 11*, **41**, 317–26.

Kowal, J. M. and Kassam, A. H. (1976) Energy load and instantaneous intensity of rainstorms at Samaru, northern Nigeria. *Trop. Agr. (Trinidad)*, **53**, 185–97.

Laflen, J. M. (1987) Effect of tillage systems on concentrated flow erosion, in *Soil Conservation and Productivity* (ed. I. Pla Sentis). Sociedad Venezolana de la Ciencia del Suelo, Maracay, pp. 798–809.

Laflen, J. M. and Colvin, T. S. (1981) Effect of crop residue on soil loss from continuous row cropping. *Trans. Am. Soc. Agric. Engnrs.*, **24**, 605–9.

Landsberg, J. J. and James, G. B. (1971) Wind profiles in plant canopies: studies on an analytical model. *J. Appl. Ecol.*, **8**, 729–41.

Leyton, L., Reynolds, E. R. C. and Thompson, F. B. (1967) Rainfall interception in forest and moorland, in *Forest Hydrology* (eds W. E. Sopper and H. W. Lull). Pergamon, Oxford, pp. 163–78.

Liddle, M. J. (1973) The effects of trampling and vehicles on natural vegetation. PhD Thesis, University College of North Wales, Bangor.

Maene, L. M. and Chong, S. P. (1979) Drop size distribution and erosivity of tropical rainstorms under the oil palm canopy. *Lapuran Penyelidikan Jabatan Sains Tanah 1977–78*, Universiti Pertanian Malaysia, Serdang, pp. 81–93.

Marshall, I. S. and Palmer, E. M. (1948) The distribution of raindrops with size. *J. Meteorology*, **5**, 165–6.

McKeague, J. A., Wang, C. and Coen, G. M. (1986) Describing and interpreting the macrostructure of mineral soils. *LRRI Contribution No. 84–50*, Agriculture Canada, Ottawa.

Merriam, R. A. (1973) Fog drip from artificial leaves in a fog wind tunnel. *Water Resources Research*, **9**, 1591–8.

Meyer, L. D. and Wischmeier, W. H. (1969) Mathematical simulation of the processes of soil erosion by water. *Trans. Am. Soc. Agric. Engnrs.*, **12**, 754–8, 762.

Monteith, J. L. (1973) *Principles of Environmental Physics*. Edward Arnold, London.

Morgan, R. P. C. (1980) Field studies of sediment transport by overland flow. *Earth Surf. Proc.*, **5**, 307–16.

Morgan, R. P. C. (1985) Effect of corn and soybean canopy on soil detachment by rainfall. *Trans. Am. Soc. Agric. Engnrs.*, **28**, 1135–40.

Morgan, R. P. C. (1989) Design of in-field shelter systems for wind erosion control, in *Soil Erosion Protection Measures in Europe* (eds U. Schwertmann, R. J. Rickson and K. Auerswald), Soil Technology Series No. 1, Catena Publications, Cremlingen-Destedt, pp. 15–23.

Morgan, R. P. C. and Finney, H. J. (1987) Drag coefficients of single crop rows and their implications for wind erosion control, in *International Geomorphology 1986*. Part II (ed. V. Gardiner). Wiley, Chichester, pp. 449–58.

Morgan, R. P. C., Finney, H. J. and Williams, J. S. (1986) Fundamental plant parameters for wind erosion control. Final Report to the UK Agriculture and Food Research Council, Research Grant No. AG63/170.

Morgan, R. P. C., Finney, H. J. and Williams, J. S. (1988) Leaf properties affecting crop drag coefficients: implications for wind erosion control, in *Land Conservation for Future Generations* (ed. S. Rimwanich). Department of Land Development, Ministry of Agriculture and Cooperatives, Bangkok, pp. 885–93.

Morin, J., Benyamini, Y. and Michaeli, A. (1981) The effect of raindrop impact on the dynamics of soil surface crusting and water movement in the profile. *J. Hydrology*, **52**, 321–6.

Mosley, M. P. (1982) The effect of a New Zealand beech forest canopy on the kinetic energy of water drops and on surface erosion. *Earth Surf. Proc. Landf.*, **7**, 103–7.

Moss, A. J. and Green, T. W. (1987) Erosive effects of the large water drops (gravity drops) that fall from plants. *Austral. J. Soil Res.*, **25**, 9–20.

Nielsen, S. A. and Styczen, M. (1986) Development of an areally distributed soil erosion model, in *Partikulaert bundet stoftransport i vand og jorderosjon* (ed. B. Hasholt). Nordisk Hydrologisk Program Rapport No. 14, pp. 293–302.

Noble, C. A. (1981) The effect of plant cover on the erosivity of rainfall. MSc Thesis, Silsoe College, Cranfield Institute of Technology.

Noble, C. A. and Morgan, R. P. C. (1983) Rainfall interception and splash detachment with a Brussels sprouts plant: a laboratory simulation. *Earth Surf. Proc. Landf.*, **8**, 569–77.

O'Loughlin, C. L. (1984) Effectiveness of introduced forest vegetation for protecting against landslides and erosion in New Zealand's steeplands. Paper presented to Symposium on effects of forest land use on erosion and slope stability, Honolulu, Hawaii.

Park, S. W. (1981) Modeling soil erosion and sedimentation on small agricultural watersheds. PhD Thesis, University of Illinois.

Park, S. W., Mitchell, J. K. and Scarborough, J. N. (1982) Soil erosion simulation on small watersheds: a modified ANSWERS model. *Trans. Am. Soc. Agric. Engnrs.*, **25**, 1581–8.

Penman, H. L. (1949) The dependence of transpiration on weather and soil conditions. *J. Soil Sci.*, **1**, 74–89.

Quansah, C. (1985) Rate of soil detachment by overland flow, with and without rain, and its relationship with discharge, slope steepness and soil type, in *Soil Erosion and Conservation* (eds S. A. El Swaify, W. C. Moldenhauer and A. Lo) Soil Conservation Society of America, Ankeny, IA, pp. 406–23.

Quinn, N. W., Morgan, R. P. C. and Smith, A. J. (1980) Simulation of soil erosion by human trampling. *J. Environmental Management*, **10**, 155–65.

Randall, J. M. (1969) Wind profiles in an orchard plantation. *Agric. Meteorol.*, **6**, 439–52.

Rauws, G. and Govers, G. (1988) Hydraulic and soil mechanical aspects of rill generation on agricultural soils. *J. Soil Sci.*, **39**, 111–24.

Rawls, W. J., Brakensiek, D. L. and Soni, B. (1983) Agricultural management effects on soil water processes. Part I. Soil water retention and Green and Ampt infiltration parameters. *Trans. Am. Soc. Agric. Engnrs.*, **26**, 1747–52.

Ree, W. O. (1949) Hydraulic characteristics of vegetation for vegetated waterways. *Agric. Engng.*, **30**, 184–7, 189.

Rose, C. W., Williams, I. R., Sander, G. C. and Barry, D. A. (1983) A mathematical model of soil erosion and deposition processes. I. Theory for a plane element. *Soil Sci. Soc. Am. J.*, **47**, 991–5.

Rutter, A. J. and Morton, A. J. (1977) A predictive model of rainfall interception in forests. III. Sensitivity of the model to stand parameters

and meteorological variables. *J. Appl. Ecol.*, **14**, 567–88.

Schoklitsch, A. (1950) *Handbuch des Wassbaues.* Springer, Wien.

Seginer, I. (1972) Windbreak drag calculated from the horizontal velocity field. *Boundary Layer Meteorology*, **3**, 87–97.

Shaw, R. H. and Pereira, A. R. (1982) Aerodynamic roughness of a plant canopy: a numerical experiment. *Agric. Met.*, **26**, 51–65.

Singer, M. J. and Blackard, J. (1978) Effect of mulching on sediment in runoff from simulated rainfall. *Soil Sci. Soc. Am. J.*, **42**, 481–6.

Skidmore, E. L. and Hagen, L. J. (1977) Reducing wind erosion with barriers. *Trans. Am. Soc. Agric. Engnrs.*, **20**, 911–15.

Styczen, M. and Høgh-Schmidt, K. (1986) A new description of the relation between drop sizes, vegetation and splash erosion, in *Partikulaert bundet stoftransport i vand og jorderosjon* (ed. B. Hasholt). Nordisk Hydrologisk Program Rapport No. 14, pp. 255–71.

Styczen, M. and Høgh-Schmidt, K. (1988) A new description of splash erosion in relation to raindrop size and vegetation, in *Erosion Assessment and Modelling* (eds R. P. C. Morgan and R. J. Rickson). Commission of the European Communities Report No. EUR 10860 EN, pp. 147–98.

Styczen, M. and Nielsen, S. A. (1989) A view of soil erosion theory, process-research and model building: possible interactions and future developments. *Quaderni di Scienza del Suolo*, **2**, 27–45.

Temple, D. M. (1982) Flow retardance of submerged grass channel linings. *Trans. Am. Soc. Agric. Engnrs.*, **25**, 1300–3.

Tengbeh, G. T. (1989) The effect of grass cover on bank erosion. PhD Thesis, Silsoe College, Cranfield Institute of Technology.

Thom, A. S. (1971) Momentum absorption by vegetation. *Quart. J. Roy. Met. Soc.*, **97**, 414–28.

Thom, A. S. (1975) Momentum, mass and heat exchange of plant communities, in *Vegetation and the Atmosphere*, Vol. 1 (ed. J. L. Monteith). Academic Press, London, pp. 57–109.

Thornes, J. B. (1988a) Competitive vegetation–erosion model for Mediterranean conditions, in *Erosion Assessment and Modelling* (eds R. P. C. Morgan and R. J. Rickson). Commission of the European Communities Report No. EUR 10860 EN, pp. 255–82.

Thornes, J. B. (1988b) Erosional equilibria under grazing, in *Conceptual Issues in Environmental Archaeology* (eds J. Bintliff, D. Davidson and

E. Grant). Edinburgh University Press, Edinburgh, pp. 193–210.

Thornes, J. B. (1990) The interaction of erosional and vegetational dynamics in land degradation: spatial outcomes, in *Vegetation and Erosion* (ed. J. B. Thornes). Wiley, Chichester, pp. 41–53.

Tollner, E. W., Barfield, B. J. and Hayes, J. C. (1982) Sedimentology of erect vegetal filters. *J. Hydraulics Division, ASCE*, **108**, 1518–31.

Torri, D., Sfalanga, M. and Del Sette, M. (1987) Splash detachment, runoff depth and soil cohesion. *Catena*, **14**, 149–55.

Tsukamoto, Y. and Kusakabe, O. (1984) Vegetative influences on debris slide occurrences on steep slopes in Japan. Paper presented to Symposium on Effects of forest land use on erosion and slope stability, Honolulu, Hawaii.

US Soil Conservation Service (1954) *Handbook of Channel Design for Soil and Water Conservation.* USDA Tech. Pub., SCS-TP-61.

Valentin, C. (1985) Effects of grazing and trampling on soil deterioration around recently drilled water holes in the Sahelian zone, in *Soil Erosion and Conservation* (S. A. El-Swaify, W. C. Moldenhauer and A. Lo). Soil Conservation Society of America, Ankeny, IA, pp. 51–65.

Van Elewijck, L. (1988) Influence of leaf and branch slope on stemflow amount. Paper presented to British Geomorphological Research Group Symposium on Vegetation and geomorphology, Bristol, UK.

Van Elewijck, L. (1989) Stemflow on maize: a stemflow equation and the influence of rainfall intensity on stemflow amount, *Soil Technology*, **2**, 41–8.

Voetberg, K. S. (1970) *Erosion on Agricultural Lands.* Agricultural University, Wageningen.

Waldron, L. J. (1977) The shear resistance of root-permeated homogeneous and stratified soil. *Soil Sci. Soc. Am. J.*, **41**, 343–9.

Waldron, L. J. and Dakessian, S. (1981) Soil reinforcement by roots: calculation of increased soil shear resistance from root properties. *Soil Sci.*, **132**, 427–35.

Wang, W. L. and Yen, B. C. (1974) Soil arching in slopes. *J. Geotechnical Engineering Division, ASCE*, **100**, 61–78.

Wiersum, K. (1985) Effects of various vegetation layers of an *Acacia auriculiformis* forest plantation on surface erosion in Java, Indonesia, in *Soil Erosion and Conservation* (eds S. A. El-Swaify, W. C. Moldenhauer and A. Lo). Soil Conservation Society of American, Ankeny, IA, pp. 79–89.

Wiersum, K., Budirijanto, P. and Rhomdoni, D. (1979) Influence of forests on erosion. Seminar on the erosion problem in the Jatiluhur area. Institute of Ecology, Padjadjaran University, Bandung, Report No. 3.

Wischmeier, W. H. (1975) Estimating the soil loss equation's cover and management factor for undisturbed areas, in *Present and Prospective Technology for Predicting Sediment Yields and Sources.* USDA Agr. Res. Serv., Pub., ARS-S-40, pp. 118–24.

Wischmeier, W. H. and Smith, D. D. (1978) Predicting rainfall erosion losses. *USDA Agr. Res. Serv. Handbook 537.*

Withers, B. and Vipond, S. (1974) *Irrigation: Design and Practice.* Batsford, London.

Wossenu Abtew, Gregory, J. M. and Borrelli, J. (1989) Wind profile: estimation of displacement height and aerodynamic roughness. *Trans. Am. Soc. Agric. Engnrs.*, **32**, 521–7.

Wright, J. L. and Brown, K. W. (1967) Comparison of momentum and energy balance methods of computing vertical transfer within a crop. *Agron. J.*, **59**, 427–32.

Yen, C. P. (1972) Study on the root system form and distribution habit of the ligneous plants for soil conservation in Taiwan (preliminary report). *J. Chinese Soil & Water Conserv.*, **3**, 179–204.

Yen, C. P. (1987) Tree root patterns and erosion control, in *Proceedings of the International Workshop on Soil Erosion and Its Countermeasures* (ed. S. Jantawat). Soil and Water Conservation Society of Thailand, Bangkok, pp. 92–111.

Ziemer, R. R. (1981) Roots and stability of forested slopes. *Int. Assoc. Hydrol. Sci., Pub. No. 132*, pp. 343–61.

Zinke, P. J. (1967) Forest interception studies in the US, in *Forest Hydrology* (eds W. E. Sopper and H. W. Lull). Pergamon, Oxford, pp. 137–61.

ECOLOGICAL PRINCIPLES FOR VEGETATION ESTABLISHMENT AND MAINTENANCE

Nick Coppin and Richard Stiles

3.1 INTRODUCTION

On very few parts of the earth's surface are the conditions so inhospitable that they do not support some kind of vegetation. Wherever it is sufficiently vigorous and remains intact, the vegetation cover forms a protective layer, reducing the potentially erosive and destabilizing effects of natural processes on the underlying soils. It is this natural protection that biological construction (bioengineering) techniques aim to exploit when they are applied to stabilize slopes and waterside situations.

Although the presence of vegetation is a consistent factor in contributing to the protection of the land surface throughout the world, the nature of the vegetation itself is by no means constant. Its structure, composition and dynamics vary dramatically in response to the widely different natural conditions that are to be found between the poles and the equator.

The aim of this chapter is to set out the principles of how to achieve a stable vegetation cover on disturbed ground, such as cultivated slopes, formed slopes and earthworks. Such vegetation will have to conform to certain functional requirements, for erosion control and slope stabilization, as set out elsewhere in this book. However, site conditions may be difficult, or even hostile, so there may be many constraints on the establishment process. Many approaches have been developed around the world to cope with different environments and circumstances, and the basis of these approaches is reviewed.

Vegetation types and their 'ecology' vary considerably around the world and the intention of this chapter is to take a global approach to vegetation establishment and maintenance. To do this in detail would require an immense amount of space, hence the emphasis is on principles, which local specialists can apply using local knowledge. Before describing these, however, it is helpful to define a few basic terms. **Establishment** involves the process of obtaining a vegetation cover using seeding and planting techniques, including a period of aftercare until the vegetation is fully established. In some situations the aftercare period has to be quite long. **Maintenance** involves the periodic inputs and management required in order to maintain the required vegetation in the required form, and to prevent unwanted effects occurring.

The full meaning and implications of these terms will become clear in the following sections. However, before considering the establishment process itself, it is necessary to understand how nature organizes vegetation, how vegetation behaves naturally, and what factors need to be taken into account when selecting the best vegetation to use. This chapter is therefore

Slope Stabilization and Erosion Control: A Bioengineering Approach. Edited by R. P. C. Morgan and R. J. Rickson. Published in 1995 by E & FN Spon, 2-6 Boundary Row, London, SE1 8HN. ISBN 0 419 15630 5.

Table 3.1 Climatic zones and their corresponding soil and vegetation types

Climax vegetation	Zonal climate	Zonal soil type
Evergreen tropical rain forest	Equatorial with diurnal climate, humid	Equatorial brown clays (ferralitic soils, latosols)
Sub-tropical deciduous forests and savannahs	Sub-tropical with summer rains	Red clays or red earths (savannah soils)
Sub-tropical desert vegetation	Sub-tropical arid (desert climate), arid	Sierozems (poorly developed rocky and sandy soils)
Sclerophyllous woody plants (Mediterranean type)	Winter rain and summer drought, arido-humid	Brown earths
Temperate evergreen forests	Warm temperate (maritime), humid	Yellow or red podzols
Broadleaf deciduous forests (bare in winter)	Typical temperate with a short period of frost	Forest brown earths and grey forest soils
Steppes and prairie grasslands	Arid temperate with a cold winter (continental)	Chernozems to sierozems
Boreal coniferous forests (taiga)	Cold temperate	Podzols (raw humus bleached earths)
Tundra vegetation (treeless)	Arctic (including Antarctic) polar	Tundra humus soils

organized into four main sections: vegetation as a natural component of the landscape, factors affecting plant selection and vegetation growth, establishment, and management.

3.2 VEGETATION AS A NATURAL COMPONENT OF THE LANDSCAPE

In order to be able to assess whether biological construction techniques are likely to be feasible in any particular area, it is important to have a broad understanding of the nature of the earth's vegetation cover and the way in which it closely reflects the interaction of natural conditions prevailing at any position on the earth's surface.

3.2.1 WORLD VEGETATION ZONES

Potential natural vegetation and climate

Climate is recognized as the major factor influencing the natural vegetation cover. Differences in climate at the global scale are normally represented in the form of a series of broad horizontal belts running around the earth, each having a generally similar set of climatic conditions. The distribution of the main zonal vegetation types

is of particular interest from the point of view of biological construction techniques in that it links climatic zones directly to corresponding soil and vegetation types. Table 3.1 illustrates this relationship and how it changes from the equator towards the poles. The climax vegetation referred to is the potential natural vegetation, without human interference. Within these zones and vegetation types there are many variations due to human modification and management, the most profound being agriculture.

Within each climatic zone, the type of soil and the natural vegetation communities associated with it are a function of the interactions between climate, the underlying geology (and thus the soils) and the indigenous flora. These interactions link vegetation, fauna, soils, hydrology and climate to form the ecosystems characteristic of each climatic zone. From the biological construction point of view the interrelationships between climate, soils and vegetation are close and important.

1. The climate, both its nature and seasonality, will have a profound influence on the potential erosion hazard to which the soils are

subjected, as well as on the ability of vegetation to flourish.

2. The soils, together with the climate, will determine the nature of the vegetation that can be supported and thereby also influence the extent to which they themselves can be protected against erosion.

3. The vegetation, in turn, provides the basic material with which to implement biological construction techniques to protect the *in situ* soils from the effects of extremes within the prevailing climatic conditions.

Boundaries between zones are far from clear cut, and intermediate zones of varying extent can be identified between each of the major zones. There is also further variation in climate, soils and vegetation within the main and intermediate climatic zones in response to changes in both oceanity (proximity to the sea) and altitude.

Mountainous regions are, of course, likely to be of considerable importance from a bioengineering point of view. As regions of both high altitude and steep slopes, their vegetation will be subject to relatively greater stress due to the more extreme conditions, while the risks from slope failure and soil erosion will be correspondingly higher.

Climatic influences and variation in vegetation

Climate is the result of the interplay of a number of factors. It is the product of, amongst other things, differing precipitation levels, variations in temperature, together with the way in which these are distributed throughout the year.

The established pattern of seasonal and diurnal variations in temperature and moisture availability in any part of the globe provides the immediate climatic conditions under which the annual growth of vegetation takes place. This pattern can be seen as one of the key evolutionary pressures that influences the types of plant species and plant community that evolve in and become adapted to a particular environment.

Two extremes of evolutionary pressure can be identified as factors which have shaped the response of vegetation in terms of contrasting climatic conditions. Higher levels of solar radiation reaching the earth, towards the equator, between the tropics, clearly lead to warmer climates, but more significantly perhaps, to climates in which there is relatively little seasonal distinction between a cooler and a warmer period, or between wetter and drier times of the year. This is, of course, most marked at the equator, where an almost perfect diurnal climate exists, with day and night remaining of equal length throughout the year, and temperatures not varying more than a few degrees from month to month. Under these conditions vegetation, not surprisingly, continues to grow rapidly throughout the year.

At the poles, in contrast, where overall levels of radiation and therefore temperatures are far lower anyway, such that there are always limitations on plant growth, the diurnal variation in temperature is also almost entirely eliminated. Thus for half the year there is almost continual daytime, when light at least is not a limiting factor for plant growth, while for the other six months continual night prevails so that no photosynthesis can take place. Here the climate can be described as completely seasonal, and vegetation can only grow during the short summer period.

Between the poles and the equator the diurnal and seasonal factors gradually vary in their effects on both the periodicity of vegetation growth and on the type of vegetation which has evolved.

The effects of latitude, however, only account for part of the variation in climatic conditions. The location on the earth's surface in relation to distribution of the landmasses and the oceans also has a significant influence. This is expressed in terms of the degree of oceanity or continentality of the climate, which in turn affects the relative seasonal variation in temperature and precipitation. The significance of this, from the point of view of vegetation, is that, in oceanic regions, it affects the extent to which there is a clearly defined growing season, or conversely, in the more continental regions, the extent to which the role of vegetation in protecting the soil is, to some extent, limited to the period during which it is not dormant.

To some extent the effects of increasing altitude on climate mirror those of increasing distance from the equator in moving from an essentially diurnally influenced, to a more seasonally influenced climate as altitude increases. At intermediate latitudes temperature decreases with altitude and this results in a shorter growing season. Close to the equator this seasonal effect is lost but the diurnal climate remains, while overall temperatures are reduced throughout the year as altitude increases.

3.2.2 CHARACTERISTICS OF WORLD VEGETATION ZONES AND THEIR IMPLICATIONS FOR BIOENGINEERING POTENTIAL

Anecdotal evidence seems to suggest that experience of the application of bioengineering techniques tends to be concentrated in temperate (e.g. Europe, the USA and New Zealand) and subtropical (e.g. Hong Kong and Malaysia) climates and the adjoining climatic zones. An understanding of the nature of the other climatic zones suggests that this is perhaps to be expected, as is discussed below.

The low productivity and sparse vegetation growth characteristic of certain natural zones, as a result of climatic and other constraints, will clearly limit the scope for using vegetation as a stabilizing tool. In some cases it will be possible to compensate artificially for natural shortcomings. However, there are obvious limitations in compensating for the low temperatures and short day lengths experienced in the Arctic and much of the boreal regions, and to some extent for the shortages of water in desert climates.

In assessing the potential for the application of bioengineering techniques within the various climatic zones it is important to consider the characteristics and scope for natural regeneration of the vegetation types concerned, as this

will decisively influence the likelihood of their successful use.

Tundra and boreal zones

Soil depths, and therefore the scope for plant growth, in tundra climates are determined by the extent to which thawing occurs during the summer months. In southern Siberia the growing season may be as long as four months. This gives way, closer to the poles, to what is described as cold desert, which supports little vegetation. The predominant vegetation form in the less extreme conditions is dwarf willow and birch scrub, while large areas are also subject to solifluction and support little vegetation.

The most favourable habitats tend to be the banks of rivers and south-facing slopes, which warm up most rapidly in the low summer sun, while on lower lying land the frozen ground means that even the low levels of rainfall give rise to extensive swamps. Even in the most favourable habitats the ability of the vegetation to regenerate after damage tends to be limited by the slow growth rates allowed by the limitations imposed by both climate and soil conditions.

Clearly the establishment of protective/ stabilizing vegetation under these conditions is not easy. However, on the whole the consistent nature of the climatic conditions, the long periods during which the ground remains frozen and the consequent low prevalence of natural hazards mean that the spread of soil erosion is likely to be limited.

Boreal climates are typified by extensive evergreen coniferous forests growing on podzolized soils. In North America and Eastern Asia there are a relatively large number of species represented, while in the European forests spruce and pine predominate. Spruce tends to be concentrated on wetter soils and is thus relatively shallow rooting, while pine is more common on drier ground and tends to root more deeply. The degree of layering within the vegetation, within the forest types, depends both on the local density of the canopy and the fertility of the soil.

Prairies and steppes

The low rainfall and short growing seasons of these relatively continental zones provide conditions insufficient to support tree cover and consequently extensive grasslands form the natural vegetation. This vegetation type is to be found across large parts of central North America (prairie), Eastern Europe and Central Asia (steppes) as well as limited areas of South America (pampas) and New Zealand. It contains a varying proportion of herbaceous species as well as a wide range of grasses.

The soils of this zone are typified by the deep, fertile black earth or chernozem soils. This is densely penetrated by the extensive root growth of the grasses, while other species tend to root more deeply. Generally the proportion of the biomass below the ground is in excess of that above. This suggests that the potential of the natural vegetation for soil reinforcement is relatively high, although the relatively low levels of above-ground biomass mean that the scope for intercepting rainfall and protecting the ground surface from erosion is more limited.

Temperate deciduous forest zones

Deciduous forest, with a potentially well-structured vegetation, characterizes this zone of cool summers with a shorter, marked cold season during which vegetation growth ceases. It is only to be found in the northern hemisphere, except for parts of New Zealand and mountainous regions of the Southern Andes. Precipitation is spread relatively evenly throughout the year and the loss of the leaves in the winter is a response to the cold season.

The cold season during which vegetation is dormant is the main constraint on the ability of vegetation to maintain its stabilizing role throughout the year. The severity of the winter period increases with increasing continentality and altitude. However, this tends to be asso-

ciated with decreasing rainfall levels and wind speeds, and therefore the potentially damaging environmental effects with regard to slope stabilization and erosion control tend to decline in line with the ability of the vegetation to counter them.

Nevertheless, the favourable growth rates achieved during the growing season in this zone provide an opportunity for the exploitation of vegetation as a means of protecting slopes and guarding against soil erosion. The vast majority of the detailed experience in the establishment and management of vegetation for slope protection has come mostly from tackling such problems within this climatic zone. These details are discussed in the sections below.

Mediterranean-type vegetation

Mediterranean-type climates are to be found on several continents, but those of the Mediterranean region in Europe, after which this vegetation zone is named, are the most extensive, extending from the Atlantic Ocean to Afghanistan. However, they are also present in more restricted areas of California, Australia, South Africa and western South America.

The vegetation is characterized by sclerophyllous woody plants (which have small rigid leaves) such as thorny shrubs, evergreen oaks and olive trees. These are seen as being the result of adaptation to the prevailing conditions of moist winters and summer drought. In contrast to the zone of deciduous forests, there is a decrease in the markedness of seasonal periodicity, making the active contribution of vegetation to soil and slope protection more consistent throughout the year.

In Mediterranean-type areas the potential for the use of plant material for soil stabilization is therefore relatively good if the minor constraints to plant growth resulting from seasonal water shortage can be overcome. The dry summer period does have another effect, however, in that it increases the risk of fire, which is regarded as a natural component of the Mediterranean-type ecosystem. This can pose a potential threat to the engineering application of vegetation as although it will tend to occur during the dry summer months, its effects are longer lasting and persist for several years before full regeneration of the vegetation cover takes place.

Desert vegetation

Several different types of desert can be recognized on the basis of the amount and seasonal distribution of rainfall. In all cases, however, the vegetation structure is similar in that the plant cover is consistently sparse. This is the straightforward result of the low rainfall in desert areas, which means that the numbers of plants the soil is able to support are very limited. Each plant effectively requires a considerable unvegetated 'catchment area' in order to collect enough rainwater to survive, although the unvegetated areas are usually penetrated by extensive root systems which may have some beneficial effects in strengthening soils.

The 'xerophytic' nature of the foliage of plants adapted to desert conditions will mean that their effect in protecting the ground surface from rainsplash erosion during the short rainfall events will be very limited. Similarly, the sparsity of top growth means that most natural vegetation does little to reduce windspeeds and protect soils against wind erosion. The application of vegetation for soil surface protection is therefore clearly limited if there is to be reliance on natural sources of water.

Where protection against erosion is to be sought through the protective action of vegetation, irrigation can change the situation for using plant material dramatically. In desert environments water is usually the only limiting factor as warmth, light and nutrients are unlikely to be a significant constraint to plant growth. But it must be remembered that any interruption of irrigation, even for relatively short periods, will probably result in the loss of the newly established vegetation and a subsequent reversion to a pattern of plant growth more characteristic of the region. This assumes that there

have been no negative side-effects of irrigation, e.g. the build-up of soil salinity. If this has taken place there may be a permanent change in the vegetation and the potential for the use of vegetation for erosion control may be lost altogether.

Subtropical vegetation and savannah

The climate of the subtropical zone is free from frosts and thus there is no period of winter dormancy affecting plant growth. However, periodicity in the vegetation persists in the form of an uneven distribution of rainfall, with marked wet and dry seasons. This takes the form of summer monsoons in India and southeast Asia. During the dry season many species lose their foliage in response to the shortage of water.

The semi-evergreen nature of the vegetation gives it the ability to respond relatively rapidly to fluctuations in rainfall. The seasonal nature of the rainfall pattern means that when it rains, the risk of erosion and the threat to slope stability are relatively great. However, the high temperatures and rainfall make the use of vegetation for soil protection a viable proposition.

In the drier areas a grassland savannah vegetation develops, usually supporting scattered woody plants. Soils of the tropical savannah zone are poor in nutrients, but the reason suggested for their inability to support forest vegetation is the presence of impervious layers within the soil, in the form of iron pans, which hinder free drainage and thereby interfere with the water balance of the soil.

Secondary savannah can also be found in the wetter tropics too, but this is the result of the destruction of the original vegetation by the local inhabitants, and is not a natural climatic related form.

Tropical forest zone

Although rainfall is higher than in any other climatic zone, it is very evenly spread throughout the year. Temperatures are high and consistent throughout the year, allowing the development of rich and luxuriant vegetation. This is characterized by a great variety of species and a complex, layered structure. The richness of the vegetation is, however, not mirrored by the fertility of the soil, most of the nutrients in the ecosystem being tied up in the considerable biomass.

Soils of equatorial regions are generally very old and may be weathered down to many metres on appropriate types of parent material. This means that if they do become exposed the potential for erosion is considerable. Furthermore both fertility and pH levels tend to be very low as nutrients and most basic ions have long been washed out of the soil. Thus the risks to soil and slope stability associated with loss of the vegetation cover are significant, as indeed are the difficulties of re-establishing vegetation if it is lost, unless considerable efforts to ameliorate the soil are undertaken.

The extremely favourable climatic conditions make vegetation growth very rapid and thus the potential for establishing a protective plant cover is very good if the nutrient deficiency problems can be overcome, although the recreation of anything like the diversity of the indigenous rain forest is notoriously difficult. Root development in tropical soils is not well researched. However, it is reasonable to assume that, despite the depth of weathering, plants do not root deeply because of low nutrient availability and the effects of rapid mineralization and mycorrhizae close to the soil surface which make this unnecessary.

3.2.3 PRESENT LAND USES AND BIOENGINEERING POTENTIAL IN THE NATURAL CLIMATIC AND VEGETATION ZONES

Although both the broader distribution and the local patterns of natural vegetation types as outlined above can be seen as sensitively reflecting the outcome of the interplay between natural climatic and physical conditions, human colonization of the landscape has, over most of the regions of the earth, brought about con-

siderable changes in the type and pattern of vegetation cover. However, while the existing vegetation will to a greater or lesser extent differ from the potential natural vegetation in most areas, the climatic and physical conditions which gave rise to the original indigenous vegetation will still continue to operate on whatever managed vegetation type has replaced it.

Given the objective of establishing forms of vegetation for slope stabilization and erosion protection which will require minimum management, the relationship between the natural vegetation and the vegetation to be established for slope protection is an important one.

Over the centuries clearance of forest for agriculture has brought about the greatest changes in the patterns of vegetation described above. Cultural landscapes based around grazing and crop cultivation have replaced natural vegetation. More recently, urbanization, the construction of transport infrastructure, mining and water resources management have played a growing part in this change.

As a result natural vegetation, where it has survived at all, has been restricted to increasingly small and isolated patches. Its ability spontaneously to re-colonize areas which are no longer being managed for agriculture or which have been disturbed by construction projects has, as a consequence, been increasingly limited. Thus, while some form of natural reversion to a type of vegetation structure similar to that which formerly existed can be expected, the time-scales involved may be long and the detailed composition of the resulting vegetation may never fully come to resemble that which was once there. This means that the use of vegetation for stabilization purposes, while ideally attempting to mimic natural structures, cannot simply rely upon natural processes for its establishment.

In any natural vegetation zone where the plant productivity is sufficient for it to be exploited for agriculture or forestry, the use of plant material for slope stabilization and erosion control can be considered a possibility. How vegetation is used, the selection of species, the ease and mode of its establishment and its management requirements will be conditioned by the zone in which a particular project lies.

Outside the developed world and the largely temperate climatic and vegetation zones, there is relatively little published experience of species selection. The willows and poplars which form the 'classic' bioengineering species of Europe and North America are absent from other vegetation zones, and substitutes must be found. Willows and poplars are pioneer species characteristic of relatively disturbed habitats and able to colonize these relatively rapidly. In searching for species to replace them in other climatic zones, the most fruitful approach is likely to be to look for equivalent species with similar growth strategies in these climatic zones. One might expect to find these in habitats such as river flood plains or in other naturally disturbed habitats, where pressures favour species able to exploit vacant niches rapidly and then overcome the stress and disturbance to which such niches are subjected.

3.3 FACTORS AFFECTING PLANT SELECTION AND VEGETATION GROWTH

3.3.1 CHOICE OF APPROACH

Whatever the climatic zone, a combination of factors affects the choice of approach to the establishment and management of vegetation. Phyto-sociological (ecological) and environmental factors and constraints have to be reconciled with biotechnical (functional) requirements. Before selecting vegetation, a basic choice has to be made between two approaches:

1. modifying the site or environmental conditions to suit the desired vegetation. This is most appropriate when the situation requires a specific type of vegetation, or when money is no object.
2. selecting appropriate species to suit the prevailing site and environmental conditions.

This choice depends on the nature of any constraints due to site conditions and the extent to which there is scope for modifying them, and on the flexibility of the desired vegetation and the functions it will be required to perform. Some site conditions can be modified fairly readily, such as soil fertility, whilst others are very difficult if not impossible to modify, such as climate. The main rule is that, in general terms, the less you modify the site conditions to suit the vegetation, the less management you will require, and the more you can utilize natural processes of vegetation development.

A balance therefore has to be struck between these approaches and in practice a combination is usually adopted. Other considerations may also be important, such as amenity, landscape and wildlife value, or the resources available for long-term management of the vegetation. The resources required and available for each step in the process: design–establishment–aftercare–management, have to be understood and allowed for if the vegetation is to fulfil the required function in the long term.

3.3.2 PHYTO-SOCIOLOGICAL CONSIDERATIONS

Vegetation is more than a collection of individuals; it is a complex, dynamic plant community with many interactions between individuals, species and the surrounding environment. The study of such plant communities is a branch of ecology known as phyto-sociology, and whilst this is a very complex subject, some basic principles need to be understood before a functioning vegetation can be designed.

The environmental preferences and behaviour of individual plants or species are often well studied, but their behaviour in dynamic communities is often less easy to predict. In practice, experimental work is often required in order to investigate how a particular set of species will develop and persist in a given situation.

Succession

The first principle to establish is that of **succession**: a sequence of developing plant commu-

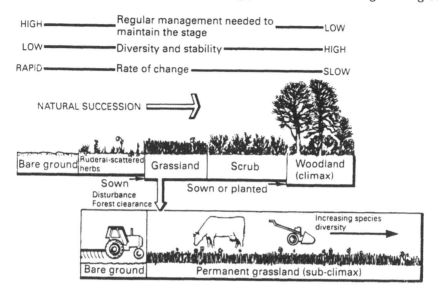

Figure 3.1 Natural succession in plant communities of temperate European climates (after Coppin and Richards, 1990).

nities from the first colonizers of bare ground, through a series of stages, until a stable natural vegetation or **climax** is reached. The direction and rate of succession depends mainly on environmental factors, particularly climate, but is also influenced greatly by the availability of plant propagules (e.g. seeds). Natural succession therefore has a large element of chance involved, though most vegetation is affected by human activity to some extent.

A typical successional sequence is illustrated in Figure 3.1. Initially, quick-establishing, mobile species colonize bare ground. This **ruderal** community then develops into a herbaceous community, usually dominated by grasses. As earlier species modify the local environment, and as the community becomes more competitive, progressive colonization by larger species, such as shrubs and trees, leads eventually to scrub, woodland and ultimately the **climatic climax**, usually high forest communities.

In tropical environments the early stages of succession are dominated more by woody species, with shrubs and trees being the early colonizers. The progressive development to high forest will then involve a large number of

forest stages, as the canopy and the complex age structure develops.

As a succession progresses, the rate of change decreases, and the diversity and stability of the communities increase greatly. There are also many outside influences, notably human, which will artificially alter a succession or maintain a particular stage by introducing a management regime. Many grassland systems are maintained for stock grazing. Similarly, natural grassland systems, such as steppe and prairie, are maintained by a combination of climate and wild grazing herds.

One of the main factors influencing the rate and direction of natural succession is the input, or 'rain', of seed from other areas. For most of the areas where human influences are major there is a decrease in the seed rain of natural species (reflecting the great reductions in natural habitat) and an increase in that of introduced species. In many places the potential natural vegetation will not be the same as what was once completely natural.

Nevertheless, for almost any area of the world it is possible to identify the natural climatic climax vegetation, or potential natural

Table 3.2 Some characteristics of competitive, stress-tolerant and ruderal plants

Characteristics	Competitors	Stress-tolerators	Ruderals
Morphology			
Life forms	Herbs, shrubs and trees	Lichens, herbs, shrubs and trees	Herbs
Morphology of shoots	High dense canopy of leaves; extensive lateral spread above and below ground	Very wide range of growth forms	Small stature, limited lateral spread
Life history			
Longevity of plant	Long or relatively short	Long – very short	Very short
Longevity of leaves and roots	Relatively short, continual new growth	Long	Short
Leaf phenology[a]	Well-defined peaks of leaf production	Evergreens, with various patterns of leaf production	Short phase of leaf production
Perennation[b]	Dormant buds and seeds	Stress-tolerant leaves and roots	Dormant seeds only
Regenerative strategy[c]	V, S, W, B_s	V, B_j	S, W, B_s
Physiology			
Potential growth rate[d]	Rapid	Slow	Rapid
Response to stress	Rapid morphogenic responses (root:shoot ratio, leaf area, root surface area) maximizing growth	Morphogenic responses slow and small in magnitude	Rapid curtailment of vegetative growth, diversion of resources into flowering

[a] Leaf phenology will also apply to roots; the descriptions apply to seasonal climates.
[b] Perennation applies only to seasonal climates where plants exhibit dormancy.
[c] Regenerative strategies: V = vegetative expansion; S = seasonal regeneration in vegetation gaps; W = numerous small wind dispersed seeds or spores; B_s = persistent seed bank; B_j = persistent seedlings or juveniles.
[d] Maximum potential relative growth rate

vegetation, allowing for human influence on the regional vegetation type. Simple observation of the local vegetation can help the bioengineer greatly, by indicating how the natural vegetation would develop and progress under the local site conditions. Selecting the most appropriate vegetation and management regime can then be easier; it is always preferable to follow nature's preferred route as much as possible.

Plant strategies

The role and success of an individual species within a community will depend on its strategy for establishment and growth. Based on three basic strategies for dealing with varying inten-

sities of environmental stress (brought about by the availability of light, water, nutrients, temperature, etc.) and disturbance (arising from the activities of humans, herbivores, pathogens, damage, erosion and fire), three types of plants may be recognized:

1. competitors, which exploit conditions of low stress and low disturbance, but where many species are competing for the available resources;
2. stress-tolerators, which exploit conditions of high stress but low disturbance;
3. ruderals, which tolerate disturbance but not high stress.

Some of the typical characteristics of these

types are described in Table 3.2. These characteristics will help the bioengineer select plants with the right strategy.

It can be seen that there is no strategy for dealing with both high stress and high disturbance, which are the conditions on many types of degraded land and in hostile environments, where vegetation establishment is most difficult. In these conditions it is necessary artificially to remove one of the constraints, i.e. relieving stress or reducing disturbance, before plants can be successfully established.

As well as the strategy that plants adopt during their established phase, there are a number of strategies for regenerating and invading new areas. Almost all plants reproduce by seeds, but the extent to which they rely on this mechanism for exploiting new areas varies considerably. There are a range of vegetative means of spreading, such as suckers, rhizomes and stolons, though seeds are usually the means of dispersing long distances. Regenerative strategies are important in bioengineering because they determine the extent to which the vegetation cover will repair itself after damage, and the species that are most likely to invade or take over if disturbance is a regular occurrence.

A practical implication of plant strategies and a plant's natural niche within a community is that mixtures of species have to be designed to take account of plant community dynamics and the characteristics of the site. Most bare sites will start off as ruderal habitats and the direction in which vegetation will develop subsequently depends on the relative stress due to climate and soil. Sites where stress is not very great, such as lowland areas where there is a good soil-forming material, will progress from ruderal to competitive communities. In contrast, where stress is greater, due to poor soil and/or harsher climate, the vegetation will progress towards a community of stress-tolerant species.

Slopes requiring erosion control and stabilization will probably fall within the latter category, i.e. with high stress. It is therefore important to select plants that will tolerate these conditions, unless a high level of management input or intervention is available. The type of management must also be compatible with the natural direction that the succession is likely to take.

Wildlife value

In many situations the 'naturalness' of the vegetation will be important, not just because natural wild species are likely to survive better, but because of wildlife considerations. Exotic species are often introduced into an area for good reasons, but they can give problems and sometimes get out of hand and disrupt the natural balance of species.

The wildlife value of vegetation is related to the diversity of species present and to the naturalness of the plant community, or how closely it represents a wild community. In some areas therefore it may be appropriate for these considerations to be included in the selection criteria for vegetation establishment.

3.3.3 ENVIRONMENTAL CRITERIA

There are few environments to which plants of one species or another have not become adapted, though as a general rule the more extreme the environment, the fewer the species that are available. Also, in harsh environments, growth rates and overall productivity are very low. Species also vary widely in the range of environments in which they will grow or thrive. The most widely used species for erosion control and slope stabilization are those with a wide 'ecological amplitude', that thrive in a wide range of situations and environments. This is because:

1. there is greater confidence that the plants will establish without the need extensively to research the site conditions;
2. longer-term survival of the plants should not be compromised by wide variations in site conditions, especially as the plant community may have little stability from diversity;

3. propagation and establishment techniques are well established for a relatively narrow range of commercially available species.

This approach tends to result in standardized species mixtures and, in most situations, little effort goes into properly developing an adaptive approach. Whilst the vegetation used might be very robust, the long-term management, development (succession) and even survival of the resulting plant communities can become a problem. In most situations a lot more imagination could be used to develop more appropriate vegetation.

The criteria that it is important to establish as a basis for selecting vegetation can be divided into three groups: soil physical, soil chemical and bioclimate. These are discussed briefly here, but reference should be made to an ecological text for further information.

Soil physical factors

Soil texture and density determine the nature of the rooting medium. The development of natural soils serves as a basis for understanding how these factors apply, but few bioengineering soils are natural. Soil physical factors are especially important in engineered soils and on formed slopes, because the handling of the soils and construction of the soil profile has a fundamental effect on the soil's properties.

The handling and placement of soils within about 1 m of the final ground surface will be crucial to the ability of the soil profile to be exploited by plant roots and to its ability to provide the right water regime. The nature of this depth is also crucial to the stability of a slope, and conflicts between geotechnical and ecological requirements can arise. Geotechnical requirements usually override other considerations of course, but it should be remembered that in the long term the stability of a slope depends on having a healthy vegetation cover on it. The enhanced stability achieved by having a vigorous vegetation cover will usually more than compensate for any loss of stability arising from the less than ideal geotechnical properties.

Perhaps the most useful parameters to control in the soil profile are the **profile available water** and the **packing density**. The packing density is a more reliable indicator of the effects of compaction than bulk density alone, by allowing for the influence of clay content.

$$\text{Packing density } (L_d) = \rho_b + (0.009 \times \%\,\text{clay})$$
$$(3.1)$$

where ρ_b is the dry bulk density (Mg/m^3).

The profile available water is the sum of the available water for each horizon in the soil profile, down to 1 m depth or to an impermeable horizon, whichever is shallower. The **available water capacity** for each horizon can be calculated or measured.

Using these parameters it should be possible to design a soil profile with the appropriate physical characteristics, or at least to modify the way it is constructed to provide the optimum rooting depth and geotechnical properties with the soil materials that are available. With an existing soil, it will be possible to predict how it will behave and the intensity of any stress on plants that might grow in it.

Similarly, it is possible to design the soil surface environment, which must be seen as critical to the establishment of vegetation. The usual method of engineering slopes, where the surface is compacted and smoothed, provides a very hostile environment for plants. The surface permeability can be very low, which not only exacerbates surface erosion but reduces infiltration and subsequent storage of water in the soil profile. Water storage in the profile is important for plants during dry periods; good water storage will reduce drought stress. Compacted layers at the surface or within the soil profile will also prevent effective plant rooting.

Soil chemical factors

The chemical factors that affect plant growth are perhaps the best understood, and there are many textbooks and research papers covering this topic. The two main factors involved are:

1. soil fertility, the presence of major nutrients (nitrogen, phosphorus and potassium) and minor nutrients (e.g. calcium, magnesium, sulphur and trace metals);
2. soil pH, the acidity or alkalinity which affects the chemical environment and the availability of nutrients or undesirable elements.

Soil fertility can be readily manipulated using lime and fertilizers, and regular application of these materials is common practice. However, in bioengineering situations, regular or intensive inputs of fertilizers over a long term will not be possible or appropriate, so it is important to build up soil fertility quickly and in a way that enables natural cycling of nutrients to provide the vegetation's requirements. The form in which nutrients are applied is therefore important, with an appropriate balance between soluble and longer-term release forms. Organic fertilizers or manures are usually the best materials.

As well as the total amount of nutrient in a soil, an important factor is the way that it is stored and released. This is determined not only by its chemical form (inorganic or organic) but by the properties of the soil. Clay minerals adsorb mineral ions on to their surface or into their structure, and the presence and nature of clay minerals, together with organic matter, affects the soil's buffering capacity, or **ion exchange capacity**. The soil texture is therefore important in the soil's fertility, as well as in its water relations. The combined effects of exchange capacity and permeability of a soil will determine the likelihood that leaching during rainfall will remove nutrients and progressively deplete the soil's fertility.

Of all the mineral nutrients, nitrogen is the most important in most situations. In most ecosystems nitrogen is the main limiting factor for plant productivity, and will exert a powerful influence on the structure and development of the ecosystem. Whilst nitrogen deficiencies in a soil can be readily corrected using fertilizers, for long-term stability reliance should be placed on building up a viable soil-nitrogen cycle, with

sufficient nitrogen capital (usually based on organic matter) and release through microbial mineralization, to support the desired plant community. A typical nitrogen cycle is illustrated as part of the nutrient cycle in Figure 3.2.

Some species, notably those of the legume family, have root nodules containing bacteria which can fix nitrogen from the air, so that it is available within the ecosystem. The relationship between the plant and the bacteria is a 'symbiotic' one, and usually dependent on specific environmental conditions. However, legumes form a very useful group of plants for use on infertile soils, and can be a vital component in the long-term build-up of soil nitrogen and organic matter. Care needs to be taken with their establishment, in particular the need to inoculate the legume seeds with the correct strain of bacteria for effective nitrogen fixation. On infertile soils the legumes may not become inoculated naturally.

Bioclimate

Bioclimate is the combination of climatic factors that determines plant growth and the survival of different species to form a particular plant community. The major climatic distinctions form the principal vegetation zones discussed in section 3.2.1 earlier. However, within these zones there are many variations and extremes determined by local climate. These variations can be discerned at two levels:

1. a regional level, usually determined by altitude and distance from the sea (oceanicity);
2. a local level, determined by topography, such as exposure, moisture regime and aspect (collectively referred to as microclimate).

Generally, bioclimate cannot be manipulated or ameliorated, so the only approach available is an adaptive one: choosing the right combination of species to thrive in the conditions. However, often the microclimate can be manipulated over the short term, improving conditions for, say, vegetation establishment. After this, the developing vegetation will itself modify

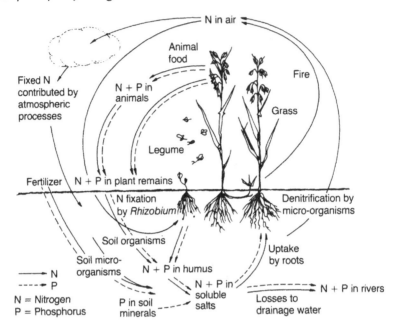

Figure 3.2 Typical nutrient cycle for N and P (after Bradshaw and Chadwick, 1980).

the local microclimate to some extent, which may influence the succession of plant communities. This factor can often be used to advantage.

The principal determinants of bioclimate operate over a yearly cycle. Their quantities/averages alone are only part of the story; their distribution over the year is equally, if not more, important. The determinants are rainfall, temperature and evaporation, which between them define the length and intensity of the plant growing season. In tropical and subtropical climates the growing season is defined by the availability of water, i.e. the distribution of rainfall. In temperate climates, the growing season is determined mainly by temperature, with summer drought being a secondary factor.

Altitude exerts an important influence on bioclimate, and conditions in mountainous or alpine environments, where bioengineering is often a valuable approach, can be difficult. The availability of water generally increases, but the growing season is reduced by decreasing temperatures. Difficulties are often exacerbated by steep slopes, thin soils and intensive rainfall.

3.3.4 BIOTECHNICAL CRITERIA

Chapter 2 of this book describes the engineering properties of vegetation, and the specific requirements for different erosion control and stabilization functions are discussed in the following chapters. Selection of the appropriate vegetation, especially the combination of species, to deliver the desired properties is therefore very important.

The biotechnical properties can be said to be determined by three aspects of plant growth:

1. the strength and architecture of the root system;
2. the nature of the top growth, i.e. shoots, leaves;
3. the annual pattern of growth and overall growth rates.

Root systems vary considerably between species, mainly in the depth and distribution of roots. Species suitable for surface erosion control would have shallow-rooting systems, with a dense surface mat of fine roots. Many grasses and herbs have root systems like this. On the other hand, stabilization of soils requires deeper-rooting plants that will anchor and bind soil layers together, and will provide a network of soil reinforcement fibres. Larger herbs, together with shrubs and small trees, will usually be the preferred choice for this role.

It should be remembered that the depth and density of the soil profile will exert a large effect on the pattern of rooting of all vegetation. It is no good selecting deep-rooted species for soil stabilization if the soil profile is not suitable for deep rooting.

Tensile strengths of individual roots have been measured, but there is a wide variation even within a species, depending on age, health and time of year. A more important factor than the strength of individual roots is the strength of the overall root-reinforced mass of soil. This will depend more on other factors such as the density and orientation of roots, and the distribution of roots within the soil profile.

For effective erosion control, the top growth of the plant community should be distributed evenly over the ground surface. Clumped or tussocky growth is undesirable (section 2.3.1).

When matching the desired biotechnical properties of species with a given situation, the annual growth pattern of the species should be taken into account. Plant growth is seasonal, and most species have a dormant period outside the growing season. Root growth in particular can be very seasonal, and a large part of the root system will usually die back during the dormant season. When selecting species for a specific biotechnical role, it is important to ensure that the plant will be able to provide the function required of it during the season in which it is most needed.

So-called **bioengineering** approaches to slope stabilization mainly involve the establishment of vegetation using mature, but usually unrooted, parts of live woody plants. As well as simply acting as unconventional propagules from which new plants are able to develop, the cuttings, branches, poles, wands, etc., also function in a number of other ways, mainly through

exploiting the mechanical and structural properties of the woody material.

A number of recognized techniques, some of which are outlined in section 3.4.7, have been developed, making use of woody material to perform such roles as reinforcing soil structure, retaining surface material and improving soil drainage. Clearly the use of dead brushwood or inert materials can have much the same superficial effect. However, the use of living woody material, capable of striking root and developing into new plants, has a number of advantages.

Dead brushwood will operate only over a relatively short time-scale, depending on the level of soil microbial activity. In the wet tropical and semi-tropical areas any stabilizing effects due to mechanical action are unlikely to last beyond a matter of weeks or, at most, months, because the breakdown of organic matter in the ground is very rapid. In temperate zones the stabilizing effects can be expected to last for a few seasons, while in boreal and Arctic climates microbial activity is so slow that the effects may be expected to last for a considerable period. In contrast to dead material, living plants are in various ways adapted to resist microbial attack.

Inert materials such as metal reinforcing rods, meshes and synthetic geotextiles are not subject to organic breakdown and can therefore be expected to persist in the soil and perform the mechanical stabilizing functions more or less indefinitely. This in itself, though, may be viewed as a disadvantage from an environmental, rather than an engineering viewpoint. The use of live plant material can similarly have the advantage of long lifespan, as a result of its capacity for self-renewal rather than any inherent longevity, but potentially it has other advantages over and above inert material.

First, by virtue of its ability to grow, the amount of stabilizing material in the ground will increase over time, as will the volume of the soil mass penetrated by the expanding root system. In woody plants these effects are also persistent and the vegetation retains much of its function outside the growing season. This would itself be an advantage over the use of inert materials, quite apart from any cost benefits. However, plant roots are not merely acting passively to provide mechanical strength. Transpiration also means that roots are operating actively to remove water from the soil and thereby further increase its strength, although this is only the case when active growth is taking place.

It is this unique combination of functions – immediate mechanical reinforcement, and removal of soil moisture together with the other protective and strengthening roles of vegetation – which the bioengineering use of woody plant material exploits. Live plant material is, in addition, often used in combination with timber, metal and/or stone in order to provide a combination of immediate protection with long-term stability. Because the material to be used is live, for a project to be successful it is important that it is harvested and used during the dormant period, where one exists. In temperate climates this will be during the winter months, while in sub-tropical areas for example, dormancy may occur during the dry season.

3.3.5 PLANT SELECTION

The selection of appropriate species and, more importantly, a complementary mixture of species, requires a careful balance of considerations. Inevitably it will not be possible to have the ideal mixture for every situation and combination of circumstances. A strategy has to be adopted that combines short-term and long-term functional requirements, site limitations and management constraints. Two strategies are possible:

1. seek to establish the ideal long-term vegetation as quickly as possible, introducing the desired 'sub-climax' or 'climax' species, with perhaps a proportion of **pioneer** species as a 'nurse'. This strategy will require a lot of management input both to achieve it in the shortest time-scale and possibly to maintain it against successional changes. There may

also be an initial shortfall in the required biotechnical properties whilst the vegetation develops to maturity;
2. establish a pioneer community which has the required biotechnical properties and which will develop into a suitable climax or subclimax by natural succession. Less management should be required, sufficient only to ensure succession in the desired direction. It may be appropriate to introduce further species at a later time in order to encourage the required succession.

Whichever strategy is adopted, it is vital to understand how the plant communities will behave in the long term. Vegetation is never static; it is continually developing, both throughout the year and progressively from year to year.

The best way of defining what is likely to be the natural vegetation and succession is to examine what is growing locally, and its relation to different soil types and bioclimate regimes. If this natural vegetation is unlikely to fulfil the biotechnical or ecological requirements, then it will be necessary to develop a strategy based on different vegetation. In these circumstances, a certain amount of experimental or trial work will be required in order to investigate how the vegetation will behave and perform in the short and long term. This will often be the only way of determining the precise management requirements.

As described above, legumes can be an important component of any vegetation mixture. There is a wide range of herbaceous forage legumes, but these are usually intolerant of hostile soils. They can be useful in the short term, to give an initial boost to soil fertility, but care should be taken as they can smother other slower-growing species. As they die out there may not be other species to take their place. However, there is a wide range of wild-type legume species, comprising herbaceous, shrub and tree forms. These are usually preferable to forage legumes, but seed is often not so readily available in commercial quantities.

Many bioengineering techniques involve the use of live woody material, that subsequently roots in the soil and grows into a full plant. Willows and poplars are widely used for this and are common in temperate climates. The ability to propagate readily in ways like this is a valuable attribute, because it means that local plant material can be used, without the need to have a separate propagation stage. However, in remote areas, local nurseries for collection and propagation of local plant material are often the best approach.

3.4 ESTABLISHMENT

3.4.1 SLOPE CHARACTERISTICS AND ESTABLISHMENT PROBLEMS

The process of plant establishment involves speeding up or by-passing the natural process of invasion by pioneering plant species. The speed of natural colonization depends in part on the hostility, or degree of stress, of the ground surface. An important part of plant establishment therefore involves first understanding and then overcoming the constraints to rapid establishment.

In many cases there will be few constraints and plants will establish quickly without help. However, most formed slopes will exhibit some kind of constraint, usually as a result of the engineering or natural process that formed them. In this respect there are differences between cut (cuttings, cliffs), fill (embankments) and natural (hillside) slopes. The important considerations are stability, erosion, access, aspect and soils. These are discussed further below.

Stability

Formed slopes will usually have been designed to be stable in the engineering sense, that is safe against deep-seated failures and slips. However, some movement can occur due to localized wet patches, and overconsolidated clays can experience problems after a few years. Instability on slopes is therefore due primarily to surface pro-

cesses, such as erosion by rainfall, runoff and gravity, after weathering has loosened the surface. It is this instability that vegetation can overcome, but which also gives rise to establishment problems.

Fill slopes and embankments consist of compacted soils in an artificial mixture or layering. This layering will have a large effect on the properties of the soils and the surface conditions, and it will take many years for the soil structure and natural drainage to develop. Engineering stability usually depends on the compaction of the soil and exclusion of water, so growing conditions can often be hostile.

Cut slopes are formed in the natural undisturbed soil, albeit relatively unweathered and deeper soil strata. Stability and compaction can therefore be less of a problem, except as a result of the natural ground characteristics. Disturbance due to engineering operations is confined to the surface, which can be compacted.

Cut slopes in rocky uncohesive material are usually steeper, being formed of more competent material. However, surface erosion due to spalling of weathered material can be a major problem until vegetation becomes established.

Erosion

The erodibility of a cut or fill slope will depend on the nature of the soil material and the infiltration capacity of the surface. Engineered soils tend to have limited infiltration, due to compression damage, unless some form of amelioration by cultivation is carried out. In general, as the slope increases in steepness above about 33°, the reduction in plan area in relation to actual surface area means that rainfall intensity effectively decreases, reducing erosion. However, at the same time the erosion due to gravity will increase.

Access

The ability to establish vegetation on a slope depends on access to the slope surface itself. Normal agricultural machinery for cultivation and ground preparation can work on a slope up to about 30%; specialized machinery can work up to about 50% (1 in 2). Steeper than this requires hand work or remote access from a flatter area. Where access for machinery is critical for site work then the slope can include benches on a regular spacing, depending on the reach of the machinery.

Aspect

Aspect in relation to the sun or prevailing winds can make a significant difference to the microclimate on a slope. This must be taken into account when selecting vegetation, and may influence the community dynamics and succession. An important factor can be fire; slopes facing the sun can dry out much more and be more prone to burning.

Soils

As referred to above, the soil conditions on cut and fill slopes will vary considerably, depending not only on the nature of the materials but on the way that they are handled. Whilst the potential for damaging soils on fill slopes is greater, there is also greater scope for amelioration by careful handling and placement. The soil depth within about 0.5–1 m of the final ground surface should receive special consideration.

Rooting conditions on cut slopes can be limited, unless soil-forming materials are spread on the surface, or the soil is sufficiently friable itself. If soil is spread, it must be keyed into the underlying surface, otherwise a slip surface will form and the soil-forming layer will be unstable.

3.4.2   SITE PREPARATION AND AMELIORATION

The first step in establishment is to prepare the ground surface. This is as important in planting slopes as it is in agricultural situations, though the scope and techniques available are much more limited. For most slopes, the best time to undertake surface preparation is when they are

being formed. This can avoid many of the problems that are inherent in the engineering process, such as compaction. There are two aspects to consider: the nature of the soil surface (its texture and microclimate) and amelioration with fertilizers and other materials.

Soil amelioration

Amelioration of a soil to improve its ability to support vegetation will involve the use of chemical fertilizers and/or bulk materials. The precise requirements will depend on the nature of the soil and the deficiencies that are identified. Fertilizers can be used to correct nutrient deficiencies, and lime to adjust the pH, but the effects of both of these materials can be short-term, unless repeated regularly until a reserve of nutrients has built up.

Bulk materials, either organic matter or mineral soil-forming material, can be used to improve both the fertility and physical properties of the soil. Organic matter, especially manures, will usually contain nutrients in an ideal form for slow release to plants. Organic matter will also enhance the structure of the soil, improving its drainage, aeration, water-holding capacity and nutrient retention properties. Mineral materials can help produce a better balance of soil texture; for example, adding fine silts to a coarse porous soil to improve its water retention.

Amelioration with bulk materials depends to a great extent on what materials are readily available in sufficient quantities and at the right price. Organic waste products can be ideal, provided they are properly investigated and conditioned before use, if necessary. For example, composting of raw wastes can improve their properties. The soil type and potential ameliorants should be carefully matched to make sure that they are complementary, and whether there are any deficiencies still to make up, such as a particular nutrient element that can be added as normal fertilizer.

Soil ameliorants should be well incorporated into the ground surface if possible, by cultivation. Mixing to depths of 150 mm to 400 mm is ideal. The deeper the ameliorants are placed the deeper the plants will root. Deep rooting for soil reinforcement can be encouraged by deep placement of nutrients, for example. Conversely, if amelioration is only to a shallow depth, then rooting could be only superficial, depending also on water relations.

Surface finishing

The texture and microclimate of the ground surface can have a fundamental effect on the establishment of plants. Seedlings are very vulnerable to extremes of temperature and moisture, and a friable surface will generally have less fluctuation. In addition, when incorporation of the seed into the soil surface is not practicable, because of access for example, the natural texture of the surface will have to provide the necessary protection. Erosion is also dependent to a large extent on the condition of the ground surface.

Loosening or light cultivation of the soil surface is therefore important. This can be done by machine or hand, depending on the access and whether hand labour is available. On long slopes that cannot be accessed by machinery, some cultivation can often be achieved by working from the top or bottom of the slope with hydraulic arms. Alternatively, harrows or chains can be dragged across the slope surface by a tractor working from the top of the slope.

The need for extensive improvement of the slope surface can be reduced by careful preparation as it is being formed. Heavy machinery will tend to compact the surface, but final trimming can remove this. The temptation to run over a slope to smooth and trim it off with earthmoving machinery should be resisted. A rough finished surface will be less inhospitable to vegetation.

3.4.3 ESTABLISHMENT AIDS

On many slopes it will not be possible to prepare the ground surface adequately. There are a number of ways that the surface microclimate can be modified to assist vegetation establishment and reduce the severity of stress that non-stress tolerant species have to endure.

Geotextiles

Chapter 4 describes the role of geotextiles in simulating a vegetation cover to modify surface erosion processes. This role can be a temporary one, which has an instant effect but which is superseded once a vigorous vegetation has developed. Biodegradable products would then gradually disappear, whereas a synthetic geotextile would remain to provide reinforcement or as a backup in case the vegetation cover degraded. Surface-laid synthetic products can give rise to problems in the longer term, however, with 'snagging' and loss of continuity with the soil surface.

As an establishment aid, geotextiles have two functions:

1. to control surface erosion and prevent the loss of soil, ameliorants and seeds before establishment;
2. to improve the surface microclimate, maintaining soil moisture and protecting the seeds against desiccation and extremes of temperature.

There are two types of geotextile suitable for this purpose:

1. woven meshes, with an aperture size of 5–15 mm, usually consisting of jute or coir;
2. mulch mats, consisting of a layer of chopped straw, shredded paper and/or coir fibre 10–50 mm thick, retained between two layers of light string or plastic mesh.

These materials would normally be laid over the prepared and ameliorated ground surface and firmly pegged into place. Seed would then be broadcast over the top of them, so as to fall down within the textile, preferably to the soil surface. If seed is sown before the material is installed, the installation process itself will disturb the seed layer and result in a patchy cover.

Geotextiles are expensive to buy and install. Their use is therefore normally confined to areas where surface conditions are particularly hostile or where the erosion risk is especially high.

Binders

Chemical 'glues' can be applied to the soil surface in a suspension with water, to bind or stabilize the soil particles. A skin or crust so formed will give some protection against erosion, but on its own will not protect the seed or improve the surface microclimate. However, binders can be useful where the seed has been buried within the soil surface and the surface is liable to erosion. They are also widely used in conjunction with a mulch, to hold it in place.

Care should be taken with selecting a chemical binder. Binders vary considerably in their ability to penetrate the soil surface and some are actually toxic to young seedlings. It is also important that the skin or crust formed does not prevent seedling emergence. In general the skin-forming materials, such as emulsions of bitumen and lignin-derived products, should be avoided. The better crust-forming products are usually based on emulsions of butadiene oils or PVAs.

Mulches

Mulches comprise bulk organic materials spread over the ground surface to protect seeds, encourage water infiltration and reduce surface erosion. Mulches are also used to suppress weeds and improve soil moisture around planted stock. They act in the same way as geotextiles, but are normally spread in a layer rather than as a prefabricated product. They are often used in conjunction with a binder, to hold them in place (see Chapter 4).

Suitable mulch materials are chopped straw, shredded paper, cellulose (wood or cotton based) fibres, coir fibre, wood fibre or chippings. The best mulches have fairly long fibres, which interlock to form an erosion-resistant layer. Short-fibre materials, such as peat and ground woodpulp, are less effective as mulches.

Granular materials, such as wood chips and bark, should have a large particle size, usually >25 mm.

3.4.4 ESTABLISHMENT METHODS

Ways of establishing plants have been developed for many different plant types and circumstances, and there are often local variations. The establishment techniques used reflect the characteristics of the species and the nature of the site and ground conditions. For erosion control and slope stabilization the main requirement is that the vegetation should establish quickly and vigorously, usually at low cost. The availability of manual labour is often a major factor.

As indicated earlier (section 3.3.2), plants naturally propagate by two methods:

1. setting seed which is dispersed into new niches and develops into new and unique individuals;
2. vegetative propagation by small plantlets which do not disperse very far and are genetically identical to the parent plant.

Man has widely adapted seed dispersal for his own purposes, harvesting and storing seed and then sowing it where required. For many species the seeds are not sown directly into the site where they are ultimately to grow, but are raised first in a 'nursery', and then transplanted into their final location after one or more growing seasons. This is done in order to increase the chances of survival of the plants, giving greater care and protection during the critical establishment stage when the plant is most vulnerable to stress, or when the plant's natural regeneration strategy is not suitable for bare ground or high-stress situations. However, transplanting can itself introduce a highly stressful step into a critical stage of the plant's life cycle, and many problems can occur.

Vegetative propagation methods are also widely utilized. Naturally occurring plantlets, such as bulbs, offshoots, suckers or rhizomes, are harvested and replanted in the required locations. Alternatively, the natural ability of some species to root spontaneously from cuttings can be a very cost-effective method of propagation.

Adapting these two basic approaches, there are eight techniques for plant propagation that are commonly used on slopes.

Seeding

Seeding is the most widely used technique for herbaceous plants, and is becoming more widely used for woody plants. Seeding is usually cheap and very versatile, and suitable seeds can often be collected from locally growing plants, though commercial seed is widely available for a wide range of species. One drawback is that germinating seeds are very vulnerable to desiccation and predation during the establishment stage.

Sprigging

Plants which spread by creeping rhizomes or stolons can be propagated by harvesting and replanting these plantlets, or even simply chopped rhizome sections, which will take root and grow. This is not a widely used technique, but can be useful for certain types of grass and herbaceous species. Harvesting suitable material is usually destructive, i.e. the donor area from which it is removed has to be revegetated. The propagative material is also very vulnerable to desiccation before it has rooted, so has to be protected after spreading.

Bare-rooted plants

Raising young plants in the nursery and transplanting them on site with bare roots is the most common method used for woody species. Depending on the age of the plant, which is usually between one and three years, transplants are cheap and can be produced in large numbers. Having bare roots, the plants are very vulnerable to drying and damage to the root system whilst they are being transplanted.

Planting of bare-rooted stock can usually only be undertaken during the dormant season, when the plant's demand for water is lowest.

Container-grown plants

Many woody species are also grown in containers, which can be planted at any time of the year and with minimal disturbance to the root system. However, the main disadvantage of this method is cost.

Tubed seedlings

This is a variation on container-grown plants, where the seedlings are raised in small containers or tubes, and are planted on-site up to a year old. They are cheap and easy to handle, and again can be planted with minimal disturbance to the root system. Establishment success is usually highest if they are planted during the dormant season. Special tubes can be used, which are long and have longitudinal grooves down the side (often called root-trainers) which encourage deep rooting and the formation of a good root system.

Cuttings

Species that root freely from live wood can be raised from cuttings. These can either be used to produce bare-rooted, container-grown or tubed stock for planting on-site, or can be planted directly on-site to root *in situ*. Species such as willow and poplar are widely used for direct slope planting in temperate regions, using techniques such as those described in section 3.4.7.

Turves

Grass and herbaceous vegetation can be pre-grown as turves for transplanting on-site. This gives an instant vegetation cover, but is very expensive. Turves can be cut from natural ground or can be especially grown using prepared soil or artificial growing media. Specially grown turf can incorporate a geotextile, which can be useful for circumstances where structural reinforcement is required.

Plant-rich soil

Natural topsoil can contain a 'seed bank' or reserve of vegetative propagules which will grow and develop if the soil is disturbed. Techniques have been developed to utilize this for establishing certain types of semi-natural vegetation where conventional techniques cannot be used, for example where the range of plants is not available commercially. It is a variation on sprigging, where the surface soil material is removed in total, not just the vegetative material, and like sprigging it is usually destructive to the donor area. Examples of communities where this technique has been used are temperate heathlands and coastal woodlands on sand. However, there are many community types for which this technique could be a useful method of establishment on slopes.

3.4.5   SEEDING ON SLOPES

Access for conventional seeding machinery is not normally possible on steep slopes. There are a number of seeding techniques developed to overcome this limitation, and also to overcome the results of the lack of good soil and surface preparation.

Broadcast seeding

Seed can be broadcast dry by hand or machine, depending on the topography and ground conditions. Hand seeding is ideal for slopes, especially for smaller areas. For large areas where conventional machinery cannot reach, aerial seeding from a helicopter is possible.

Unless it is incorporated into the surface by raking or harrowing, broadcast seed will rely on the natural soil surface texture for protection. A coarse surface texture into which the seed will fall will give the best results. As an alternative, the surface can be covered with a dry mulch, spread by hand or machine.

Hydroseeding

Hydroseeding is a special technique developed for quickly and cheaply seeding slopes. All the seeding constituents: seed, fertilizers, mulch and binder, are mixed into a slurry with water, and sprayed on to the slope surface under pressure. Special hydroseeding machinery is used, consisting of a large tank (up to 5000 l capacity), a paddle system for mixing, a pump (usually centrifugal) and a spray hose or monitor controlled by an operator. Hydroseeding machines can reach about 15–20 m, depending on the wind direction, though larger machines can reach up to 25 m. For longer distances a hose system can be used, the distance only being limited by the capacity of the pump.

Most hydroseeding machinery can cope with a slurry up to about 10% solids content, and would spread the slurry at a level of between 1 and 2 l/m². Slurry application levels at the higher end of this range will give a more even coverage of the surface, as the operator can make a more even application. If the amount of material (such as mulch) required to be added is large and cannot be applied in one application, then it will have to be split into a multiple application in two (or more) passes. If this is the case, then the seed, fertilizer and some of the mulch should be applied in the first pass, and the remaining mulch in the second or subsequent passes.

Organic ameliorants and mulches can also be applied by hydroseeding machinery, though the amount of material involved may require multiple passes. One drawback to using this method is that the ameliorant is not incorporated into the ground surface.

When seeds and fertilizers are applied to the soil surface together in hydroseeding, problems can occur due to the proximity of the seed to high concentrations of dissolved fertilizer chemicals. These high concentrations can be damaging to the fragile seedling, unless they have been diluted by sufficient rainfall. There are two ways to overcome this problem:

1. by using insoluble, slow-release fertilizers, keeping the soluble fertilizers to the minimum required to give vigorous initial plant growth;
2. by delaying the application of most of the fertilizer until the plants have germinated and become established. Timing is critical, to avoid applying fertilizer too soon yet soon enough to stimulate vigorous vegetation growth as quickly as possible. Two to six weeks is the normal delay period.

Even with a mulch, hydroseeded seeds may have limited protection and are vulnerable to surface drying. Seeding should therefore be timed so that the likelihood of an extended period of drought is unlikely once the seeds have begun to germinate. However, in dry climates, it would be quite normal to seed towards the end of a dry season, allowing the seeds and protective mulch to remain on the surface for a while until rainfall commences. The most important thing to avoid is alternating wet and dry spells during the establishment period.

Spot seeding

Spot seeding is a method used when seed is in short supply, or when overall seeding is impractical. It is often used for larger seeded trees and shrubs. The technique involves sowing a few seeds of a single species in a small area, usually by hand. The seed spot can be ameliorated and lightly cultivated, the seeds sown at the appropriate depth (usually two to three times the seed diameter), and any additional protection such as a mulch added.

The density of spots depends on the available seed material and the density of plants required. Spacing would usually be between 1 and 5 m, though at the closer end of this range on slopes. Spots of different species can be mixed in the required proportion, in the same way as for planted species.

Seed spots can be successfully used to introduce new species into an established vegetation cover, such as a sward. The area around the spot must be kept free of vegetation to avoid the risk of the young seedlings being swamped or out-competed by other species before they become fully established.

Dry mulch seeding

This technique is a variation on other seeding methods, such as broadcasting, where a heavy mulch is required for protection against either extreme exposure/cold or drought during establishment. High mulch rates, such as 3–5 t/ha of chopped straw, may be required in such circumstances.

After seeding, the mulch is spread either by hand or using a dry-mulching machine which throws the mulch over the soil surface. To hold the mulch in place, a binder or tackifier is either sprayed on afterwards or applied to the mulch as it is blown on.

As an alternative to a loose mulch, mulch mats can be laid over the ground surface, in a similar way to a geotextile. These will have a similar effect to a normal dry mulch, and may have the advantage in some circumstances that spreading machinery is not required.

Dry mulching should be routinely used for tree and shrub seeding, to improve the seed-bed conditions and assist establishment. When fine seeds are also to be applied, it is usual to sow these over the top of a heavy mulch, so that they fall down within it and are not smothered by the mulch.

Seeding trees and shrubs

The techniques described above apply as much to trees and shrubs as to herbaceous species. However, when seeding trees and shrubs a few further points need to be borne in mind.

The seed-bed conditions for most trees and shrubs are generally much more important than for herbaceous species. Pioneer species are able to cope with poor ground conditions, but most species are only able to establish successfully if the conditions are ideal. Careful ground preparation and the use of ameliorants and mulches are therefore important.

Tree and shrub seeds vary widely in size, so for any mixture of species the appropriate planting depth for each species has to be considered. The usual approach is to sow the larger-seeded species first, and incorporate them to the appropriate depth, then to sow the smaller-seeded species. Alternatively, the larger-seeded species can be sown last of all, in spots.

Many tree and shrub species exhibit various levels of dormancy, enabling them to establish progressively over a period of time, or to spend an inhospitable season as a dormant seed until the beginning of the next growing season. When artificially seeding such species it is usually desirable to break this dormancy, in order to get a rapid establishment and recruitment of individuals. Various methods of breaking dormancy can be employed, depending on the species and seed type.

Seeding in cold environments

Environments where the growing season is very short, due to latitude or altitude, often require special approaches to seeding. The short season may be accompanied by highly erosive conditions as well, making plant establishment very difficult.

Good ground preparation and soil amelioration are clearly important to give the plants the best start. Increasing fertilizer rates can stimulate vegetation growth to a large extent. However, care is required with this approach. Species that are best adapted to cold environments are stress tolerators (see section 3.3.2), which have low potential growth rates and do not respond well to higher fertility. By increasing fertility the less well-adapted species may be stimulated at the expense of the desired long-term species, resulting in a sward that is vulnerable to cold and exposure stress.

Extensive use of mulches and other establishment aids (section 3.4.3) will help in hostile environments. Erosion control netting, for example, can help bridge the gap between seeding and the vegetation cover being sufficiently robust to resist erosion and instability unaided, even if this gap is more than one growing season.

In order to make maximum use of the growing season, seeding should be undertaken as early as possible. This can even mean seeding on to a snow or frost cover, leaving the seed to remain dormant until the snow melts and the ground becomes warm enough for the seed to germinate. This approach can have the added advantage that access to the ground by machinery is best when the ground is frozen hard, because less damage is then caused to fragile alpine vegetation.

Seeding in hot, dry environments

The first requirement in hot, dry environments is to maximize water infiltration and storage in the soil by proper ground preparation. Soil ameliorants (section 3.4.2) can improve water-holding capacity, but their use is often impractical on a large scale or in remote areas.

Heavy mulches over the soil surface after seeding will help conserve soil moisture, though care needs to be taken with very thick layers, which can smother some seeds (section 4.5). In areas where the rainfall season involves erosive rains, temporary or permanent erosion control netting may be appropriate (section 4.6).

Successful seeding in dry climates usually involves utilization of the wet growing season. This may require seeding at the very beginning of the season, or even at the end of a dry season so that the seeds remain dormant on the surface until conditions are right for germination.

3.4.6   PLANTING ON SLOPES

Planting is the usual method for establishing woody vegetation or herbaceous species where the site conditions are not suitable for germination and early growth. As a general rule, the younger the plant when it is planted out, the better are its chances of vigorous growth though, if planted out too young, predation or competition from other species can reduce or prevent establishment. Most planting on a large scale uses plants between one and three years old.

The period between being lifted in the nursery and becoming established on site is very critical for the young plants, and is when most problems with establishment occur. During this period they are vulnerable to mishandling and drying out, and may sustain damage from which they never recover. Later death or dieback might be attributed to hostile site conditions, when in fact more often than not the cause is damage sustained during this period. Careful handling of young plant material, especially if it is bare-rooted, involves keeping physical damage to a minimum and preventing exposure to a drying atmosphere. The fine adventitious roots can dry out very quickly when exposed, but usually they can be maintained in a good condition in a plastic bag out of direct sunlight, for example.

There are a number of basic methods of planting young trees and shrubs, and local horticultural practice should be followed as much as possible. Amelioration of the soil around the plant with fertilizers or water-holding material is achieved by incorporating the ameliorant to the full root depth or more with a hand tool before the plant is inserted. Care should always be taken with bare roots, to ensure that they are well spread out and in good contact with the soil. In good soil the plants can be notch-planted directly, without prior amelioration or cultivation of the planting spot. In poorer soils loosening of compacted layers and cultivation of the rooting area are usually necessary.

In general, the younger the plant, the quicker and cheaper the planting process. Small tubed seedlings can be very quickly planted over large areas, and damage to the root systems is minimized by the tube. The plant should be removed from the tube before planting. Small plants, however, will also be vulnerable to swamping by larger plants or existing vegetation.

Planting spacing on slopes would normally be 1–3 m between plants, depending on species and the initial density required. Closer spacings will provide a complete cover and closed canopy of vegetation in a shorter time, but will require management to thin them out much sooner. When using mixtures, plants should be randomly mixed, though when an even coverage is not so important small species-groups can be easier to manage in the long term.

When planting any young trees or shrubs, existing vegetation around the planting positions must be removed, either by hand cultivation or using herbicides. An area about 1 m in diameter is usually sufficient. In densely planted areas it may be easier to remove all the existing groundcover vegetation. Removal of existing vegetation, especially persistent growth such as rhizomatous species, will prevent overcrowding and competition for nutrients and water at a critical stage in the life of new plants.

After the tree or shrub is planted, a mulch can be applied to the surface around it, in order to suppress weed growth and to help conserve soil moisture (section 4.5). A mulch can consist of:

1. granular material such as chopped bark or wood chippings; fine material should be avoided as this will encourage other vegetation growth;
2. plastic (black) or bitumenized felt laid on the ground surface and pinned in position;
3. a mulch mat (section 3.4.3).

Other local materials may be suitable provided they conform to the basic requirements of reducing surface evaporation and preventing weed seed germination. Mulching an area about 1 m diameter around the plant is usually sufficient.

Planting in dry environments

A number of measures can be combined to help establishment in dry environments:

1. amelioration of the planting area with water-holding material such as a polymer, which should be incorporated to more than the depth of planting, in an area 0.5 m in diameter;

2. use of a good mulch around the planting location – granular material should be spread to at least 75 mm depth;
3. planting on a small ledge which is graded back into the slope, which will help to increase infiltration and water storage in the vicinity of the plant;
4. careful ground preparation to maximize water infiltration and soil moisture storage as a whole;
5. planting at the very end of the dormant season, immediately before growth starts in response to increased rainfall.

Irrigation can also assist establishment, but care has to be taken on slopes not to reduce the stability of the soils. High moisture contents can lead to slippage, and irrigation is very difficult to control carefully enough. Localized wet areas around the plant could lead to small slips. A further factor is that if plants are established with the aid of irrigation, they become dependent on the irrigation and they do not adapt to the prevailing dry conditions. Nevertheless, some watering immediately after planting can be beneficial, providing that the water infiltrates sufficiently to reach the roots.

3.4.7 COMBINED BIOENGINEERING APPROACHES

The specific techniques outlined below are the result of the systematic refinement of traditional stabilization and erosion control methods which have evolved over a long time, in some cases centuries, in areas of the world susceptible to particular stability problems, especially mountainous areas and wetlands. The traditionally small-scale approaches have been developed for use in slope construction, stabilization and repair in an engineering context, usually in areas where conventional or recent engineering methods utilizing stone and concrete have been impractical. The common techniques outlined below are branch layering, fascines, live crib walling and slope grids.

Branch layering

This is a means of establishing vegetation to stabilize slopes involving, as the name implies, the use of reasonably large vegetative material. Two variants of the technique have been developed; one for the stabilization of existing slopes and the other for the construction of new slopes from fill material.

The technique for stabilizing cut slopes involves cutting shallow ledges between 0.5 and 1 m deep, and between 2 and 4 m apart, in the new slope on to which the branch material is laid. The surface of the ledges should slope gently back towards the main slope and the live branches should be placed in a criss-cross fashion, such that only a few centimetres of the tips protrude from the slope, when they are covered with soil. Work is carried out from the bottom of the slope upwards, so that each shelf can easily be filled in again using the material excavated from the one above.

When constructing new slopes of fill material, much longer live branches of 1–5 m can be used. These are placed into the slope as it is built up and compacted layer by layer. They should also be laid criss-cross and at an angle of at least 10° so that the branches are inclined slightly upwards and their tips protrude.

The branches will have an immediate effect in both reinforcing the slope and preventing uninterrupted surface runoff leading to gullying. Once they begin to root and throw out shoots these two effects will be considerably enhanced.

Fascines

Fascines are bundles of freshly cut, woody material which are built into the surface of a newly constructed cut slope to help prevent loss of freshly spread topsoil or other applied growing medium, and to control and enhance surface and subsurface drainage of the slope. Up to four or five branches are wired together in bundles to form long continuous cylinders of 100–300 mm or so in diameter. The exact dimensions will be governed by the available plant material.

The fascine bundles are placed in trenches just deep enough to cover them, cut at a shallow angle to the slope. Trenches are constructed every few metres up the slope. The shallow angle allows them to intercept surface water runoff and lead it gently down the slope so that it causes minimal erosion. Working from the bottom to the top of the slope allows each trench to be refilled using the material excavated from the one above. Before the trenches are filled the fascines should be fixed into place by driving steel pins or wooden stakes through them and into the slope behind.

Fascines do not provide the initial depth of slope reinforcement afforded by branch layering, but their tubular nature means that they are very effective in controlling and channelling surface water runoff. The extent to which lengths of fascine can be prefabricated means that the amount of intensive labour needed to install them can be minimized, in comparison to wickerwork or wattle fencing techniques. Eventually the material should root into the slope and produce aerial shoots, and as these develop both the depth and degree of permanence of the slope protection will increase.

Live crib walls

This technique also provides a combination of the benefits of immediate protection with the long-term advantages of vegetation for stabilization. In simple terms, crib walls are just a specialized form of gravity-retaining structure using on-site fill material, held within a constructed framework, in order to provide most of the necessary mass to resist overturning by the weight of the slope. Crib retaining walls are usually constructed to stabilize the base of slopes which have failed. They can either be of single or double construction and made from purpose-made pre-cast concrete elements or built using sawn or roundwood poles. Single construction is possible using timber, but with concrete units it is usually necessary to construct a double wall.

By planting live cuttings or branches into the structure as it is built, in a similar fashion to brush layering, additional reinforcement is achieved. As these root and grow, the stability of the structure and the repaired slope it is supporting can be expected to increase.

The crib wall should be built with a batter of at least 10°. Horizontal timbers or elements are laid at the base of the slope and these are fixed in position by nailing to sharpened timbers or by interlocking with concrete header elements running back into the slope. How closely these are spaced will depend on the slope which is to be retained. The next horizontal course is then placed on top of the headers, and the space behind is filled with earth and a layer of live wood. As far as is possible the base of the live wood should make contact with the *in situ* soil of the existing slope, while the tips should protrude from the wall approximately 10 cm. The wall is built up to the required height layer by layer, and is back-filled and planted as it is constructed.

Slope grids

Slope grids are usually larger structures which take the form of a grid of horizontal and vertical timbers. This is constructed and fixed to lean up against the slope which it is intended to support. The voids in the grid are then filled with soil or appropriate available fill material and then planted with woody cuttings and, in some cases, rooted plants.

Slope grids are usually used to repair limited areas of slope failure. The timber grid provides a support framework which operates to retain the soil of the reconstructed slope until the plant material can become established and root into both the newly filled soil as well as into the *in situ* material forming the existing slope.

The grid is usually constructed on a base provided by a horizontal timber member which in turn rests on a small concrete or stone foundation. Depending on the depth of the slope failure to be repaired, either single vertical timbers can be used or prefabricated ladder-shaped elements. These vertical elements are linked together with further horizontal timbers and the whole structure is then back-filled with earth and planted. Ideally the live timber poles will be driven through the fill material of the repaired slope and into the undisturbed earth behind. The idea is that when these strike root, they will bind together the new fill material and the existing slope.

Further planting of the fill material with rooted plants can also be undertaken at this stage. As a temporary measure the slope grid should in any case be anchored to the hillside with steel pegs. Small slope grids can also be completely constructed using live wood, but to ensure that this has any chance of rooting it must be completely buried in soil.

3.4.8 AFTERCARE DURING ESTABLISHMENT

The transplanting of plants on to a site by seeding or planting is only the beginning of the establishment process. Aftercare is the process of assisting the initial vegetation to become fully established and stable, such that a less intensive regime of management can take over. The objectives of aftercare, which can last for between two and five years, are:

1. to ensure that the plants become established and overcome the initial constraints on growth;
2. to establish a viable soil–plant system, with sufficient nutrient 'capital' and turnover to support the vegetation.

A fundamental aspect of aftercare is **monitoring**, i.e. keeping track of what is happening to the soil and the plant community, enabling the proper response to be devised for any situation, rather than relying on prescriptions according to a predefined set of rules. Monitoring involves, amongst other things:

1. examining the vegetation cover, i.e. the species present and their relative abundance, and the overall density and distribution of plant cover;
2. examining the soil development, develop-

ment of root systems, re-establishment of soil structure and profile water relations; periodic soil analysis for pH and nutrient content.

Monitored conditions should be compared to what is required or expected, and also to the functions required of the vegetation. Trends in the development of the plant community will give an indication of the development of the whole system, and how it might behave in the future.

Aftercare for herbaceous swards

Apart from the need to overseed any bare patches that have failed to establish, the main aftercare requirement is to maintain the soil fertility. If the initial soil is already sufficiently fertile then the sward will develop rapidly, and competitive species will quickly dominate. Aftercare in this case involves controlling the density and structure of the sward, such as preventing the development of tussocky species.

If the soil is initially infertile then the vegetation will consist of mainly stress-tolerators and competitive-stress-tolerators (section 3.3.2). The fertility will need to be built up progressively, using fertilizers/manures and nitrogen-fixing species such as legumes, until such time as the system is self-sustaining and competitive species begin to dominate.

A self-sustaining system is one that has a pool of nutrients held in the soil, mostly bound up with accumulated organic matter. Mineralization from this pool by microbial action releases soluble nutrients available for plant growth. This 'turnover' depends on the balance of nutrients, especially nitrogen, the soil pH and the temperature (see the cycle illustrated in Figure 3.2). In infertile soils the cycle can become blocked, leading to poor growth and a 'moribund' sward. The usual causes of this are nitrogen deficiency and/or low pH, which reduce microbial activity.

Defoliation of the sward by mowing or grazing can be beneficial in a number of ways:

1. it stimulates lateral vegetative growth and tillering in grasses, thereby thickening the sward;
2. it reduces the vigour of tussock-forming species;
3. it returns organic matter to the soil and stimulates nitrogen cycling.

Conversely, mowing or grazing can cause damage to the fragile soil, especially through poaching by animal feet. Defoliation may also reduce the root development, which is in balance with the shoots. The abundance of flowering herbaceous plants, which are not adapted to defoliation, may also decrease if they cannot flower and set seed, so the timing of defoliation may be important. Flowering plants may be an important component of the community, providing a deeper-rooting function.

Aftercare for planted trees and shrubs

The same principles for the development of soil fertility apply to trees and shrubs as to herbaceous swards described above. However, trees and shrubs can be more tolerant of low soil fertility, though rapid development of a dense canopy depends on a good nutrient supply. For woody species, periodic fertilizing in the root zone is usually the most cost-effective approach, and this avoids over-stimulating the herbaceous ground cover between the plants.

Planted or seeded trees and shrubs will be vulnerable to competition from other vegetation for moisture and nutrients for about three years. During this time it is therefore important to keep the area around each plant free of competing growth, and maintain a mulch if necessary.

The survival and growth of trees and shrubs should be monitored during each dormant period and any plants that have failed or have died back should be replaced as required. Up to 10% losses are usually tolerated in tree plantations, though this depends on the distribution of the losses. Planting losses may have to be made good for up to three years, before the effects of transplanting have ceased.

Coppicing of selected species can be a useful method of manipulating the structure of the vegetation in the first few years. Coppicing will stimulate a dense thicket of growth closer to ground level, which can provide more shelter to other species and greater protection to the ground surface. Coppicing, or other pruning operations, would normally be carried out during the dormant season.

3.5 MANAGEMENT

Once established, if left to its own devices vegetation will continue to change and develop, and the results of this change may, or may not, be beneficial with regard to the desired biotechnical objectives that were the reason for its initial establishment. Consequently, in order to ensure that the vegetation develops in such a way as to be able to fulfil the functional expectations placed upon it, and to continue to be able to do so over the course of time, some form of continued monitoring and management may be required.

3.5.1 OBJECTIVES FOR MANAGEMENT

Management objectives can be conveniently classed under two main headings: ecological and biotechnical. Although it is convenient to separate the objectives from the point of view of analysing their role, it is important to remember that these functions are inseparably interlinked as far as the development of the vegetation and its management are concerned.

Ecological objectives

The first aim of management from an ecological point of view is to encourage the transition from an artificially established collection of plants, or propagules, into something which begins to resemble and function as near-natural vegetation. This objective is based on the premise that, as discussed previously, vegetation of the appropriate kind forms the natural protective layer for the ground surface over the majority of the land surface and that, at least as far as indigenous vegetation is concerned, is relatively self-sustaining requiring little human intervention to maintain its continued existence.

As well as a straightforward increase in size of the individual plants, vegetation development will involve an increase in complexity within the planting as a whole. The development of complexity will operate at a number of levels, but these can best be summarized in terms of floristic and structural complexity.

For obvious practical reasons the initial planting will be likely to consist of large numbers of individuals of largely similar age and size, representing relatively few species. The gradual invasion of this species-poor new planting by species from the locality will, by definition, affect both the floristic complexity as well as, in most cases, the structural complexity. The invasion of new species to take over niches left vacant at the establishment stage, such as the ground and herb layers in a planting of woody species, will increase both kinds of complexity and be of particular importance in helping to provide long-term protection of the ground surface from the erosive effects of rainfall and runoff.

Other woody species, some with different growth habits, can also be expected to colonize the planting and these will add further to its floristic complexity as well as increasing the age, form and size range of plants. In time, structural and age diversity are also likely to increase, even in the absence of other species, due to differential development of the individuals which formed the initial planting. The result of this maturing will be expressed in the form of the development of layering in the vegetation.

The development of vegetation will enhance the self-sustaining characteristics of a protective planting. As well as the implied benefits of a reduction in the levels of intervention that are likely to be necessary to sustain the planting, there are also advantages from a landscape and nature conservation point of view. Whether or not it is composed of largely indigenous species, vegetation which has developed a degree of internal complexity will be more visually stimu-

lating and thus more attractive to the observer than a monotonous monoculture. By the same token, structural complexity will mean a greater scope for the number of physical niches for further colonization by plants and animals.

The encouragement of the transition from a species-poor structurally simple planting scheme towards a complex near-natural plant community is therefore an important goal to be taken into account when managing the new stabilizing vegetation. This is considered under the next four headings.

Maintenance of vegetation structure

Having achieved the transition to a more complex form of vegetation, it will be important to manage this in such a way as to maintain the newly diversified structure in a form that will continue to have maximum effectiveness in protecting and stabilizing the underlying soil. Natural development of the vegetation will lead eventually to greater structural diversity. This will be represented to some extent by the development of patches and gaps in the vegetation, and also by an increasing range in the age and size of individual plants.

Gaps in the vegetation may represent areas of weakness in the protective layer at which erosion and instability could start and so must be avoided. Large individuals also provide potential problems in that they will be subject to greater danger of windblow. The resulting gaps in vegetation cover and damage to slopes will be an equally difficult problem. For this reason, whatever the species diversity and visual interest arguments in favour of highly structured stands of vegetation, from the point of view of slope protection and erosion control a minimum degree of uniformity in the vegetation cover ought to be a management aim.

Maintenance of stability of vegetation

Long-term stabilization of the ground surface through the medium of vegetation requires the maintenance of long-term stability in the vegetation itself. Ecological stability is conventionally seen as being synonymous with diversity. Stability can also be seen as being broadly synonymous with resilience to change in the face of external pressures or in response to internal cycles in the vegetation.

External pressures may take the form of attack by pests or disease. As pests and diseases tend to be largely specific to single species or to groups of closely related species, vegetation composed of a diverse range of species will be likely to suffer limited damage at any one time in the face of an attack by a specific pest or disease. Resistant species, not affected, can respond to losses by expanding into the space occupied by those affected and replacing them in their protective role.

Internal instability can also result from species-poor communities or monocultures where, for example, the senescent stage in the life history of the dominant species is reached by most individuals at the same time, causing large gaps in the vegetation cover. Diverse vegetation which has a broad age structure will clearly be able to recover relatively rapidly from the effects of such cycles. The effects themselves will also be much reduced in their intensity and consequent threat to soil or slope stability.

Management measures should therefore also be directed towards the maximization of stability in the vegetation. This will involve intervening where necessary to further and maintain species and age diversity.

Direction of successional development

The establishment and maintenance of a floristically diverse and well-structured vegetation in order to gain the most from the stabilization potential of the vegetation cover will require some intervention to influence the course and extent of the succession process. The aim of intervention is likely to involve attempting to strike a balance between:

1. allowing or encouraging succession to progress as far as possible in the direction of increasing species diversity, in order to maximize the potential stability of the vegetation, while

2. keeping a rein on development so that the diversity, especially of vegetation size, does not present a risk to the continued integrity of the vegetation cover.

No management case: nutrient cycle development or blockage

In situations where no management of vegetation established for the purposes of stabilization takes place, and the process of succession is allowed to take its course, nutrients are gradually accumulated in the vegetation and the fertility of the soil can become depleted. In situations where this development has been allowed to continue unhindered for exceptionally long periods of time, such as in tropical rain forests, this state of affairs becomes very extreme and almost all the nutrients in the system become locked in the vegetation.

Biotechnical objectives

Biotechnical objectives for management are concerned with the maintenance of a continued, extensive and vigorous root system for the purpose of soil reinforcement, together with the upkeep of a continuous cover of foliage to protect the ground surface from the effects of erosion.

The increasing biomass and diversity of the vegetation resulting from the process of succession will be reflected in both the underground and aerial components of the vegetation cover. Competition for nutrients as the vegetation develops will result in an increasingly dense root penetration of the soil mass by woody species.

A largely consistent structure can be seen as providing a consistent degree of protection. The structure exhibited by the aerial parts of the plant is likely, in relative terms, to be reflected in the below-ground parts. This is unlikely to be true in strict physical terms, i.e. a 2 m high densely-branched shrub is unlikely to have a

2 m deep densely-branched root system but, in general terms, a vigorously growing plant will have a relatively vigorous root system.

3.5.2 MANAGING HERBACEOUS VEGETATION

The approaches adopted for managing herbaceous vegetation are likely to be related to the specific reasons for selecting it as a means of slope protection. If these require its maintenance as herbaceous vegetation, which will often be a plant community dominated by grasses, the main objective of maintenance will be to prevent the succession taking place from grass into scrub and then woodland. An additional aim is likely to be the stimulation of vigorous root growth and the maintenance of a close sward to protect the soil surface from erosion. Maintenance operations should also be aimed at helping to maximize the diversity of the vegetation within the sward.

As discussed above, there is a relationship between the above- and below-ground plant biomass. The usual response to defoliation is for a plant correspondingly to reduce the amount of root growth. The requirements for a vigorous root system together with a close, short sward are therefore to some extent incompatible.

There are two main approaches to achieving the various objectives. These involve cropping the vegetation through either mowing or grazing.

Mowing

By their nature, areas of herbaceous vegetation established for soil or slope stabilization purposes are likely to be in locations which are difficult to reach, in particular on slopes. This brings into question the feasibility of mowing on steep slopes, and the consequent need for specialist equipment. Tractor-mounted mowers can be safely operated on slopes of 33% or less or on steeper slopes close to level areas and within reach of hydraulic arms. Various types of pedestrian-held mowers, strimmers or scythes

will be needed for steeper and less accessible areas.

Depending on the type of herbaceous vegetation which has been established and the role it is to fulfil, it may be desirable to cut it more or less frequently. Less frequent cutting will allow root development to be maximized to improve soil reinforcement, while more frequent cutting will encourage a closer sward for better surface erosion protection, but at the expense of root development. All herbaceous vegetation should be cut at least once in a growing season, to remove invading woody plants and to discourage tussocky, uneven sward development.

Harvesting the cut grass and herbaceous species in a sward over time will remove nutrients from the soil and thereby gradually reduce the vigour of the sward. This may, in turn, affect the ability of the vegetation to perform its required functions. Nutrients can be replaced by fertilizer applications or by leaving the cuttings on the slope. One possible disadvantage of not removing cut material is that, while it may contribute marginally to protection against surface erosion in the short term, in the longer term it may well suppress the growth of many of the less vigorous species which go to make up the sward, resulting in gaps and possible areas of weakness.

Grazing regimes

In some circumstances grazing may provide an economic solution to the management of herbaceous vegetation. It is important, however, that it is managed in a controlled manner, and does not merely take the form of sporadic attention from passing wild herbivores. Although this might have the apparent attraction of requiring no outside management input, it cannot guarantee that the necessary integrity of the vegetation, and thus the slope it is stabilizing, will be maintained.

Grazing animals will certainly have none of the access problems to slopes faced by conventional mowing machines, but it is also possible that they may have some detrimental effects which mowers do not. These arise from the trampling effect of the feet of grazing animals, which in some circumstances may destroy areas of protective vegetation as well as cause physical damage to the surface of the slope. For this reason it is important that small, light animals such as sheep or goats should be used in preference to large ungulates such as cattle or buffalo.

The possible selectivity of grazing is another factor which may have to be considered. By concentrating on the more palatable species, grazing may result in only some of the species being grazed, with the consequence that the sward may develop unevenly and bare patches may result. Nutrient loss is less of a problem with grazing animals than with mowing.

Under some conditions controlled burning may be an option for managing herbaceous vegetation, but this needs careful planning and can only be undertaken at a time of year when there is no immediate risk of surface erosion, and when sufficient time is available for the sward to re-grow before surface erosion control again becomes a critical issue.

3.5.3 MANAGING WOODY VEGETATION

As in the case of herbaceous vegetation, the approach chosen for the management of woody vegetation will have to take account of the specific stabilization objectives for which it has been established. The most common management activity is likely to involve the periodic removal of woody shoots and branches in order to promote the development of a stable vegetation community and to maintain a well structured and even coverage. This can be achieved by pruning, thinning or coppicing.

Pruning involves the removal of individual branches or parts of branches, and will probably be undertaken where small, horticultural-scale adjustments to the balance between plants is required. For large-scale schemes management measures are more likely to focus on thinning and/or coppicing.

Thinning is the term used for the selective removal of individual plants to allow other individuals more room to develop. It may be necessary to reduce initial planting densities, which may have been high, to ensure the achievement of a good initial ground cover. The removal of fast-growing but short-lived nurse species, planted initially to ensure quick protection, may be another reason for thinning.

Coppicing is a form of management involving the regular cutting of some or all of the aerial shoots of a tree or shrub back to ground level. It is only appropriate for species that will re-grow from cut stumps or stools, and many species will not do this. Coppicing was formerly done in order to harvest small timber, but it also has the effect of continually rejuvenating the plant and preventing it from developing into a large crowned tree or shrub.

The advantage of coppicing from a slope stabilization point of view is that the growth of many young shoots is stimulated which increases the protection of the soil surface by providing good interception of precipitation above ground, and promotes a young, vigorous and well-balanced root system below ground. The wind resistance of a regularly coppiced plant will also be relatively low, thus lowering the risk of windthrow and subsequent damage to the slope.

As with coppicing, thinning and pruning also provide some scope for obtaining young live woody material which can itself be used in bio-engineering projects. If this is one of the objectives of managing an existing area of woody vegetation, then the timing of the operation needs to coincide with the proposed season for establishing plants on a new stabilization project. From the point of view of the existing vegetation, all of these management operations are usually best carried out during the dormant season in those areas where there is one. However, in temperate areas at least, this may also be the season when the woody vegetation may be playing its most important role in stabilizing the slopes from which it is to be harvested. For this reason, whatever is to happen to the harvested material, considerable thought needs to be given to the timing of management operations.

3.5.4 OPERATIONAL IMPLICATIONS AND SOLVING SPECIFIC PROBLEMS

Relying on vegetation to provide an essential contribution to slope stability and to help control erosion means that it must be viewed as an integral component of the structural system providing slope stability. As such, in the same way that a reinforced-concrete retaining wall or similar inert engineering structure needs to be the subject of a programme of regular structural inspections to ensure its continued stability, there is a need for continuous supervision and maintenance of vegetation.

Regular inspections should be designed to monitor the condition of the vegetation. Information on its condition should be fed into the preparation of management schedules and be linked to the monitoring of management operations. The timing of vegetation inspections clearly needs to be tied into the cycles of vegetation growth and dormancy where these exist.

Any damage to the integrity of the vegetation will clearly present a possible threat to the stability of the slope. Damage to vegetation can result from a number of factors, both natural and man-made. These include attack by pests and diseases, the results of fires (both spontaneous and those started by man), the effects of other forms of vandalism, and the impact of climatic extremes. Both the initial design and the planning of management measures should, as far as this is possible, aim to minimize the possible risks which these factors might cause.

The effects of attack by pests and diseases and of climatic extremes will differ from species to species, and the creation and encouragement by management of a varied and well-structured mix of species will provide the best insurance against any 'natural disasters' to the vegetation, which will also have devastating effect on slope stability. Maintenance measures should include inspections for any signs or symptoms of the early stages of pest or disease attack, so that appropriate remedial measures can be taken.

BIBLIOGRAPHY

Alscher, R. G. and Cumming, J. R. (1990) *Stress Responses in Plants: Adaptation and Acclimatization Mechanisms.* Wiley-Liss.

Bohm, W. (1979) *Methods of Surveying Root Systems.* Springer Verlag, Berlin.

Bradshaw, A. D. and Chadwick, M. J. (1980) *The Restoration of Land.* Blackwell, Oxford.

Buckley, P. (ed.) (1989) *Biological Habitat Reconstruction.* Belhaven Press, London.

Coppin, N. J. and Bradshaw, A. D. (1982) *Quarry Reclamation – The Establishment of Vegetation in Quarries and Open Pit Non-metal Mines.* Mining Journal Books, London.

Coppin, N. J. and Richards, I. G. (eds), (1990) *Use of Vegetation in Civil Engineering.* Construction Industry Research and Information Association/Butterworths, London.

Daubenmire, R. F. (1974) *Plants and Environment. A Textbook of Autecology,* 3rd edn, Wiley, New York.

Doorenbos, J. and Pruitt, W. O. (1984) *Guidelines for Predicting Crop Water Requirements.* FAO Irrigation and Drainage Paper 24. Food and Agriculture Organisation of the United Nations, Rome.

Etherington, J. R. (1982) *Enviroment and Plant Ecology.* Chichester, Wiley.

Fenner, M. (1985) *Seed Ecology.* Chapman and Hall, London.

Fitter, A. H. and Hay, R. K. M. (1989) *Environmental Physiology of Plants,* 2nd edn, Academic Press, London.

Grace, J. (1983) *Plant Atmosphere Relationships.* Chapman and Hall, London.

Grace, J., Ford, E. D. and Jarvis, P. G. (1981) *Plants and Their Atmospheric Environment.* Blackwell, Oxford.

Gray, D. H. and Leiser, A. T. (1982) *Biotechnical Slope Protection and Erosion Control.* Van Nostrand Reinhold, New York.

Grime, J. P. (1979) *Plant Strategies and Vegetation Processes.* Wiley, Chichester.

Grime, J. P., Hodgson, J. G. and Hunt, R. (1988) *Comparative Plant Ecology.* Allen and Unwin, London.

Hall, D. G. M., Reeve, M. J., Thomasson, A. J. and Wright, V. F. (1977) Water retention, porosity and density of field soils. *Soil Survey of England and*

Wales, Technical Monograph No. 9. Harpenden, UK.

Hammitt, W. E. and Cole, D. M. (1987) *Wildland Recreation, Ecology and Management.* Wiley, Chichester.

Harper, J. L. (1977) *The Population Biology of Plants.* Academic Press, London.

Hartmann, H. T. and Kester, D. (1975) *Plant Propagation: Principles and Practices.* Prentice Hall, Englewood Cliffs.

Haslam, S. M. (1978) *River Plants.* Cambridge University Press, Cambridge.

Haslam, S. M. and Wolseley, P. A. (1981) *River Vegetation: Its Identification, Assessment and Management.* Cambridge University Press, Cambridge.

Hubbard, C. E. (1968) *Grasses.* Pelican, London.

Jarvis, M. G. and Mackney, D. (1979) Soil survey applications. *Soil Survey of England and Wales, Technical Monograph No. 13.* Harpenden, UK.

Jones, H. G. (1992) *Plants and Microclimate; A Qualitative Approach to Environmental Plant Physiology,* 2nd edn, Cambridge University Press, Cambridge.

Kellman, M. C. (1975) *Plant Geography.* Methuen, London.

Kershaw, K. A. (1973) *Quantitative and Dynamic Plant Ecology,* 2nd edn, Edward Arnold, London.

Kozlowski, T. T. (ed.) (1981) *Water Deficits and Plant Growth,* Vols I–VI (1968–1981). Academic Press, London.

Levitt, J. (1980) *Responses of Plant to Environmental Stresses,* 2nd edn, Vols I and II. Academic Press, New York and London.

Monteith, J. L. (1975) *Vegetation and the Atmosphere. Volume I – Principles.* Academic Press, London.

Moore, P. D. and Chapman, S. B. (1986) *Methods in Plant Ecology.* Blackwell, Oxford.

Rosenberg, N. J., Blad, B. L. and Verma, S. B. (1983) *Microclimate: The Biological Environment.* Wiley, New York.

Russell, E. W. (1973) *Soil Conditions and Plant Growth.* Longman, London.

Russell, G., Marshall, B. and Jarvis, P. G. (eds) (1989) *Plant Canopies: Their Growth, Form and Function.* Cambridge University Press, Cambridge.

Russell, R. S. (1977) *Plant Root Systems: Their Function and Interaction with the Soil.* McGraw Hill, New York.

Shimwell, D. W. (1971) *The Description and Classification of Vegetation.* Sidgwick and Jackson, London.

Silvertown, J. W. (1982) *Introduction to Plant Population Ecology.* Longman, London.

Treshow, M. (1970) *Environment and Plant Response.* McGraw Hill, New York.

Walter, H. (1973) *Vegetation of the Earth.* English Universities Press Ltd.

Wild, A. (1988) *Russell's Soil Conditions and Plant Growth.* Longman, Harlow.

SIMULATED VEGETATION AND GEOTEXTILES

4

R. J. Rickson

4.1 THE NEED FOR SIMULATED VEGETATION

The engineering properties of vegetation that can control soil erosion and enhance slope stability have been considered in Chapter 2. The effects of vegetation are only fully realized once it has reached maturity. During the critical stage of plant establishment the beneficial engineering properties of the vegetation may not be apparent and a site is still highly susceptible to soil erosion. This fact applies to agricultural crops as well as to the vegetation used for civil engineering projects such as roadbanks, side-slopes of borrow pits, dam backslopes, ditch banks and cut and fill slopes. Here, annual erosion rates in excess of 480 t/ha have been recorded (Diseker and Richardson, 1962). The vulnerable period when no protection by vegetation exists may be extended in time by detrimental climatic or site conditions such as extreme temperatures or high rainfall intensities, soil toxicity or acidity, or excessive trafficking by humans or machinery, which can all hinder rapid establishment and growth of healthy vegetation. Delays in design project implementation may also increase the length of time the slope is prone to erosion, which may cause particular difficulties if seeding is postponed beyond the end of the growing season.

Without immediate, appropriate and adequate protection, slopes can suffer severe soil erosion and instability, which in turn make any further attempts at vegetation establishment extremely difficult, if not impossible. Erosion of seeds and seedlings from unprotected sites by surface runoff and high winds incurs costs in time and money as all previous attempts to establish vegetation on the slope have to be repeated. For example, in agriculture, sugar beet seedlings are extremely susceptible to scour by windborne sediment, and the damage done to the establishing vegetation is so severe that expensive re-drilling is required (Rickard, 1985).

Simulated vegetation, in the form of mulches and erosion control geotextiles, however, needs no time for establishment; the benefits for erosion control and slope stability are immediate. Simulated vegetation mimics the salient properties of natural vegetation which control erosion, namely the canopy, stem and root effects (Chapter 2). Another advantage of simulated vegetation is that the materials used can combine with the establishing and ultimately mature vegetation to give 'composite' erosion control. This effect can be long term, or even permanent, if nondegradable mulch or geotextile materials are used. Such a combination of live and inert materials may lead to 'synergistic' relationships, where the control of erosion from the two combined is greater than that of the sum of the two approaches used in isolation.

In some cases, simulated vegetation can be used to modify microclimatic and soil conditions on-site, so affecting the rate of vegetation establishment – sometimes positively, sometimes not, depending on the specific site. Finally, many of the materials used to simulate vegetation are readily available. Many are

Slope Stabilization and Erosion Control: A Bioengineering Approach. Edited by R. P. C. Morgan and R. J. Rickson. Published in 1995 by E & FN Spon, 2-6 Boundary Row, London, SE1 8HN. ISBN 0 419 15630 5.

by-products from agriculture and industry and can be relatively inexpensive to use. Kay (1978) states that the price of agricultural mulches is a reflection of the transport and handling costs, rather than the intrinsic value of the product. Meyer, Johnson and Foster (1972) show that mulching at rates costing less than conventional sodding techniques can be successful in stabilizing slopes.

4.2   THE USE OF MULCHES

Mulches are used to protect soil surfaces from the erosive agents of rainfall, runoff and wind. They also help to reduce intense solar radiation, suppress extreme fluctuations of soil temperatures, reduce water loss through evaporation and increase soil moisture, which can assist in creating ideal conditions for plant growth in many circumstances (Sprague and Triplett, 1986). Mulches can also be used to suppress weed growth by preventing unwanted seeds settling on, and germinating in, the unprotected soil. The benefits of mulching are proportional to the adversity of the environment in which they are applied (Jackobs *et al.*, 1967).

Mulch materials are diverse, ranging from crop residues such as cereal straw (Meyer and Mannering, 1963; Lattanzi, Meyer and Baumgardner, 1974; Laflen and Colvin, 1981), corn stalks (Worku and Thomas, 1992), hay, oat straw (Singer, Matsuda and Blackard, 1981), rice straw (Lal, 1976), sugar cane residue (Ruiz and Valentin, 1987), fresh grass cuttings and grass straw (Kay, 1978), leaves (Singer and Blackard, 1978), tree bark, wood shavings (Meyer, Johnson and Foster, 1972), wood pulp (Kill and Foote, 1971), paper, crushed stones (Meyer, Johnson and Foster, 1972), gravel (Seginer, Morin and Sachori, 1962; Adams, 1966), to the non-natural mulches such as glass fibre rovings and plastics. Other by-products used as mulches include cotton gin trash (Fryrear and Koshi, 1971). A combination of different mulches such as wheat straw and hay can be effective in erosion control and plant establishment (Kay, 1978).

The quantity of residue produced varies with different crops. For example, corn and small grains yield higher quantities of residue than soybean, cotton or tobacco (Meyer and Mannering, 1963). This is also determined by climatic, soil and management factors.

The effectiveness of different mulch materials in soil erosion control and vegetation establishment will be dependent on:

1. type of mulch material used (Meyer, Johnson and Foster, 1972 – comparing straw, stone and wood chip mulches; Kay, 1978 – comparing wood residues with straw);
2. mulch morphology, e.g. corn stalks are more effective in erosion control than simulated corn leaves (Okwach, Palis and Rose, 1992);
3. application rate;
4. method of application (surface versus incorporated, Roose and Asseline, 1978; Poesen, 1986);
5. soil type;
6. slope;
7. climatic characteristics (e.g. Barnett, Diseker and Richardson (1967) compared the effectiveness of mulches under a 1 year and 10 year frequency storm event).

These factors will also affect the critical slope length at which unanchored mulches will fail either by movement of individual mulch elements, *en masse* movement of the mulch or by development of rills beneath the mulch (Kramer and Meyer, 1969; Foster, Johnson and Moldenhauer, 1982a; Cadena-Zapata, 1987). Decisions as to what type of mulch is to be used are usually based on local availability and costs. Often crop residues are used for livestock feeding, bedding, fuel or thatching, so they may not be available for mulching. Also, in areas of high fire hazard, certain combustible mulches may not be appropriate. Other problems associated with mulches include harbouring of diseases and pests, and the creation of favourable habitats for rodents.

The durability of different mulch materials is important as this will affect their effective lifespan. Mulches composed of residues with low carbon/nitrogen (C:N) ratios such as legumes will decompose quickly, whereas straw and cornstalks are longer lasting as they have relatively higher C:N ratios. Decomposition rates are also affected by whether the mulches are surface laid, incorporated or covered with soil (Unger and Parker, 1968; Douglas *et al.*, 1980), and by climate.

Mulches can be applied as a surface treatment, where they simulate the role of ground cover on soil erosion processes by intercepting rainfall and retarding runoff velocities (sections 2.2.2 and 2.3.1). On agricultural land, often the mulch is the previous season's crop residue left after harvesting. If the residue is left standing, it can be killed with a herbicide, but will still provide excellent soil protection as the roots remain anchored in the soil (Meyer and Mannering, 1963). However, there are increasing concerns that excessive use of such chemicals may have detrimental effects on the environment, if they are leached into groundwater sources or surface water bodies. Large-stalked residue is best mechanically shredded and spread to provide more effective soil cover. Application can be by hand or by specially designed straw blowers, which can spread mulch materials at a rate of 15 t/h over a distance of 30 m. Blown straw mulches tend to have better contact than mulches spread by hand (Kay, 1978).

Surface mulches can be applied hydraulically, where the mulch is sprayed on to the slope in the form of a slurry containing the mulch, seeds, fertilizers, chemical soil binders, humectants and fibre tackifiers. This process, known as 'hydromulching' is most effective when good quality wood chips are used as the mulch (Kay, 1978). One problem of the technique is that about 60–70% of the seed sticks to the mulch elements and has little or no chance of getting its primary roots into the soil. The technique has limited application on agricultural land, although is more common on engineered slopes, such as road cuttings and embankments.

Optimum protection by surface mulches is given by 65–75% cover (Morgan, 1986). Lower percentage covers were thought to give insuffi-

cient protection, although recent studies have shown the effectiveness of lower application rates. Mannering and Meyer (1963) found that the percentage reduction in erosion per increment of mulch applied was greater at lower application rates. This reiterates Jackobs *et al.*'s (1967) statement that mulches perform best under the more adverse conditions (i.e. where there is a low level of soil protection). Denser applications may inhibit vegetation emergence beneath the mulch, through interception of light, impedance of rainfall to irrigate the soil or as a physical barrier to emergence. Usually surface mulches are applied uniformly over the site, but the use of 'vertical mulches' or trenches filled with mulching material are an exception to this (Brown and Kemper, 1987). In this case, vegetation establishment within the trenches is not the objective, so higher application rates can be justified.

Estimates of percentage soil cover per unit weight for different mulches can be calculated based on average length, diameter and weight of a random sample of clean, oven-dried mulch material. Greb (1967) found that the most efficient soil cover/unit weight relationship was given by a mulch of spring barley followed equally by winter wheat and spring oats, then sudan grass, grain millet, and finally grain sorghum, which had the lowest soil cover/unit weight efficiency of the materials tested.

Applying the mulch on the surface can lead to practical problems, particularly on cultivated lands (Abrahim and Rickson, 1989). Conventional farming operations such as drilling are difficult through the thick layer of residue. This problem is often held to be the reason why many land managers are reluctant to use mulches. Other problems include the lack of anchorage and contact of the mulch with the soil surface when the site is subject to strong winds or erosive runoff. One solution to this is to 'tack' the mulch material to the soil with asphalt (Swanson *et al.*, 1965; Kay, 1978; Pla, Florentino and Lobo, 1987), but this can inhibit germination beneath the mulch (Sheldon and Bradshaw, 1977). Galvanized wire netting has also been used extensively to fix straw mulches. Plastic, light-degradable meshes have been used to secure straw, wood chip and other mulch materials. These are reviewed more extensively in section 4.3.

Weed control under surface mulching relies on increased herbicide applications, rather than using cultivations to control the weeds. Again, this has attracted a great deal of criticism as to the environmental impacts of excessive use and leakage of the chemicals into the surrounding water sources

These disadvantages have encouraged the development of subsurface mulching techniques, involving the incorporation or 'trashing' of the mulch material (commonly crop residues). Whilst percentage cover is reduced (for the same application rate when applied as a surface mulch), the effect of the plant stems and roots on erosion may be simulated more effectively. Incorporation is usually carried out by conventional operations such as disc or chisel ploughing, with better erosion protection given by across the slope operations rather than up and down. Other techniques of incorporating the mulch including 'crimping', where the mulch is left in vertical tufts by a disc packer, usually at an application rate of 2 t/ha, and 'punching' where the residue (4 t/ha) is rolled into the soil. Details of the various machines used for mulch incorporation are found in Kay (1978). Several Highway Departments in the United States use the technique of 'whisker dams', where a straw mulch is embedded or pressed into the soil (Barnett, Diseker and Richardson, 1967). However, with all of the various machines used, there are slope steepness limits as to their safe use.

Swanson *et al.* (1965) found no difference in the erosion control afforded by a loose mulch of prairie hay compared with when it was incorporated in the soil, although differences due to the mode of application were observed for wood chips. Where these had been incorporated by discing, protection was less effective than for the surface treatment.

From the work outlined above on the mode of installation, the success or otherwise of either surface- or subsurface-applied mulches appears to be dependent on the unique characteristics of the site on which they are to be used.

4.3 THE ROLE OF MULCHES IN WATER EROSION CONTROL

Most research into the use of mulches for erosion control has concentrated on how surface mulches (Wischmeier, 1973; Foster and Meyer, 1975; Lal, 1977) can reduce the rates of soil erosion by water. Lal (1976) states that mulches reduce soil erosion by:

1. reducing raindrop impact;
2. increasing soil infiltration;
3. increasing surface storage;
4. decreasing runoff velocity;
5. improving soil structure and porosity;
6. improving the biological activity in the soil so improving soil structure and porosity.

Lattanzi, Meyer and Baumgardner (1974) show that soil erosion on a plot mulched with wheat straw at an application rate of 0.5 t/ha (= 25% cover) was reduced by 35–40% of that observed for an unmulched plot. At an application rate of 2 t/h (61% cover) the losses were reduced by 75–80%. These results agree with those under similar conditions by Meyer, Wischmeier and Foster (1970). Kramer and Meyer (1969) also found that mulch application rates as low as 0.5–1 t/ha would reduce soil erosion greatly, compared with an unmulched plot. Hussein and Laflen (1982) found that rill erodibility was directly related to the amount of surface residue cover. Duley and Russel (1943) showed that sandy loam plots with incorporated residue yielded only 9.9 t/ha of sediment, whereas the bare plot yielded 35.7 t/ha.

Much of the research on surface mulches illustrates that different types of mulch have different degrees of effectiveness in controlling soil erosion. Singer and Blackard (1978) show that a 40% cover of oat straw reduces sediment in runoff significantly, but application rates of 75% and 70% are required for mulches of redwood litter and oak leaves respectively, if soil losses are to be reduced significantly compared with the unmulched condition. Meyer, Johnson and Foster (1972) found that twice the application rate was needed for wood residue mulches as for a straw mulch to give the same erosion control. Swanson *et al.* (1965) found this application rate ratio between the two products to be even higher at 6:1.

In order to explain how mulches control soil erosion, it is necessary to analyse their effect on the various factors affecting erosion.

4.3.1 THE ROLE OF MULCHES IN CONTROLLING RAINDROP IMPACT

Rainsplash erosion can be controlled using mulches because they simulate the effect of ground cover. An exponential relationship exists between the percentage area covered by the mulch and the detachment rate of soil by raindrop impact (Laflen and Colvin, 1981). The kinetic energy of the raindrops is intercepted and reduced by the mulch components, so that lower energies act on the soil surface. In other words, the incident rainfall is less erosive. This arises because the mass of the drops is reduced as the drops shatter on impact with the mulch, and drop velocities approach zero as there is insignificant fall height from the mulch components to the ground. As raindrops are the most efficient and effective means of soil detachment (Morgan, 1986), control of their erosive energy dramatically reduces total soil losses from the site. Splash transport rates can also be reduced under a mulch cover of 96% to less than 8% of that observed for bare soil (Singer, Matsuda and Blackard, 1981).

Mulch thickness can determine the degree of raindrop interception. Adams (1966) found a gravel mulch of 5 cm depth reduced soil losses significantly compared with a mulch of 2.5 cm depth. Light rain falling on the mulch may help to increase its effective weight, so improving the contact between mulch and soil.

4.3.2 THE ROLE OF MULCHES IN INCREASING SOIL INFILTRATION

Many workers have shown the role of mulches in increasing infiltration rates and encouraging the movement of excess water at depth (Adams, 1966; Lal, De Vleeschauwer and Malafa Nganje, 1980). Lattanzi, Meyer and Baumgardner (1974) found that with low rates of mulch application (0–2 t/ha), infiltration rates were only 20% of those observed under an application rate of 8 t/ha. Mulches help prevent the process of crusting or sealing on the soil surface, which can severely restrict soil infiltration rates, particularly on fine textured soils. As soil detachment rates are curbed under mulching (see section 4.3.1), fewer detached particles are available to seal the soil surface; sealing occurs less rapidly, so that soil infiltration capacities remain high, reducing runoff generation (Cadena-Zapata, 1987).

Compaction of the soil by raindrop impact can also reduce soil infiltration rates. Covering the soil with a mulch will protect the soil from this process. Whether increases in infiltration rates also increase the susceptibility of certain soils to slaking (i.e. building up of soil porewater pressures followed by their sudden release, so destabilizing soil structure) does not appear to have been investigated in the literature.

With subsurface mulches the incorporation procedure helps to loosen soil particles, reducing bulk densities, so enhancing soil infiltration rates and capacities. Whilst soil disturbance may enhance detachability of soil particles, this appears to be more than compensated for by the increases in the ability of the soil surface to absorb rainfall. Subsurface mulches will also provide preferential lines for rainfall to infiltrate the soil profile, thereby reducing surface runoff volume.

The role of mulches on infiltration is graphically shown by Aldefer and Merkle (1943), where runoff hydrographs for surface-applied and incorporated mulches are compared with the infiltration and runoff characteristics of

Table 4.1 Effect of surface mulching on runoff coefficient and peak runoff rate (after Ruiz and Valentin, 1987)

	Rainfall rate (mm/h)	Bare surface	Mulched surface
Runoff coefficient (%)[a]	30.0	57.9	11.4
	120.0	84.3	21.2
Peak runoff rate (mm/h)[b]	30.0	20.0	11.0
	120.0	108.0	65.0

[a] Runoff coefficient = runoff volume/rainfall volume.
[b] Peak runoff rate = maximum value of runoff rate recorded during the four simulated rainfalls.

an unmulched control plot. Infiltration capacities equal to or in excess of 76 mm/h were measured for the surface mulched plots, and of 63 mm/h for the incorporated plots. These figures compare with only 25 mm/h for the control plot.

4.3.3 THE ROLE OF MULCHES IN INCREASING SURFACE STORAGE

Surface storage reduces runoff volumes and peak runoff rates (Ruiz and Valentin, 1987; Table 4.1). Singer and Blackard (1978) and Adams (1966) note that surface storage with different mulches leads to different rates of runoff, times to onset of runoff, and times between the end of a rainfall event and the cessation of runoff. Mulch elements increase surface water storage by forming miniature dams on the soil surface behind which water is ponded (Bonsu, 1980). This layer of water may protect the soil from further raindrop impact. The static water will infiltrate the soil profile rather than running off the slope (Meyer and Mannering, 1963). This helps to reduce both soil and water losses (Valentin and Roose, 1980). The shape of the mulch elements affects the contact between the mulch and the soil, so affecting surface storage behind the mulch. Njoroge (1985) compared a wheat straw and a broad-leaved mulch, concluding that the former was more effective at storing potentially erosive runoff than the latter at equivalent application rates.

Abrahim and Rickson (1989) investigated the effect of various incorporated mulches on runoff volumes. The materials used were equal application rates of hay, wheat straw and maize stalks, which were compared with a control plot, with no incorporated residue. All the mulches reduced runoff volume, particularly the wheat straw. Observations showed that this was due to the excellent contact between the straw and the soil, and the higher percentage cover at the given application rate (measured as a weight of material) for the wheat compared with the corn stalks. Both these factors helped to pond surface water. It was found that the wheat straw was effectively incorporated, unlike the larger corn stalks, which resisted incorporation and clogged up the machinery. The hay treatment performed relatively badly, as it tended to be over-incorporated, thus losing vital surface cover. The reductions in runoff volume explained the observed soil losses for the different treatments.

However, using a simulated stone mulch cover, Poesen (1986) noted the generation of 'rock flow', whereby the mulch elements did not reduce runoff volume, but actually increased it as all the rainfall intercepted by the non-absorbent rocks became runoff. Where the stones were placed on the soil surface, this runoff could infiltrate. However, for the same cover percentage, but with the stones embedded (particularly on soil susceptible to capping), the formation of a continuous crust prevented this 'rock flow' from infiltrating, so increasing the erosion risk on these slopes.

4.3.4 THE ROLE OF MULCHES IN CONTROLLING RUNOFF VELOCITY

Even if soil particles are detached, mulches help to control the efficiency with which runoff transports them away. Morgan (1986) states that sediment transport capacity is more sensitive to runoff velocity (transport capacity varies with runoff velocity raised to the fifth power), than to runoff volume (transport capacity varies with the square of the runoff volume). Any reduction in flow velocity will have a dramatic effect on the ability of the runoff to transport detached particles. Many workers have investigated the degree to which mulches can reduce runoff velocities (Hines, Lillard and Edminster, 1947; Meyer and Mannering, 1963). Lattanzi, Meyer and Baumgardner (1974) showed that interrill runoff velocities on unmulched plots were 2.5 cm/s, whereas for wheat straw mulched plots the velocities were 1.3 cm/s and 0.8 cm/s for application rates of 0.5 t/ha and 2 t/ha, respectively. Runoff velocity was halved with a straw mulch rate of 0.56 t/ha compared with bare soil (Meyer, Wischmeier and Foster, 1970). Such a reduction results in a dramatic decline in the transporting capacity of the flow (Singer, Matsuda and Blackard, 1981).

Foster, Johnson and Moldenhauer (1982b) use basic hydraulic theory to explain how mulches reduce runoff velocity. The shear stress of water flow is subdivided into that acting on the soil (equivalent to the grain roughness), and that acting on the mulch (equivalent to the form roughness) (Graf, 1971). This roughness can be expressed as a Manning's *n* value for different types of mulch (Table 4.2). Any roughness acting on the flow will reduce runoff velocity.

Flow shear stress is imparted to the mulch rather than to the soil. At heavy mulch rates shear stress on the soil was less than that observed for the unmulched control plot, even though the total shear stress exerted by the flow on the mulched plots was greater. The mulch elements resist the stress which slows the velocity of runoff, in turn reducing the shear stress exerted on the mulch elements. Hence there is a positive feedback of reduced velocity leading to reduced shear stress. The mulch will only fail when the shear stress acting on it exceeds a critical threshold, which is dependent on the type of mulch and its application rate.

Van Liew and Saxton (1983) found similar processes operating when the mulch is incorporated into the soil. The subsurface mulch had a significant effect on flow shear stress so that the incidence of rilling was reduced by 60% under the medium application rate (4.7 t/ha) and by 85% with the high application rate (9.4 t/ha), compared with no residue. They also found that incorporated wheat straw acted as a binding agent with the soil, so higher runoff velocities and energies were required to erode the soil.

Runoff becomes less efficient with the longer flow paths caused by the mulch elements (Singer and Blackard, 1978). Mulch shape affects the roughness imparted by the mulch to overland flow, thus affecting runoff velocity (Singer and Blackard, 1978). Length of the mulch fibres is also important, with long fibres performing better than short fibres (Meyer and Mannering, 1963; Kill and Foote, 1971; Kay, 1978; Pickles, 1984). This is because longer fibres exert more resistance to the flow as their anchorage is greater. Swanson *et al.* (1965) estimated that an

Table 4.2 Manning's *n* values for different mulch types and application rates (after Foster, Johnson and Moldenhauer 1982b)

Mulch type	Rate (kg/m^2)	Manning's n (due to mulch)
Corn stalks	0	–
	0.22	0.0102–0.0196
	0.45	0.0221–0.0296
	0.66	0.0393
	0.70	0.0477
	0.90	0.0655–0.0594
	1.3	0.112
Wheat straw	0	–
	0.11	0.0029
	0.22	0.0418
	0.45	0.131

application rate of 13 t/ha of short-fibred wood-chips was required to attain the same degree of erosion control as 2 t/ha of long-fibred mulch such as hay or straw.

Mulch elements act as miniature grade control structures, behind which runoff velocities are locally reduced and deposition of sediment occurs, decreasing local hydraulic gradients and reducing runoff velocity still further. This feedback principle can be seen working in practice with the use of 'whisker dams' by US Highway Departments.

Despite these effects, Stallings (1949) illustrates that crop residue is more efficient at absorbing raindrop impact energies than impeding runoff in the control of soil erosion processes.

4.3.5 EFFECT OF MULCHES ON SOIL STRUCTURE AND POROSITY

The effect of soil structure on erosion is usually linked with the degree of aggregation in the soil, and the stability of the soil aggregates when subjected to the forces of rainfall impact and runoff. In a classic paper by Aldefer and Merkle (1943), indices of soil stability were calculated for a number of different mulch treatments, three and four years after their initial application (Table 4.3). Analysis showed that none of the surface nor incorporated treatments produced statistically significant changes in soil structural stability when compared with the fallow check plot. However, qualitative analysis showed that the maximum differential effect of the surface-applied mulches on soil stability was found just below (<2.5 cm) the soil surface. Of the ten surface mulches, manure and pine needle mulched soils had the highest structural stability. The sawdust, corn stover, oak leaves, bluegrass clippings and glasswool mulches gave similar indices of structural stability as for the control plot. The straw, charcoal and sand and gravel mulched plots showed a reduced index of structural stability, compared with the control.

Lal, De Vleeschauwer and Malafa Nganje (1980), using different mulch rates rather than different mulch types, found a very strong posi-tive relationship ($r = 0.98$) between percentage water-stable aggregates (>0.5 mm) and mulch rate. Other workers found that soil mulched with sugar cane residues for three years had a higher proportion of water-stable aggregates than an unmulched plot (Prove and Truong, 1988). Since many workers have shown a direct relationship between aggregate stability and soil erodibility (Bryan, 1968), a mulch can make the soil more resistant to water erosion, albeit indirectly.

Organic matter levels, also known to influence soil structure and aggregation, increase when mulches are applied, whether on the surface (Aldefer and Merkle, 1943) or incorporated (Biggerstaff and Moore, 1982; Table 4.3). However, disturbance during mulch incorporation can destroy stable aggregates and have a negative effect on soil structure, so increasing soil susceptibility to erosion. Indeed, Meyer and Mannering (1963) suggested this soil disturbance gave higher soil losses for plots where residues had been incorporated, than where the mulch had been applied on the surface. However, it is likely that in some situations, enhanced infiltration through mulch incorporation will counterbalance the effect of disturbance to soil structure.

The effect of mulches on soil porosity was investigated by Lal, De Vleeschauwer and Malafa Nganje (1980), using rice straw at application rates of 0, 2, 4, 6 and 12 t/ha. Total porosity for these treatments was 48, 50, 55, 55 and 59% respectively, with related increases in the number of macropores, hydraulic conductivity, field infiltration rates and moisture retention at 0.1 and 0.3 bar suctions.

Aldefer and Merkle (1943) expressed soil porosity as 'probable permeability', which was taken as the percentage of the soil which consists of either primary or secondary particles having diameters giving optimum moisture permeability. They found that both surface and incorporated mulches had more effect on porosity than they had on soil aggregation. For the surface mulches, the numbers and size of larger granules decreased with increasing depth. Most

Table 4.3 Stability index, probable permeability, and organic matter values for soils 3 and 4 years after the surface application or incorporation of various mulching materials (after Aldefer and Merkle, 1943)

Plot symbol	Treatment	Soil depth (mm)	Stability index 1941	Stability index 1942	Probable permeability 1941	Probable permeability 1942	Organic matter (%) 1941	Organic matter (%) 1942
A	Check: fallow, uncultivated; weeds removed by scraping			56.6[a]		25.3[a]		3.09[a]
		0–75	49.5	57.6	27.8	27.3	2.85	3.19
		75–150	49.2	57.3	27.4	27.5	2.83	2.98
	Check: fallow, cultivated			54.3		26.3		2.96
		0–75	51.1	53.6	26.2	27.0	2.94	2.92
		75–150	51.3	56.1	20.4	28.4	2.75	2.83
H	Manure mulch			61.1		61.1		5.40
		0–75	54.9	62.7	42.0	50.3	3.67	4.07
		75–150	52.8	64.6	38.6	47.4	3.49	3.27
	Manure incorporated			54.2		30.8		3.21
		0–75	51.0	56.1	32.7	36.6	3.29	3.87
		75–150	53.7	62.7	32.7	39.1	3.85	3.59
B	Straw mulch			54.1		32.2		3.61
		0–75	52.9	56.2	20.7	31.1	2.93	3.12
		75–150	58.9	61.0	27.8	28.9	2.84	3.03
	Straw incorporated			53.0		31.8		2.80
		0–75	58.7	56.1	21.3	31.1	2.59	2.91
		75–150	56.1	61.8	25.8	30.2	2.75	2.61
C	Sawdust mulch			58.6		51.6		3.91
		0–75	63.7	57.8	24.3	37.8	3.02	3.14
		75–150	60.7	62.7	24.6	27.9	2.96	2.88
	Sawdust incorporated			54.3		26.9		3.05
		0–75	51.1	57.1	29.9	27.0	3.00	3.19
		75–150	59.6	59.6	31.7	27.1	3.66	3.67
E	Corn fodder mulch			57.9		42.8		3.44
		0–75	52.1	60.7	26.0	38.7	2.79	3.07
		75–150	49.1	60.5	29.8	36.7	2.44	2.35
E	Corn fodder incorporated			54.7		28.4		2.71
		0–75	49.0	56.7	26.4	28.0	2.31	2.78
		75–150	54.9	57.7	20.4	24.5	2.40	2.17
F	Charcoal mulch			51.9		28.4		4.39
		0–75	49.1	56.6	24.8	27.2	3.63	2.88
		75–150	49.2	59.7	26.3	32.1	2.61	2.70
	Charcoal incorporated			50.2		25.4		4.75
		0–75	42.0	49.7	27.3	28.3	4.01	5.18
		75–150	48.4	45.6	25.7	28.0	5.86	5.62
G	Peat incorporated	0–75	54.4	–	32.9	–	4.98	–
		75–150	55.9	–	31.2	–	3.50	–
I	Oak leaves mulch			55.9		45.6		3.72
		0–75	51.8	58.5	22.6	42.7	3.03	3.06
		75–150	53.8	60.7	31.0	41.6	3.00	2.84

Table 4.3 (cont.)

Plot symbol	Treatment	Soil depth (mm)	Stability index		Probable permeability		Organic matter (%)	
			1941	1942	1941	1942	1941	1942
	Oak leaves incorporated			54.4		26.0		2.60
		0–75	47.2	56.2	27.4	28.7	2.63	2.79
		75–150	53.3	57.5	26.4	33.9	2.86	2.48
J	Pine needles mulch			62.2		39.5		3.50
		0–75	52.9	59.1	25.5	33.2	2.89	2.94
		75–150	50.0	60.0	27.4	29.1	2.80	2.65
	Pine needles incorporated			55.3		25.3		2.62
		0–75	53.0	55.2	26.8	26.6	2.67	2.73
		75–150	47.4	57.2	28.2	31.2	2.94	2.53
T	Grass clippings mulch			56.1		31.8		2.66
		0–75	43.8	57.0	26.0	34.4	2.76	2.47
		75–150	46.2	57.6	24.6	27.4	2.36	2.38
	Grass clippings incorporated			53.4		25.6		2.29
		0–75	43.2	54.8	23.5	28.0	2.32	2.28
		75–150	49.6	55.6	22.0	27.8	2.30	1.94
D	Sand and gravel mulch			50.9		31.5		2.66
		0–75	52.9	58.4	24.2	29.4	2.66	2.60
		75–150	55.7	59.6	23.0	26.9	2.76	2.49
	Sand and gravel incorporated			47.1		34.2		2.49
		0–75	47.4	42.9	30.6	37.2	2.43	2.45
		75–150	48.4	48.4	29.2	35.6	2.37	2.15
M	Glasswool mulch			56.2		28.2		2.45
		0–75	48.4	56.3	25.6	26.7	2.47	2.49
		75–150	50.8	59.1	21.7	27.7	2.52	2.28
L	Complete fertilizer, 4–12–8			54.0		26.4		2.59
		0–75	50.9	57.1	20.5	28.9	2.74	2.64
		75–150	48.1	59.8	20.6	27.7	2.88	2.50
O	Nitrate of soda			50.4		20.8		2.35
		0–75	47.9	54.0	24.2	27.0	2.39	2.57
		75–150	51.7	58.3	21.4	26.5	2.66	2.23
R	Muriate of potash			50.5		26.9		2.60
		0–75	52.6	52.8	21.1	27.3	2.58	2.66
		75–150	50.1	57.8	27.4	˙30.2	2.74	2.69

[a] The first value is for a 0–25 mm depth in each case; the second value is for a 25–75 mm depth.

effective was manure, followed by oak leaves, corn stover, sawdust, pine needles, bluegrass clippings and finally straw. Surprisingly, incorporating the mulches did not improve the soil porosity to the same extent as the surface-applied mulches, although there was a slight increase in porosity within a soil layer 8–15 cm beneath the soil surface.

Despite the results for both soil structure and permeability, Aldefer and Merkle concluded that the chief value of mulches in controlling runoff lies in their protective effect, rather than in any fundamental structural change, even four years after they were applied. It may be that any structural changes occur within this period of time, but that after four years the effects may have been nullified.

4.3.6   EFFECT OF MULCHES ON BIOLOGICAL ACTIVITY

Studies of earthworm activity beneath mulches by Lal, De Vleeschauwer and Malafa Nganje (1980) found an increase in worm casts (measured in metres per month!) with an increase in mulch rate ($r = 0.98$, significant at the 5% level).

4.3.7   SUMMARY OF THE EFFECT OF MULCHES ON EROSION PROCESSES

The various effects of mulches on the factors affecting erosion have been combined and expressed in the following equation (Laflen and Colvin, 1981):

$$\text{Erosion} = A\,e^{b \cdot RC},$$

where A and b are constants, and RC is the percentage residue cover.

The b value is dependent on the shape of the curve expressing the relationship between the mulch factor and the crop residue. This relationship is affected by soil and slope conditions as illustrated by differences in the particle size distributions of the eroded sediments, but surprisingly the variables of crop canopy, type of residue or crop rotation did not seem to have any effect on the relationship, according to these researchers. The conclusion drawn was that a single mulch factor/crop residue relationship will be valid for various row crops at different canopy levels, under specific soil and slope conditions. Typical b values for soils high in sand content on low slopes are given as <-0.075, and for soils high in silt content on steep slopes as >-0.04. More specific values than these depend on individual site conditions.

Laflen and Colvin's equation is then used to calculate a 'mulch factor', by dividing it by its intercept, A to give:

$$\text{Mulch factor} = \frac{e^{b \cdot RC}}{A}.$$

This mulch factor will account for the interactions between the mulch and soil and/or slope conditions, which are often ignored in other indices of erosion control with vegetation or simulated vegetation (Laflen and Colvin, 1981). The mulch factor/crop residue relationship changes over time, as evidenced in changing eroded sediment size distributions. When plotted (Figure 4.1) the relationship gives an exponential decay function between the mulch factor and the type of residue cover. Similar relationships have been found by other workers (Lal, 1976). Morgan (1982) found a curvilinear relationship between soil loss and percentage cover, in the following equation:

$$\text{Soil loss} = K\,e^{y \cdot PC},$$

where $K =$ constant; y ranges between -0.03 and -0.07; and $PC =$ percentage cover or percentage rainfall interception.

This curvilinear relationship between percentage cover and soil loss appears to apply to other mulch materials such as redwood litter, oak leaves and oat straw (Singer and Blackard, 1978).

4.4   THE ROLE OF MULCHES IN WIND EROSION CONTROL

The mechanisms of wind erosion control with mulches are the same as those using low-growing vegetation (section 6.6.3). Residue roughness and the amount of soil covered by the residue, rather than the weight of the mulch, determine the effectiveness of a mulch in controlling wind erosion (Fryrear and Koshi, 1971). Surface residue will reduce the direct wind force acting on the soil from 99% to as low as 5%, thereby substantially reducing wind erosion (Zingg, 1954). Air-flow velocities and therefore wind erosivity are also reduced with

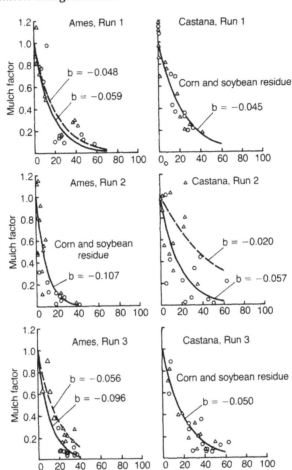

Figure 4.1 Mulch factor–residue cover relationships for each run at two locations. O ———, corn residue; △———, soybean residue (Laflen and Colvin, 1981).

increasing levels of surface mulching (Chepil, 1944; Englehorn, Zingg and Woodruff, 1952). By reducing wind velocities, the mulches help to trap any eroded sediment, leading to deposition. Also, damage to plants caused by the scouring action of the eroding sediment can be minimized (Pierson, Lewis and Birklid, 1988). This is important with some agricultural crops that are especially prone to scour damage during early stages of growth. The damage can be quantified in terms of costs incurred as loss of

crop yield and/or quality at harvest, or, alternatively re-drilling costs.

Fryrear and Koshi (1971) used cotton gin trash, and wheat and sorghum residue to reduce wind erosion rates on a fine loamy sand to below a given tolerable level, assumed to be 9 t/ha per year. Surface covers required were 76, 52 and 35 % for the three mulch types respectively. These differences were due to the texture and density of the different materials. The effectiveness of the mulches was increased with

Table 4.4 Effect of tackifier products on wind stability of barley straw broadcast at 2242 kg/ha (2000 lb/acre) (after Kay, 1978)

| Product | Rate/ha | | | Wind speed (km/h) at which 50% of straw was blown away |
	Chemical	Fibre (kg)	Water (l)	
None				14
SS-1 asphalt	1870 l			64
SS-1 asphalt	3740 l			128
SS-1 asphalt	5610 l			134
Fibre only		542		75
Fibre only		824		134
Fibre only		1104		134
Terratack I	50 kg	168	7014	109
Terratack II	100 kg	336	14 028	134
Ecology Control M-Binder	112 kg	168	6546	134
Styrene butadiene copolymer emulsion	560 l	84	3740	134
Polyvinyl acetate	935 l	280	9352	86
Copolymer of methacrylates and acrylates	935 l	280	9352	122

appropriate tillage techniques such as chiselling or listing.

Siddoway, Chepil and Armbrust (1965) studied the interactions of mulch type (wheat stubble and fine- and coarse-textured sorghum stubble), residue orientation (flat, leaning or standing), application rate, wind velocity and soil cloddiness on wind erosion rates. They concluded that wind erosion varies exponentially with the quantity of residue and that fine-textured mulches were more effective than coarse-textured ones. Any orientation of the residues tended to reduce erosion to a greater extent than when the stubble was laid flat.

As in water erosion control, longer mulch elements are more effective in reducing wind erosion than equivalent application rates of short-fibred mulches. Whilst incorporated mulches give less percentage cover, they are less subject to removal by high winds. Best protection is given by combining incorporated stubble and surface-applied straw (Chepil, 1944). Another alternative to reduce mulch failure is to spray asphalt onto the mulch to hold it in place (Kay, 1978; Table 4.4).

4.5   THE ROLE OF MULCHES IN VEGETATION ESTABLISHMENT

Mulches play an important role in vegetation establishment in three ways. First, erosion is controlled so that seeds, soil, fertilizer and nutrients are not washed away, resulting in maintenance of soil fertility and hence a higher and more uniform vegetation cover (Bungolo, Lenvain and Lungo, 1989). Control of rainsplash erosion also minimizes crust formation, which can hinder seedling emergence after germination. Kill and Foote (1971) claim that such erosion control is the most influential factor enabling successful vegetation establishment. Second, the microclimate of the site, such as the temperature, evaporation and light penetration, will be modified by the use of mulches. It must be remembered that these effects may not always be beneficial in terms of establishing a vegetation cover on a site. Third, mulches can improve soil characteristics that enhance vegetation establishment and growth.

Much of the research on the role of mulches on vegetation establishment uses measurements of biomass or crop yields as indicators of the positive or negative effect of the mulch on vegetation growth.

Pla, Florentino and Lobo (1987) report how peanut yields in semi-arid to subhumid regions doubled under an asphalt mulch due to the better efficiency of water use, with fewer days of soil water stress and reduced evaporation losses. Other workers have also found a positive relationship between mulch rate and crop yield (Box and Walker, 1965; Smika and Wicks, 1968; Greb, Smika and Black, 1970). Meyer, Wischmeier and Daniel (1971) obtained fescue-bluegrass establishment rates of 3, 28 and 42% with surface straw mulch rates of 0, 2.24 and 4.48 t/ha, respectively. Significant differences (at the 5% level) were observed in seedling establishment on a fill slope, for eight out of 11 different mulch treatments, compared with an unmulched plot (Dudeck *et al.*, 1970).

These positive relationships are site- and mulch-specific. There are instances where vegetation does not thrive under mulch conditions (Fryrear and Koshi, 1971). However, in turn these apparently negative effects may have other benefits, such as the suppression of weeds. This would be advantageous for sites where the mode of vegetation installation was by direct planting of already established shrubs and/or trees.

4.5.1 THE ROLE OF MULCHES IN MODIFYING MICROCLIMATE

Temperature

Soil temperatures reflect the energy exchange between the soil and the atmosphere. Thick mulches will suppress the energy transfer between these two media, thus suppressing fluctuations in temperatures beneath the soil surface. Different mulches have different degrees of radiation transmission or thermal efficiency (Othieno, Stighler and Mwampaja, 1985), giving different temperature ranges beneath the mulch (Dudeck *et al.*, 1970). Mulches will also intercept the warming effect of the sun's rays on the soil.

Analysis of the role of mulches on temperature is rather confusing because mulches can both increase and reduce soil temperatures, depending on the climatic environment, particularly the diurnal and seasonal characteristics in which they are applied. In turn, any modification of temperature by the mulch can be either beneficial or detrimental to vegetation establishment and growth, again depending on the site conditions.

Adams (1965), on plots in Texas, USA, compared soil temperatures under gravel and straw mulches, with those on bare soil. He found highest soil temperatures on the bare soil, followed by the 2.5 cm thick gravel mulch, then the 5 cm thick gravel, with the lowest temperatures under the 5 cm thick straw mulch. Using mulches of green and burnt sugar cane residues, Prove and Truong (1988) found soil temperatures were 2–4 °C lower than under conventionally managed plots in Queensland, Australia. However, this decrease only caused a temporary delay in the ratooning of the following crop.

The effect of suppressed temperatures under different mulch application rates on crop growth has been investigated by Burrows and Larson (1962). Whilst they point that out that the results will only be valid for north-central United States, they found increasing levels of chopped corn stalk mulch gave lower temperatures, which in turn affected plant height and dry matter production. They concluded that for each ton applied over the range of 0–4 ton/acre (0–9 t/ha) there was a decrease in the soil temperature during May–June at a depth of 10 cm of approximately 0.7 °F (0.2 °C). Whether this is significant in terms of vegetation establishment and subsequent growth would depend on the vegetation type being grown and the adversity of the site before a mulch was used.

Rates of mulch application affected soil temperatures in Western Nigeria, (Lal, De Vleeschauwer and Malafa Nganje, 1980). A rice straw mulch applied at 2, 4, 6 and 12 t/ha reduced maximum temperatures by 3.3, 4.1, 4.5

and 5.4 °C, respectively, compared with the unmulched control.

The effect of mulches on soil temperatures under clonal tea plants in Kenya was only significant before the plant canopy had reached 40% cover (Othieno and Ahn, 1980). However, the lower temperatures under grass mulches had detrimental effects on the growth of the young tea plants, from which they never recovered.

Conversely, Sheldon and Bradshaw (1977) and Barkley, Blaser and Schmidt (1965) found higher soil temperatures under mulch treatments in England and Kentucky respectively, giving better germination and establishment rates than where a mulch had not been used. The insulating effect of mulches also meant frost disappeared one to three weeks earlier on corn mulched fields compared with autumn ploughed or chiselled fields (Benoit *et al.*, 1986). This led to earlier drainage and soil warming for the mulched fields, and consequent improvements in the conditions for vegetation establishment and growth.

Benoit and Lindstrom (1987) found the time of application affected the thermal conductivity of the mulch, so influencing soil/mulch temperatures. Mulch applied early in the winter became dark on decomposition, which reduced its reflectance, and with time the mulch became more compact, so that by springtime its thermal conductivity had increased.

Evaporation

Evaporation from the soil is minimized under mulches, and this can result in higher soil moisture contents (section 4.5.2). Control of evaporation by mulches will also help to reduce water loss from seeds (Sheldon and Bradshaw, 1977). This is important for the critical process of imbibition to take place. Unger and Parker (1968) found evaporation was reduced by 300% over 16 weeks under a wheat straw mulch, applied at a rate of 11 t/ha compared with an unmulched soil. Adams (1966) also found that during a hot, rainless 10-day period, evaporation from the top 15 cm of the soil profile was

significantly reduced when mulching was applied. Greb, Smika and Black (1967) found lower evaporation losses were reflected by soil water gains of 70% below the surface 61 cm of the soil profile under a heavy mulch.

Rates of evaporation decrease with increasing mulch rate (Bond and Willis, 1969). Applications of wheat straw at 30, 60 and 90% cover reduced water losses by solar distillation from a wet soil to 16, 33 and 49% respectively, compared with no straw (Greb, 1966). An application rate of 180% cover (i.e. 6720 kg/ha) only reduced water losses slightly compared with the 90% cover rate. However, Unger and Parker (1968) found that incorporated mulches with relatively low percentage cover can also reduce evaporation, compared with unmulched plots.

With only limited quantities of residue, total evaporation may be less when the residue is placed in small trenches, usually on the contour, rather than spread uniformly (Bond and Willis, 1969). This technique, known as 'vertical mulching', is often used as a water conservation technique. Brown and Kemper (1987) found soils treated in this way had beneficially higher water-holding capacities, which were reflected in higher bean yields.

The relationship between evaporation rate and mulch application rate is not always straightforward. Brun *et al.* (1986) found that differences in cumulative evaporation from a bare soil and a wheat straw covered surface were affected by precipitation frequency and amount. For relatively infrequent and small precipitation events, there was little or no difference in cumulative evaporation. Larger, more frequent events resulted in less cumulative evaporation from soil protected by the mulch.

Light penetration

Mulches can reduce light penetration which is vital for the establishment and growth of young seedlings. Meyer, Wischmeier and Foster (1970) question the practicality of the often recommended, but apparently excessively high mulch rate of 5 t/ha, as such a cover may adversely

Table 4.5 Mean effects of mulch treatments on surface-soil water, seedling stands, and grass yields on a 2:1 northeast-facing fill slope near Lincoln, Nebraska, in 1966[a] (after Dudeck *et al.*, 1970)

Treatments	% soil water	Seedlings per 10 dm^2	Dry matter g/10 dm^2
Excelsior mat	24.7^a	128.5^a	7.8^a
Excelsior	23.3^{ab}	101.0^{abc}	5.2^b
Asphalt	22.4^{abc}	40.1^e	2.6^{cd}
Prairie hay and asphalt	22.0^{abcd}	98.8^{abc}	4.3^{bc}
Jute net	20.8^{bcde}	111.8^{ab}	5.4^b
Corncobs and asphalt	20.0^{cdef}	75.8^{cd}	3.9^{bcd}
Woodchips and asphalt	19.4^{def}	96.3^{abc}	5.3^b
Fibreglass and asphalt	19.0^{ef}	90.5^{bc}	4.2^{bc}
Wood cellulose B	18.4^{ef}	69.3^{cde}	3.6^{bcd}
Wood cellulose A	18.1^{ef}	70.3^{cde}	3.2^{bcd}
Excelsior and wood cellulose	17.7^f	47.8^{de}	2.4^{cd}
No mulch	13.2^g	49.1^{de}	1.5^d

[a]Means within a column followed by the same letter do not differ significantly at the 5% level.

affect crop growth by reducing the incidence of light beneath the mulch.

4.5.2 EFFECTS OF MULCHES ON SOIL CHARACTERISTICS FOR VEGETATION ESTABLISHMENT AND GROWTH

Soil moisture

Soil moisture is essential for seed germination, seedling growth and the establishment of turves. Unlike cover crops or green manures, mulches have no water demand for evapotranspiration, so conserving soil water and increasing soil moisture contents. This point is important to farmers in areas where crop yields are highly dependent on water availability. Here, farmers are often reluctant to use cover crops for fear of moisture competition with the main crop, even if this lack of cover increases erosion rates. This is because water, not soil, is regarded as the limiting factor to higher crop yields. Different types of mulch gave significantly (5% level) higher soil water contents compared with an unmulched treatment in extensive tests carried out by Dudeck *et al.* (1970) (Table 4.5). These results are compatible with those of Aldefer

and Merkle (1943), which showed that various surface mulches caused a very pronounced increase in soil moisture contents, particularly during dry periods. Sheldon and Bradshaw (1977) found that peat mulches were especially effective in retaining soil moisture under dry conditions.

Increases in soil moisture contents under mulches are often linked to lower evaporation rates. Although incorporated mulches have reduced evaporation losses (section 4.5.1), increases in soil moisture tend to be less marked than when the mulch is applied on the surface (Aldefer and Merkle, 1943).

The ability of mulches to retain soil moisture can be detrimental. The combination of excessive soil moisture and low evaporation in some climates may lead to waterlogging in some soils, inducing gleying and anaerobic conditions, which will not support plant growth. It could be argued, however, that if mulches are used for the dual purpose of erosion control and vegetation establishment, they are likely to be applied on sloping land, where hydraulic gradients would adequately drain the soil of excess water.

Ironically, the negative effects of increased

soil moisture can also be seen on dry sites (Kay, 1978). Due to enhanced moisture retention by an overlying mulch, seeds will germinate even after the first minor rainfall event following a dry spell. However, these prematurely germinated seeds often die if the soil moisture is not sustained. A solution to this is 'soil mulching', where the seeds are covered with a soil mulch. Germination will only occur once the soil has been sufficiently moist for a long period, so ensuring sustained plant growth. This technique would not be advisable where the primary purpose of the mulch was erosion control.

A similar effect has been observed under mulched young tea plants (Othieno, 1980). After a prolonged drought, soil moisture contents and water-holding capacities were higher under three different grass mulches when compared with the unmulched soil. Highest soil moisture was recorded for the Napier grass (*Pennisetum purpureum*) mulch. However, these increased moisture contents resulted in shallow rooting, so that the tea plants were in fact more prone to drought in the long term, especially once the mulch had decayed, and thus the increase in moisture content due to the mulch was lost.

Soil acidity

Agusta, Sessler and Kahnt (1988) found mulches of hydrolysed wood and straw added NH_4^+ nitrogen to the soil, which suppressed the growth of weeds. On a relatively acidic soil, Unger and Parker (1968) observed that pH was reduced under a surface mulch, but that incorporated mulch had little effect on the soil pH level.

Organic matter content

Organic matter levels increase when mulches are applied, whether on the surface (Aldefer and Merkle, 1943) or incorporated (Biggerstaff and Moore, 1982). There is little quantification as to what extent this affects the soil fertility, or over what time-scale this effect applies.

Nutrient status

On sites of low soil nutrient status, it may be necessary to add N to compensate for the tie up of nitrogen in the process of decomposition of organic mulches. This competition is a very strong argument against the use of mulches on agricultural land. The tillage practices used in stubble mulching may adversely affect the availability of some plant nutrients (Hines, Lillard and Edminster, 1947).

Toxicity

Kay (1978) found that high application rates of a grass straw mulch (*Lolium multiflora*) became toxic, leading to inhibition of plant growth. Wheat, oat, corn and sorghum residues can also contain water-soluble materials which are phytotoxic to wheat seedlings (Guenzi, McCalla and Norstadt, 1967). The amount of toxicity is dependent on the degree of decomposition of the residue. Under field conditions, the wheat and oat residues were no longer toxic after eight weeks, but the toxic materials in the corn and sorghum residues (which were relatively more toxic at harvest) remained for 22–28 weeks. The fact that surface mulches may take longer to decompose than incorporated residue has implications for the length of time newly emerging plants may be affected by these potential toxins. There is little research reported as to the extent of damage caused by these toxins to different types of establishing vegetation.

4.6   THE USE OF GEOTEXTILES

Geotextiles are defined as permeable textiles used in conjunction with soil, foundation, rock, earth or any geotechnical engineering related material as an integral part of a man-made project (John, 1987).

Geotextiles have six functions:

1. separation of two distinct ground materials;
2. filtration, where transfer of fluids but not solids takes place through the geotextile;
3. drainage, where the geotextile may increase local hydraulic conductivities, so increasing

flow to a subsurface drain for example;

4. surface erosion control;
5. slope stability and reinforcement;
6. amelioration of site conditions for vegetation establishment and growth.

An overview of the diverse engineering applications is given in John (1987), but in this chapter emphasis is given to the way some geotextiles are used to simulate vegetation, predominantly to control soil erosion. Most of the geotextiles designed specifically for surface erosion control are in the form of either three-dimensional

erosion meshes, erosion blankets and mats, or honeycomb-shaped webs, known as geocells (Figure 4.2). In addition to erosion control, some of these geotextiles have a role in modifying microclimatic and soil conditions for vegetation establishment and plant growth.

Geotextiles used for soil erosion control can be classified by their composition (natural or synthetic, which in turn affects their durability on site) and by their mode of installation (surface or buried). Whether a temporary or permanent geotextile is used for erosion control will depend on the required function. If fully estab-

(a)

(b)

(c)

(d)

Figure 4.2 Types of erosion control geotextiie. (a) Geojute (jute woven mesh); (b) fine Geojute (jute woven mesh); (c) Bachbettgewebe (coir woven mesh); (d) Cocomat (100% coir mulch mat); (e) Enviromat (100% wood chips mulch mat); (f) 100% straw mulch mat; (g) Eromat (80% straw, 20% coir mulch mat); (h) Covamat ((i) mulch mat comprising straw, cotton waste, coir, seeds with (ii) paper backing); (i) Enkamat (synthetic, three-dimensional mesh); (j) Tensarmat (synthetic, three-dimensional mesh); (k) Geoweb (synthetic web); (l) Armater (synthetic web).

lished vegetation will control erosion eventually, temporary geotextiles are sufficient. If, however, a composite of geotextile and vegetation is required to keep erosion down to a tolerable level, then permanent geotextiles will be selected. In this case, the aim is to achieve a 'synergistic' relationship between geotextile and vegetation.

Geotextiles made from natural, often vegetative materials will bio- or light-degrade in time, so their effectiveness is only temporary. In theory, as they decompose so the natural vegetation will establish and develop sufficiently to control the processes of erosion. In reality the timing of geotextile degradation and vegetation succession may not coincide as required. There have been very few studies to quantify this, because of the necessary duration of such tests and the site-specific nature of geotextile degradation and vegetation establishment rates. Jute, a commonly used raw material in surface-applied geotextiles will rot in about two years in temperate climates (Thomson and Ingold, 1986).

(e)

(f)

(g)

(h) (i)

(h) (ii)

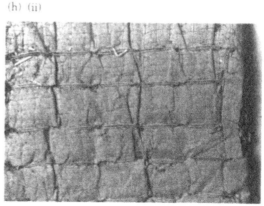

Figure 4.2 (cont.)

(i)

(k)

(j)

(l)

Figure 4.2 (cont.)

Some artificial tests into geotextile durability were carried out at the German Bundesanstalt für Materialprufung (Schürholz, 1992). The fibres tested were cotton, jute, sisal and coir. After one year of extreme tests in incubators, the woven coir textile had degraded least of all the treatments. Schürholz claims that the coir took 15 times longer than the cotton and seven times longer than the jute to degrade, especially in a wet environment, with very fertile soils. The results for the sisal are not quoted.

Temporary geotextiles, usually in the form of erosion blankets or mats are made from various materials, including woven fibres such as jute (current products available include Geojute, Soil Saver, Anti-wash) and coir (Bachbettgewebe). Others comprise loose mulch materials (paper strips, straw, wood (usually poplar or pine) chippings or shavings (Enviro-mat, Excelsior), coconut fibres (Cocomat), cotton waste, or a combination of these, as in Eromat (coir and straw) and Covamat (coir, straw and cotton waste), held together within a lightweight, often light-degradable polypropylene mesh. It has been estimated that $500\,000\,m^2$ of coir geotextiles alone are used annually in Europe (John, 1987), and a staggering $5 \times 10^6\,m^2$ of erosion control geotextile are used annually in the USA (Thomson and Ingold, 1986). Details of these products can be found in Table 4.6. The products are easy to install; after seeding the slope, the rolls of geotextile are simply laid down the slope and kept in place with wooden pegs or 11-gauge steel staples. There are some limitations when using temporary geotextiles made from natural, vegetative materials. Once installed, they may be eaten by animals or

Table 4.6 Selected geotextiles used in soil erosion control

Geotextile	Type	Composition	Natural/synthetic	Surface applied/ buried
A: TEMPORARY GEOTEXTILES				
Geojute/Soil saver/Anti-wash	B	Open-weave jute mat	Natural	Surface
Fine geojute	B	Open-weave fine jute mat	Natural	Surface
Enviromat/Excelsior	B	Mat of wood chips in light-sensitive mesh	Natural/synthetic	Surface
Bachbettgewebe	B	Open-weave coir mat	Natural	Surface
Covamat	B	Pre-seeded, coir, straw and cotton waste in light-sensitive mesh	Natural/synthetic	Surface
Eromat	B	Coir fibre and straw in a light-sensitive mesh	Natural/synthetic	Surface
Cocomat	B	Coir fibres in light-sensitive mesh	Natural/synthetic	Surface
B: PERMANENT GEOTEXTILES				
Enkamat	3D	Polymer mesh	Synthetic	Surface (US); buried (Europe)
Tensarmat	3D	Polyester	Synthetic	Buried
Geoweb	GC	High-density polyethylene	Synthetic	Buried
Armater	GC	Stiffened non-woven Bidim fabric	Synthetic	Buried

B – blanket; 3D – three-dimensional mesh; GC – geocell

present a fire hazard. Some fibres can be treated to become 'smoulder-free'. Their effect on microclimate, especially moisture status (section 4.9.2), may encourage pests and diseases.

Synthetic geotextiles are usually not degradable and will provide permanent erosion protection on site. Often carbon black is incorporated in the manufacture of synthetic geotextiles to protect against the degrading effects of sunlight. This in turn is often not aesthetically pleasing, particularly an public access sites, as well as making the geotextile conspicuous to vandals. Also, permanent geotextiles are not practical on sites where machinery is used for maintenance (such as grass cutting), as the geotextile may be snared in the cutting mechanism. There are also concerns as to the damage done by grazing animals and the effects on wildlife habitats.

Permanent geotextiles are usually made from polyethylene, and include most three-dimensional meshes, such as Enkamat (randomly welded polyamide, nylon filaments) and Tensarmat (a complex structure of multiple layers of black polyethylene mesh), and geocells, such as Armater (stiffened nonwoven Bidim fabric) and Geoweb (high-density polyethylene). The three-dimensional meshes are available in different thicknesses depending on the end use. On installation, they are rolled down the slope, pegged, seeded and then filled with topsoil. These products can be modified to incorporate ready grown turf or bitumen-bound gravel chippings for applications where erosion hazard is severe. The geocells are delivered on site as collapsed honeycombs, which are drawn out down the slope and then pegged into position. The cells are then filled with topsoil and seeded. For all buried geotextiles, the process of removing and then replacing top soil adds to the practical difficulties facing contractors when installing these products. These additional steps in geotextile installation may incur substantial economic costs relative to the use of surface-laid products. However, as almost all the buried products are permanent, these costs may be discounted in the long term.

Table 4.7 Vegetation properties simulated by geotextiles

Types of geotextile	Vegetation property[a]			
	Canopy	Stems	Roots	Litter
Woven meshes (often natural fibres)	**	**	X	*
Two-dimensional mats (often natural materials)	**	*	X	**
Three-dimensional meshes (synthetic – polypropylene or nylon)	X	X	**	X
Geocells or geowebs (synthetic)	X	X	*	X

[a] ** = Performs identically to role of vegetation; * = performs similarly to role of vegetation; X = does not simulate vegetation effect.

Reynolds (1976) compared the degradation rates of both natural and synthetic meshes used for erosion control and vegetation establishment, concluding that after three months' exposure the geotextiles made from woven jute, woven paper, woven polyethylene and extruded woven polyethylene had weight changes of −47, −53, −6 and +3% respectively. He also quantified the resistance to mechanical damage afforded by each product after this time.

Geotextiles should be resistant to strong acids or alkalis and should tolerate soils in the range pH 3–13. This is especially important if the geotextiles are to be used to reclaim potentially toxic industrial sites, such as mining spoil or landfill sites.

4.7   THE ROLE OF GEOTEXTILES IN WATER EROSION CONTROL

Like mulches, geotextiles control soil erosion by mimicking the salient properties of vegetation (Rickson, 1990; Table 4.7). Thus the principal ways in which both mulches and geotextiles control soil loss are similar. However, there is very little published research on how geotextiles specifically control erosion. Rickson and Vella (in press) give an overview of some of the research carried out in Europe and the United States into the effectiveness of natural and synthetic geotextiles for the control of soil erosion.

Early field trials monitored the relative effectiveness of different products in terms of total erosion rates. This research included testing of geotextile products on highway embank-

ments by Armstrong and Wall (1991, 1992) who attempted to use erosion prediction models such as the Universal Soil Loss Equation (Wischmeier and Smith, 1978) to explain their test results. Other field trials in North America include work by Martin (1985), Godfrey and Landphair (1991), Jacobsen and Potter (1991) and Sanders, Abt and Clopper (1990). In other parts of the world similar trials have been set up by Natarajan and Gupta (1977) and Schürholz (1991) on railway cuttings and embankments, Ingold and Thomson (undated) on a road embankment in the UK and Forth and Leung (1989) on slopes in Hong Kong.

However, these studies do not explain the mechanisms by which erosion processes are controlled by geotextiles. Thus, their results have limited application in the formulation of design procedures. This can lead to a lack of confidence by engineers in the ability of geotextiles to control soil loss (Fifield, 1987). Most product selection procedures are based on previous, *ad hoc* experience (both good and bad) and intuition. Selection is often made on subjective and cost-oriented (as opposed to performance-related) decisions. Recently, however, there has been an increase in the scientific quantification of geotextile effectiveness (for examples, see International Erosion Control Association, 1989, 1990, 1991, 1992).

The following section looks in more detail at the ways in which geotextiles affect specific soil erosion processes, through soil protection, modifications in runoff volume and velocity, and effects on soil shear strength.

4.7.1 EFFECTS OF GEOTEXTILES ON SOIL PROTECTION

Geotextiles used in soil erosion control are often very effective in controlling rates of soil detachment by raindrop impact. Surface-laid geotextiles often have a high percentage cover, which simulates canopy interception and storage effects (Chapter 2). Thomson and Ingold (1986) state that the percentage cover afforded by the geotextile or erosion mat effectively increases as the inclination of the protected slope increases above 45°.

The effect of various erosion-controlling geotextiles on soil detachment by raindrop impact without runoff has been quantified (Rickson, 1987, 1988) in experiments using small Ellison splash cups to isolate the process. Each geotextile provided a different degree of soil protection; their performance was also dependent on soil type (clay loam and sandy loam soils were used) and the intensity of the rainfall (35 mm/h and 115 mm/h; Figure 4.3). The conclusion drawn was that, when compared with unprotected control plots, the geotextiles reduced rainsplash to a greater extent on the more erodible soil, and were generally more effective under high intensity rainfall compared with the low intensity rainfall. One important finding of this research was the poor performance of the buried geotextiles in controlling rainsplash-induced detachment. In some cases the buried geotextiles actually gave higher detachment rates than those observed for the control plot. This was due to the backfilling of the geotextile with top soil, which effectively sieves the soil, breaking up any soil structure and aggregation. The loose, uncompacted backfill is highly erodible, unless the soil has a high clay content, in which case the backfill can be compacted to form soil clods. Even then, these 'temporary' aggregates will disintegrate rapidly when subjected to further raindrop impact. It must be noted that the loose unconsolidated backfill may be effective in controlling potentially erosive runoff, through increased infiltration (section 4.7.2). This is similar to the comparison of how surface- and subsurface-applied mulches act in controlling erosion processes.

In conclusion, the results of these tests illustrate the desirable characteristics of a geotextile required to reduce splash erosion. These are: high percentage cover; natural fibres for high water absorption and good soil contact; thick fibres, to intercept any splashed particles; and an installation procedure which does not involve backfilling. In particular, the jute geotextiles performed well in this set of trials.

It should be remembered, however, that the results are restricted to the process of rainsplash erosion. Reynolds (1976) states that small plot studies such as these tend to overestimate the value of interception cover of the geotextile, whereas the real erosion control ability is dependent on the area of contact and firm attachment to the soil, which affects the hydrological impact of the geotextile (section 4.7.2). This is illustrated by the fact that in his studies a jute geotextile with an open weave suffered less erosion than a woven polythene mesh which had a higher percentage cover. This difference may be explained by the different water-holding capacities of the two products, however.

Another set of experiments used larger plots over which runoff was generated from simulated rainfall. Thus the combined erosion processes of rainsplash and runoff were simulated (Rickson, 1992a). As with the rainsplash-alone experiments, the most effective geotextiles at controlling sediment production were those with high percentage cover and high water-holding capacity, so that when wetted, there was good contact between geotextile and soil surface.

The susceptibility of the soil is partly dependent on its organic content (section 2.4.2), as aggregation of soil particles and soil structure are enhanced by the presence of organic material. Biodegradable geotextiles will contribute some organic matter to the soil once they begin to break down. At present there is no scientific evidence that quantifies the amount of organic matter contributed, whether this

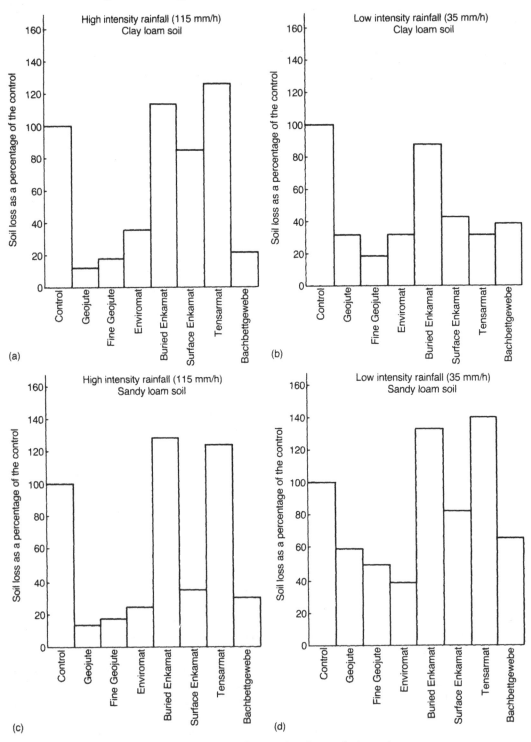

Figure 4.3 Results of geotextile performance under simulated rainsplash erosion.

amount is significant in affecting soil erodibility over any given set period of time, or the decay of this organic matter over time. Evidence from research on the addition of organic matter from decomposing mulches would suggest the quantities are likely to be insignificant, and thus have little effect on ameliorating soil susceptibility to erosion.

4.7.2   EFFECTS OF GEOTEXTILES ON HYDROLOGY

Geotextiles can control soil erosion processes by affecting the quantity or volume of runoff that can transport detached soil particles. Natural geotextiles in particular have very high water-holding capacities (Rickson, 1988), so that the fibres absorb incoming rainfall before it contributes to or generates surface runoff. By intercepting and absorbing rainfall, the weight of geotextile increases over time, giving excellent contact with the soil beneath (Reynolds, 1976; Rickson, 1992b). This retards flow on the soil/geotextile interface, so limiting erosion and undercutting beneath the geotextile which can ultimately lead to geotextile failure. Natural geotextiles tend to have better 'drapability' (i.e. the ability of the geotextile to conform to the microtopography of the slope form and profile), especially when wet, compared with the more rigid synthetic products (Thomson and Ingold, 1986), although the latter are often backfilled with soil, so increasing their weight to ensure good contact between geotextile and soil surface.

Open-weave products can store runoff water behind their thick weft fibres, which act as 'mini-dams', so reducing runoff volume. Thomson and Ingold (1986) explain this effect as similar to terracing used for soil conservation, where across-slope barriers not only reduce slope lengths over which runoff can be generated, but also provide increased storage of surface water. This ponding of water behind the geotextile fibres may protect the soil further from raindrop impact, through a 'cushioning' effect.

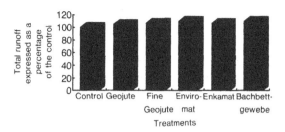

Figure 4.4 Total runoff expressed as a percentage of the control.

However, when simulated runoff was applied to a slope with five different geotextiles (four natural products, one synthetic) by Rickson (1990), there was no significant difference in the runoff volume generated from the geotextile treatments compared with that from an unprotected control ($p < 0.05$) (Figure 4.4). However, the jute product, which has the highest water-holding capacity, did produce least runoff in the test. Also, water was transmitted downslope through the fibres of this geotextile rather than on the soil surface. The geotextiles did help to delay the onset of runoff as measured at the bottom of the experimental plots used, especially if the fibres were oriented across the slope. Random orientation of fibres also slowed the onset of runoff, presumably because flow path routeing had been increased around the fibres, rather than straight down the slope, as has been observed with mulches (section 4.3.4). These observations have implications in the field. The delay in the onset of runoff represents an increase in time to peak flow on site. Peak volume of runoff is therefore decreased and time for infiltration is increased as the flow is intercepted by the geotextiles.

Although there is little quantification in the literature, buried geotextiles should reduce runoff volumes as their loose uncompacted backfill will have high infiltration rates. However, the backfill will also be very susceptible to rainsplash (section 4.7.1), which can increase soil capping, especially on soils with a high silt or very fine sand content, thus reducing infiltration and increasing runoff. Also, the interface between the uncompacted geotextile/soil mass and the relatively compacted undisturbed soil may lead to subsurface flow concentrations or oversaturation in the geotextile/soil mat. The latter may lead to a localized solifluction effect as lobes of saturated soil flow downslope.

These results show that geotextiles do not simulate vegetation in controlling runoff volumes. With increasing percentage cover, runoff volumes under live vegetation are reduced exponentially (Rickson and Morgan, 1988; section 5.2.1). However, when the same data are plotted for surface-laid geotextiles, there is no relationship between percentage cover and reduction in runoff volume. The conclusion that geotextiles are not effective in controlling runoff volumes may be important to the design of drainage ditches and soakaway areas at the bottom of slopes protected with geotextiles. From the results of the soil loss from the different geotextile treatments (Figure 4.5), this inability to reduce runoff volume does not appear to be relevant to their performance in controlling sediment production.

4.7.3 EFFECTS OF GEOTEXTILES ON HYDRAULICS

Geotextiles may change the hydraulic properties of flow, so changing its detaching and transporting capacity. This comes about from the simulation of a 'plant-induced roughness' by the geotextile elements. Meyer and Wischmeier (1969) state that the detachment and transport capacity of flow vary respectively with the square and fifth power of the runoff velocity, thus any reduction in runoff velocity by the geotextile has significant influence on the ability of flow to erode particles.

Rough geotextile fibres simulate stem effects, exerting drag forces on water flow through the geotextile. The random roughness of fibres also contributes to retarding runoff velocities (Rickson, 1992b), thereby reducing transport capacity and leading to deposition of particles within the geotextile fibres. Geotextile fibres oriented across a slope can also act as a sieve,

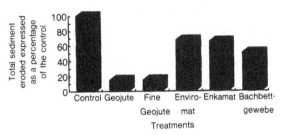

Figure 4.5 Total sediment eroded expressed as a percentage of the control.

filtering out sediment carried in the flow. In turn, deposited sediment affects the local hydraulic gradients and flow velocities, leading to a positive feedback of sedimentation, reduction in hydraulic gradient, lower flow velocity and increased sedimentation, as observed for field grass strips (section 5.3.2). This principle has been applied in the use of catchment or silt fences made from geotextiles. These are placed downstream of eroding areas to trap any sediment being transported off-site. This approach to control of sediment transport should only be used where the on-site damage of soil erosion is unimportant, and the off-site consequences, such as sediment acting as a pollutant and sedimentation in water bodies are more detrimental.

Geotextiles which comprise mulch elements in a plastic mesh, such as Enviromat and Excelsior mats, will perform in the same way as mulches in reducing runoff velocities (section 4.3.4). However, they are also susceptible to the failure mechanisms of mulches. One example is when high volumes of runoff shift mulch elements from being randomly placed relative to the slope to an up–down orientation. The mulch elements are then more susceptible to being carried away with the flow (Rickson, 1990). Flow paths are shortened, so flow tends to speed up. This orientation can also lead to microrilling as flow is directed along preferential flow paths.

Ideally, quantified roughness parameters such as Manning's n should be calculated for geotextiles used in erosion control in order to predict a 'geotextile-induced roughness to flow' parameter. This has already been done for mulches

Table 4.8 Qualitative assessment of factors affecting geotextile-induced roughness (GIR) to flow

Treatment	Depth of flow[a] (mm)	Slope (°)	Velocity[b] (m/s)	Manning's n[c]	GIR[d]	Rank	Soil loss rank
Control	0.68	10	0.0577	0.055	N/A	6	6
Geojute	3.00	10	0.034	0.258	0.203	2	1
Fine geojute	2.00	10	0.024	0.278	0.223	1	2
Enviromat	2.00	10	0.036	0.186	0.130	3	5
Enkamat	0.68	10	0.033	0.098	0.043	5	4
Bachbettgewebe	2.00	10	0.0535	0.124	0.068	4	3

[a] Depth of flow was calculated for the control plot as 0.68 mm. Relative values (based on visual observations) were then assigned to the geotextile treatments to the nearest millimetre.
[b] Velocity was calculated as the time it took for discharge to reach the bottom of the slope.
[c] Manning's n was calculated by substituting the estimated flow depths and measured velocities into the Manning equation.
[d] GIR = Geotextile Induced Roughness, taken as the Manning's n value minus the soil induced roughness, as calculated for the control (0.055).

(Foster, Johnson and Moldenhauer, 1982b). Such calculations are difficult, however, as they require runoff velocity as an input, which cannot easily be measured beneath geotextiles with high percentage cover. It is also important to distinguish between roughness imparted by the soil (which in turn is affected by the mode of installation of the geotextiles) and that exerted by the geotextile itself. There may also be interaction effects between the two forms of roughness.

However, attempts have been made to calculate an equivalent geotextile roughness parameter based on slope, hydraulic radius (assumed to be equal to flow depth), and flow velocity (Rickson, 1990; Table 4.8). The estimates of 'geotextile-induced roughness' were closely related to the observed soil losses under the different geotextile treatments.

The roughness imparted by geotextiles has implications for structures such as grassed waterways, which require protection from erosive runoff, especially during the period of vegetation establishment (Rickson, 1992b). In theory (and this requires field validation), parameters such as the geotextile induced roughness could be substituted into waterway design formulae to calculate the critical physical dimensions of a waterway, in much the same way as Manning's n values are used in design procedures at present. Many of the currently used design procedures assume a roughness factor based on fully established vegetation, which obviously results in excessive underdesign of channel dimensions during the period of vegetation establishment.

In surface erosion control, the effect of geotextiles in controlling runoff velocity is seen to be more significant than the control of runoff volume (Rickson, 1990). This is evidenced by the strong positive relationship between reduction in runoff velocity and soil loss, and yet the lack of any relationship between runoff volume and soil loss.

Most research has been concentrated on the effect of geotextiles on surface runoff hydraulics. However, geotextiles have also been used to control the hydraulics of tidal and deep-sea currents. Seabed scour can cause problems in marine pipeline installation. The velocity of seabed currents is reduced by using specialized geotextiles which simulate seaweed fronds. Bundles of plastic fibres are set into a filter geotextile sheet which is laid on the seabed. The fibres impose a viscous drag which reduces the velocity of the water flowing over the seabed. Erosion by scour is controlled, and any eroded sediment entering the zone of reduced velocity is deposited (Brashears and Dartnell, 1967).

4.7.4 EFFECTS OF GEOTEXTILES ON SOIL STRENGTH

Buried geotextiles will not enhance the shear strength of the soil itself; in fact the backfilling technique may actually reduce soil shear strength. However, this can be compensated by the additional shear strength of the geotextile itself, in the same way that plant roots add strength to a soil (sections 2.4.1 and 7.5). Hence the geotextile/soil matrix overall may have a higher shear strength than the soil alone.

Surface-laid, biodegradable geotextiles do not add strength to the soil itself, although they may have an inherent strength. However, when they begin to degrade, their fibres are often incorporated into the soil, which may add to its strength, playing much the same role as decaying roots, except that by this stage the inherent strength of the fibres is extremely low.

Many geotextile manufacturers quote tensile strength properties of their products, which for erosion control products is not extremely helpful, as it is unclear how these properties affect the erosion processes the geotextile is attempting to control. This reflects the fact that many manufacturers and specifiers, when speaking to engineers, attempt to use familar and traditional terminology to describe the geotextiles used in erosion control. However, these relatively novel products are best evaluated using totally different criteria to those commonly used by many engineers.

The potential of geotextiles to add soil shear strength has been applied in areas where soil strength has been severely reduced under high recreational pressure, such as on the Pennine Way in England. Natural and artificial geotextiles are being used extensively by organizations such as the National Trust, National Parks and the Nature Conservancy Council to strengthen heavily worn sections of footpaths (Bayfield, 1986; Rose, 1989; Coppin and Richards, 1990).

4.8   THE ROLE OF GEOTEXTILES IN WIND EROSION CONTROL

There has been very little work on the role of biodegradable geotextiles in wind erosion control. However, many of the principles of mulching in this context apply to geotextiles too. The high percentage cover afforded by some geotextiles will protect soil from erosive winds. The main determinant of wind erosion rates is the velocity of the moving air, just as water erosion rates are heavily dependent on runoff velocity. Geotextiles can interrupt the air flow close to the ground surface, especially if rough, coarse-fibred products are used, because they will impart a roughness to the flow, effectively reducing wind speed. These principles have been applied in the use of jute geotextiles for stabilization of sand dunes in the Gower Peninsula, Wales, for example.

By reducing the initial detachment of soil particles in this way, further erosion is limited as there are fewer airborne particles available, which, on falling back to earth under gravity, bombard undetached particles, causing them to saltate (section 6.4.1). Soil creep initiated by the wind will be intercepted by the fibres of open-weave geotextiles. Finally, the moisture-holding characteristics of many surface-laid geotextiles will ensure the soil surface remains relatively moist. This means the soil is less susceptible to wind erosion. Wind shear forces required for erosion of particles are higher when the soil is wet, due to the increased cohesion (section 6.4.1).

Some synthetic geotextiles have been manufactured primarily for the control of wind erosion. These products are used as artificial windbreaks and are erected as porous fences. The ideal porosity is quoted at 30–50%. Lower porosity creates a solid barrier to air flow, and hence an increase in turbulence, whilst a more open configuration has little effect on interrupting air flow. The use of such fences for erosion control is limited. They are mostly used in horticulture, as in the protection of high value crops (such as kiwi fruit in New Zealand) and film-clad tunnel greenhouses from wind damage. Other applications have been in the control of snow drifts at the sides of upland roads, and reduction in wind turbulence on golf courses.

To date, there have been few quantified studies into these specialist applications, although

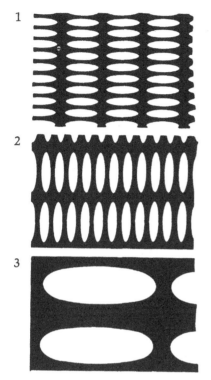

fence number	vertical × horizontal dimensions of openings	porosity	drag (CD) coefficient
1	23 × 5 mm	0.38	1.53
2	7 × 28 mm	0.42	1.43
3	82 × 25 mm	0.42	1.43

Note: All mesh openings are oval and sharp edged.
All meshes are approximately 1 mm thick.

Figure 4.6 Geotextiles used for wind erosion control (after Neal, 1988).

some manufacturers (Tensar) quote 50% reductions in wind velocities downwind, up to a distance of four times the fence height. ICI, who manufacture 'Paraweb', claim a reduction of 60% at a distance of 20 times fence height. Neal (1988) tested a number of wind erosion control geotextiles, with porosities between 38 and 42% (Figure 4.6).

In terms of vegetation establishment for wind erosion control, certain species such as marram grass have been used extensively. Mixtures of selected seed species could be incorporated

within blanket geotextiles to provide a composite material for immediate erosion control.

4.9 THE ROLE OF GEOTEXTILES IN VEGETATION ESTABLISHMENT

By controlling soil erosion, geotextiles create a stable, non-eroding environment in which vegetation can establish and grow, without the risk of wash-out of seeds or seedlings, or of damage to new shoots by the scouring action of eroded particles. As well as these physical benefits, the agronomic effects of geotextiles for plant growth are reflected by the increasing use of geotextiles in horticulture (Harper, 1990).

Geotextiles aid vegetation establishment by modifying microclimate and improving soil conditions for plant growth. After extensive field trials, Fifield (1992) and Fifield and Malnor (1990) concluded that most erosion control blankets helped increase vegetative biomass relative to that on unprotected plots under harsh, semi-arid conditions. The beneficial effects are optimized when seeds are sown directly into the geotextile mat, as with Covamat. However, not all geotextiles will benefit plant growth to the same extent. Thomson and Ingold (1986) state that geocells, for example, do not encourage the growth of vegetation *per se*, and it may be necessary to provide secondary measures to promulgate growth.

4.9.1 EFFECTS OF GEOTEXTILES ON MICROCLIMATE

Temperature

Temperatures are modified by geotextiles in the same way as they are by mulches (section 4.5.1). Geotextiles can isolate soil and plant roots from the changes in air temperatures, which are likely to be detrimental to healthy vegetation establishment and growth. Fifield *et al.* (1987) refer to this as the 'greenhouse effect'. Dudeck *et al.* (1970), working on experimental plots near Lincoln, Nebraska, showed that the daily temperature range 1.3 cm below the soil surface was 12.7 °C and 15.7 °C for Excelsior mat and jute netting respectively, compared with 21.8 °C for the bare plot (a significant difference). The colour of the geotextile affects its light and heat transmission properties (Reynolds, 1976). Dark-coloured products absorb heat and this can increase soil temperatures.

Evaporation

Whilst natural geotextiles, such as jute, absorb and retain a great deal of moisture, this may actually reduce the effective rainfall on site because of higher evaporation losses. Reynolds (1976) reports that over a 13-week growth period the effective rainfall was reduced by approximately 40 mm, which would be crucial in areas where moisture was a limiting factor to vegetation establishment and growth.

Light

Reynolds (1976) found that geotextile mesh size affected light penetration. The 1-cm mesh size of a woven paper geotextile had only 50% light penetration, whereas the larger mesh size of a woven jute product (2 cm) had 70% light penetration. However, light penetration was not seen as a limiting factor to plant growth, unless penetration of the seedlings through the geotextile was a problem. Indeed, in some cases, the meshes did form a physical barrier to seedling emergence. Fifield *et al.* (1987) found that blanket geotextiles exclude light to such an extent as to cause establishing vegetation to turn pale and sickly.

Geotextile colour will also affect the degree of light penetration. Whilst black geotextiles help increase soil temperatures, they allow very little light on to the soil. Light reflectance is affected by the colour of the geotextile. Yields and growth rates of various plants are increased under different colour geotextiles used for horticultural purposes. For example, crop yields of tomatoes were increased under a red geotextile, healthier trees and shrubs were established on a green geotextile, and irises thrived on a blue geotextile (Harper, 1990).

4.9.2 EFFECTS OF GEOTEXTILES ON SOIL CONDITIONS FOR VEGETATION ESTABLISHMENT AND GROWTH

Soil moisture

Soil moisture is essential for the establishment of vegetation, particularly if the technique of seeding is used. The composition and thickness of the geotextile will affect the amount of rainfall or surface water that can permeate the geotextile and pass into the soil. Ironically it is the thicker geotextiles which also protect the soil from high evaporation losses, so maintaining soil moisture levels. Thickness can also affect the water-holding capacity of the geotextile itself. Resultant higher moisture contents mean any irrigation requirements can be less for vegetation grown in combination with geotextiles. In turn, leaching of fertilizers and other nutrients is reduced.

Dudeck *et al.* (1970) illustrate the significant differences in soil water percentage with and without geotextiles, on a silty clay loam in Nebraska, under adequate precipitation for grass establishment. For the Excelsior mat and the jute geotextile, soil water percentages were 23.3% and 20.8% respectively. However, for a bare plot without geotextiles, the soil water percentage was only 13.2%.

As with mulches, the ability to retain water is not always wholly beneficial. Fifield (1987) observed that for some geotextiles soil moisture contents were increased, so initially encouraging high rates of grass seed germination. However, during dry periods, the competition for moisture from the resulting lush vegetation was so great that it led to substantial areas of grass being killed off.

Organic matter and nutrients

The soil fertility status of any site will affect the success of vegetation establishment and growth. There are claims that jute geotextiles can add 5 t/ha of nutrients on decomposition, which may be sufficient to reduce or even eliminate the need for top soil (Jute Promotion Project,

1985), but these are said to be rather overstated (Thomson and Ingold, 1986) as the figure appears to be derived by extrapolating the weight of the mesh per square metre to weight per hectare.

In areas where soil moisture and nutrient content are limiting, geotextiles with high percentage cover are used to suppress competition from weeds for moisture, nutrients and light. First, the geotextiles prevent foreign seeds from landing and germinating on the soil beneath the mat. This then allows the selected sown species to establish and grow, thus becoming dominant over any foreign species. Unfortunately, the early geotextile products used in horticulture were not only detrimental to weed growth, but to the seeded species too. Nonpermeable polyethylene sheets were used which restricted exchange of CO_2 and O_2. Microbiological activity was thus affected, resulting in poor vegetation development and growth. During the last decade, however, more permeable fabrics have been used, which allow the transfer of water, air and nutrients to the soil and to the plant roots. Whilst the geotextiles are permeable, they still prevent foreign seeds reaching the soil and germinating. Cuts in the geotextile will allow trees and shrubs to emerge from beneath the mat. Fewer chemical applications are required to control weeds if geotextiles are used in this way, so that the costs and benefits are often balanced out.

Soil trafficking

Vegetation establishment is extremely difficult on soils seriously affected by degradation resulting from erosion or heavy trafficking by humans and/or animals. Geotextiles have been used to help establishment of protective vegetation under such problematic conditions. Jute geotextiles were used in Arctic Norway to protect a footpath on which vegetation was being established (Parkinson, 1989). Low soil and air temperatures and more especially excessive trafficking from scientific expeditions at Lyngen had severely hindered vegetation growth. The

jute geotextile gave significant protection against the denudation of the newly emerging vegetation by walkers. Similar work has been undertaken by Scruby (1991) in the Brecon Beacons, using combinations of jute geotextile, fertilizer applications and stone scatter to aid establishment of vegetation on eroded upland areas. The inhospitable natural conditions of climate and soil were aggravated by excessive trafficking of the fragile soil by walkers. Ironically, the effectiveness of the different treatments was obscured by the fact that the stone scatter kept walkers off the newly seeded areas, whereas the jute netting provided a comfortable surface, which the walkers would then use as a preferential pathway!

Recently there has been increasing alarm at the environmental degradation occurring in the European Alps, due to excessive pressure on the land from recreation, especially downhill skiing. Oehler (1986) reports on the use of jute geotextiles to establish vegetation in the Bavarian Alps, on slopes severely disturbed and consequently degraded by mass tourism (notably hikers and skiers), by the extension of access roads to serve the tourists, and by the depletion of forest to enlarge pistes. The use of geotextiles for vegetation restoration has also been reported for plots near San Sicario in the Alps (Thomson and Sembenelli, 1987).

4.10 COMPARISONS BETWEEN MULCHES AND GEOTEXTILES

The effectiveness of both mulches and geotextiles is site-specific, both in terms of erosion control and establishment of vegetation. Likewise comparisons of the two techniques can only be valid where site conditions (soil, climate and slope) are identical. Some workers have investigated the relative performance of the two techniques.

Kay (1978) compared the performance of mulches with that of erosion mats and geotextiles for erosion control and the establishment of vegetation in California, USA. He found geotextiles more limited in practical application, because of high costs and high labour inputs required for their installation (four times higher than for tacked straw). Their effectiveness in controlling erosion and establishing vegetation was also poorer, as they had poor contact with the soil in rough or rocky areas. This lack of contact led to erosion occurring underneath the erosion mats.

Dudeck *et al.* (1970) carried out similar comparative tests in Nebraska, USA. For one set of trials, 11 different treatments of various mulches and erosion mats were compared. The other trial consisted of eight different treatments. They found the best grass establishment occurred under the Excelsior mat (also known as Enviromat), which produced healthy grass seedlings.

To the site contractor, the choice between mulching and geotextiles is likely to be determined by the availability of the materials, providing performance is reasonably good for either treatment. Geotextiles are distributed throughout the world, with extensive publicity brochures and high profile at exhibitions and conferences. Mulch materials are less accessible to the site contractor, unless a source of mulching material is known personally. Geotextiles have the advantage that they are 'ready-made' and designed for a purpose, unlike mulches which are often perceived as a by-product or even a waste product.

REFERENCES

Abrahim, Y. B. and Rickson, R. J. (1989) The effectiveness of stubble mulching in soil erosion control, in *Soil Erosion Protection Measures in Europe* (eds U. Schwertmann, R. J. Rickson and K. Auerswald). Soil Technology Series 1. Catena Publications, Cremlingen–Destedt, pp. 115–26.

Adams, J. E. (1965) Effect of mulches on soil temperature and grain sorghum development. *Agron. J.*, 57, 471–4.

Adams, J. E.(1966) Influence of mulches on runoff, erosion and soil moisture. *Soil Sci. Soc. Am. Proc.*, 30, 110–14.

Agusta, H., Sessler, B. and Kahnt, G. (1988) Studying the roles of partly hydrolised waste materials from straw and wood for soil covering and mulching in the context of soil melioration, in *Land Conservation for Future Generations* (ed. S. Rimwanich). Ministry of Agriculture and Cooperatives, Bangkok, Thailand, pp. 805–14.

Aldefer, R. B. and Merkle, F. G. (1943) The comparative effects of surface application versus incorporation of various mulching materials on structure, permeability, runoff and other soil properties. *Soil Sci. Soc. Am. Proc.*, 7, 79–86.

Armstrong, J. J. and Wall, G. J. (1991) Quantified evaluation of effectiveness of erosion control mats, in *Erosion Control: A Global Perspective*, Proceedings of the XXII IECA Conference, Florida, US, pp. 165–80.

Armstrong, J. J. and Wall, G. J. (1992) Comparative evaluation of the effectiveness of erosion control materials, in *The Environment is Our Future*, Proceedings of the XXIII IECA Annual Conference, Reno, Nevada, pp. 77–92.

Barkley, D. G., Blaser, R. E. and Schmidt, R. E. (1965) Effect of mulches on microclimate and turf establishment. *Agron. J.*, 57, 189–91.

Barnett, A. P., Diseker, E. G. and Richardson, E. C. (1967) Evaluation of mulching methods for erosion control on newly prepared and seeded highway backslopes. *Agron. J.*, 59, 83–7.

Bayfield, N. (1986) Approaches to reinstatement of damaged footpaths in the Three Peaks area of the Yorkshire Dales National Park, in *Agriculture and Conservation in the Hills and Uplands* (eds M. Bell and R. G. H. Bunce). ITE Symposium, NERC, Institute of Terrestrial Ecology.

Benoit, G. R. and Lindstrom, M. J. (1987) Interpretation of tillage residue management effects. *J. Soil and Water Conserv.*, 42(2), 87–90.

Benoit, G. R., Mostaghimi, S., Young, R. A. and Lindstrom, M. J. (1986) Tillage residue effects on snow cover, soil water, temperature and frost. *Trans. Am. Soc. Agric. Engrs.*, 29, 473–9.

Biggerstaff, S. D. and Moore, I. D. (1982) Effect of surface condition on infiltration, runoff and erosion on reconstructed soils. *Trans. Am. Soc. Agric. Engrs.*, 25, 82 2586 (microfiche).

Bond, J. J. and Willis, W. O. (1969) Soil water evaporation: surface residue rate and placement effects. *Soil Sci. Soc. Am. Proc.*, 33, 445–8.

Bonsu, M. (1980) Assessment of erosion under different cultural practices on a savannah soil in the northern region of Ghana, in *Soil Conservation: Problems and Prospects* (ed. R. P. C. Morgan). Wiley, Chichester, pp. 247–53.

Box, J. and Walker, H. J. (1965) Cotton burs for

soil improvement. *Tex. Agri. Ext. Serv.*, MP 476.

Brashears, R. L. and Dartnell, . (1967) Development of the artificial seaweed concept. *Shore and Beach*, October 1967, pp. 35–42.

Brown, M. J. and Kemper, W. D. (1987) Using straw in steep furrows to reduce soil erosion and increase dry bean yields. *J. Soil and Water Conserv.*, 42(3), 187–91.

Brun, L. J., Enz, J. W., Larsen, J. K. and Fanniny, C. (1986) Springtime evaporation from bare and stubble covered soil. *J. Soil and Water Conserv.*, 41(2), 120–2.

Bryan, R. B. (1968) The development, use and efficiency of indices of soil erodibility. *Geoderma*, 2, 5–26.

Bungolo, A. M., Lenvain, J. S. and Lungu, O. I. (1989) Soil erosion in Zambia with particular emphasis on the benefits of minimum tillage and mulching. Paper presented to the 6th International Soil Conservation Conference, Addis Ababa, Ethiopia, November 1989.

Burrows, W. C. and Larson, W. E. (1962) Effect of amount of mulch on soil temperature and early growth of corn. *Agron. J.*, 54, 19–23.

Cadena-Zapata, M. (1987) Slope limitations in the use of mulches. MSc Thesis, Silsoe College, Cranfield Institute of Technology.

Chepil, W. S. (1944) Utilisation of crop residues for wind erosion control. *Scientific Agric.*, 24(7), 307–19.

Coppin, N. J. and Richards, I. G. (1990) *Use of Vegetation in Civil Engineering*. CIRIA, Butterworths.

Diseker, E. G. and Richardson, E. C. (1962) Erosion rates and control methods on highway cuts. *Trans. Am. Soc. Agr. Engrs.*, 5, 153–5.

Douglas, C. L., Allmaras, R. R., Rasmussen, P. E., Ramig, R. E. and Roager, N. C. (1980) Wheat straw corporation and placement effects on decomposition in dryland agriculture of the Pacific Northwest. *Soil Sci. Soc. Am. J.*, 44, 833–7.

Dudeck, A. E., Swanson, N. P., Mielke, L. N. and Dedrick, A. R. (1970) Mulches for grass establishment on fill slopes. *Agron. J.*, 62, 810–12.

Duley, F. L. and Russel, J. C. (1943) Effect of stubble mulching on soil erosion and runoff. *Soil Sci. Soc. Am. Proc.*, 7, 77–81.

Englehorn, C. L., Zingg, A. W. and Woodruff, N. P. (1952) The effects of plant residue cover and clod structure on soil losses by wind. *Soil Sci. Soc. Am. Proc.*, 16, 29–33.

Fifield, J. S. (1987) Preliminary report: Testing of erosion control products. *International Erosion Control Association Report*, 19(3), 4–6.

Fifield, J. S. (1992) How effective are erosion control products in assisting with dry land grass establishment with no irrigation? in *The Environment is Our Future*, Proceedings of the XXIII IECA Annual Conference, Reno, Nevada. International Erosion Control Assocn, Steamboat Springs, CO, pp. 321–34.

Fifield, J. S. and Malnor, L. K. (1990) Erosion control materials versus a semi-arid environment. What has been learnt from 3 years of testing? in *Erosion Control: Technology in Transition*, Proceedings XXI IECA Annual Conference, Washington, D.C. International Erosion Control Assocn, Steamboat Springs, CO.

Fifield, J. S., Malnor, L. K., Richter, B. and Dezman, L. E. (1987) Field testing erosion control products to control sediment and to establish dryland grasses under arid conditions. HydroDynamics Incorp., Parker, CO.

Forth, R. A. and Leung, K. W. (1989) Use of geotextiles to prevent erosion of steep slopes in Hong Kong. in *Proc. Symp. on Applications of Geosynthetics and Geofabrics in SE Asia*, 1–2 August 1989, Selangor, Malaysia.

Foster, G. R. and Meyer, L. D. (1975) Mathematical simulation of upland erosion by fundamental erosion mechanics in present and prospective technology for predicting sediment yields and sources. *USDA Agr. Res. Service Pub. ARS-S-40*, pp. 190–207.

Foster, G. R., Johnson, C. B. and Moldenhauer, W. C. (1982a) Critical slope lengths for unanchored cornstalk and wheat straw residue. *Trans. Am. Soc. Agric. Engrs.*, 25, 935–9, 947.

Foster, G. R. and Johnson, C. B. and Moldenhauer, W. C. (1982b) Hydraulics of failure of unanchored cornstalk and wheat straw mulches for erosion control. *Trans. Am. Soc. Agric. Engrs.*, 25, 940–7.

Fryrear, D. W. and Koshi, P. T. (1971) Conservation of sandy soils with a surface mulch. *Trans. Am. Soc. Agric. Engrs.*, 14, 492–5.

Godfrey, S. H. and Landphair, H. C. (1991) Temporary erosion control materials testing, in *Erosion Control: A Global Perspective*, Proceedings of the XXII IECA Conference, Florida, US, International Erosion Control Assocn, Steamboat Springs, CO, pp. 105–20.

Graf, W. H. (1971) *Hydraulics of Sediment Transport*. McGraw-Hill, New York, p. 544.

Greb, B. W. (1966) Effect of surface applied wheat straw on soil water losses by solar distillation. *Soil Sci. Soc. Am. Proc.*, 30, 786–8.

Greb, B. W. (1967) Percentage soil cover by six vegetative mulches. *Agron. J.*, **59**, 610–11.

Greb, B. W., Smika, D. E. and Black, A. L. (1967) Effect of straw mulch rates on soil water storage during summer fallow in the Great Plains. *Soil Sci. Soc. Am. Proc.*, **31**, 556–9.

Greb, B. W., Smika, D. E. and Black, A. L. (1970) Water conservation with stubble mulch fallow. *J. Soil and Water Conserv.*, **25**(2), 58–62.

Guenzi, W. D., McCalla, T. M. and Norstadt, F. A. (1967) Presence and persistence of phytotoxic substances in wheat, oat, corn and sorghum residues. *Agron. J.*, **59**, 163–5.

Harper, D. (1990) Geotextiles in horticulture. *Geosynthetics World*, **1**(1).

Hines, R. C., Lillard, J. H. and Edminster, T. W. (1947) Applying stubble mulch tillage in Virginia. *Agric. Eng.*, **28**, 507–8.

Hussein, M. H. and Laflen, J. M. (1982) Effects of crop canopy and residue on rill and interrill soil erosion. *Trans. Am. Soc. Agric. Engrs.*, **25**, 1310–15.

Ingold, T. S. and Thomson, J. C. (Undated) Results of current research of synthetic and natural fiber erosion control systems. Paper prepared for ITC, UNCTAD/GATT, Geneva.

International Erosion Control Association (1989) *Erosion Knows No Boundaries*, Proceedings XX IECA Annual Conference, Vancouver, Canada. IECA, Steamboat Springs, CO.

International Erosion Control Association (1990) *Erosion Control: Technology in Transition*, Proceedings XXI IECA Annual Conference, Washington, DC. IECA, Steamboat Springs, CO.

International Erosion Control Association (1991) *Erosion Control: A Global Perspective*, Proceedings of the XXII IECA Conference, Florida, US. IECA, Steamboat Springs, CO.

International Erosion Control Association (1992) *The Environment is Our Future*, Proceedings of the XXIII IECA Annual Conference, Reno, Nevada. IECA, Steamboat Springs, CO.

Jackobs, J. A., Andrews Jnr., O. N., Murdoch, C. L. and Foote, L. E. (1967) Turf establishment on highway right of way slopes – a review. *Highway Res. Rec. No. 161, Highway Res. Bd. Publ. 1439*, Washington, DC, pp. 71–103.

Jacobsen, W. and Potter, P. (1991) Erosion control mats and blankets: a new generation of construction and performance. *Erosion Control: A Global Perspective*, Proceedings of the XXII IECA Conference, Florida, US, pp. 159–64.

John, N. W. M. (1987) *Geotextiles*. Blackie and Son, Glasgow.

Jute Promotion Project (1985) News about jute. Press release 10/85.

Kay, B. L. (1978) Mulches for erosion control and plant establishment on disturbed sites, in *Reclamation of Drastically Disturbed Lands* (eds F. W. Schaller and P. Sutton). ASA–SCSA–SSSA, Madison, WI, pp. 182–204.

Kill, D. L. and Foote, L. E. (1971) Comparisons of long and short fibred mulches. *Trans. Am. Soc. Agric. Engrs.*, **14**, 942–4.

Kramer, L. A. and Meyer, L. D. (1969) Small amounts of surface mulch residue reduce soil erosion and runoff velocity. *Trans. Am. Soc. Agric. Engrs.*, **12**, 638–41, 645.

Laflen, J. M. and Colvin, T. S. (1981) Effect of crop residue on soil loss from continuous row cropping. *Trans. Am. Soc. Agric. Engrs.*, **24**, 605–9.

Lal, R. (1976) *Soil Erosion Problems on an Alfisol in Western Nigeria and Their Control: Mulching Effect on Runoff and Soil Loss*. I.I.T.A. Monograph No. 1, International Institute for Tropical Agriculture, Ibadan, Nigeria.

Lal, R. (1977) Soil conserving versus soil degrading crops and soil management for erosion control, in *Soil Conservation and Management in the Humid Tropics* (eds D. J. Greenland and R. Lal). Wiley, Chichester, pp. 81–6.

Lal, R., De Vleeschauwer, D. and Malafa Nganje, R. (1980) Changes in properties of a newly cleared tropical alfisol as affected by mulching. *Soil Sci. Soc. Am. J.*, **44**, 827–33.

Lattanzi, A. R., Meyer, L. D. and Baumgardner, M. F. (1974) Influence of mulch rate and slope steepness on interrill erosion. *Soil Sci. Soc. Am. Proc.*, **38**, 946–50.

Mannering, J. V. and Meyer, L. D. (1963) The effects of various rates of surface mulch on infiltration and erosion. *Soil Sci. Soc. Am. Proc.*, **27**(1), 84–6.

Martin, J. S. (1985) Erosion control and revegetation mats: a cost-effective approach. *Public Works*, March 1985.

Meyer, L. D. and Mannering, J. V. (1963) Crop residues as surface mulches for controlling erosion on sloping land under intensive cropping. *Trans. Am. Soc. Agric. Engrs.*, **6**, 322–327.

Meyer, L. D. and Wischmeier, W. H. (1969) Mathematical simulation of the process of soil erosion by water. *Trans. Am. Soc. Agric. Engrs.*, **12**, 754–8, 762.

Meyer, L. D., Wischmeier, W. H. and Foster, G. R. (1970) Mulch rates required for erosion control on steep slopes. *Soil Sci. Soc. Am. Proc.*, **34**, 928–31.

Meyer, L. D., Wischmeier, W. H. and Daniel, W. H.

(1971) Erosion, runoff and revegetation of denuded construction sites. *Trans. Am. Soc. Agric. Engrs.*, 14, 138–41.

Meyer, L. D., Johnson, C. B. and Foster, G. R. (1972) Stone and woodchip mulches for erosion control on construction sites. *J. Soil and Water Conserv.*, 27(6), 264–9.

Morgan, R. P. C. (1982) Splash detachment under plant covers: results and implications of a field study. *Trans. Am. Soc. Agric. Engrs.*, 25, 987–91.

Morgan, R. P. C. (1986) *Soil Erosion and Conservation*. Longman, London.

Natarajan, T. K. and Gupta, S. C. (1977) Large scale field trials on the use of coir netting for erosion control of embankments and hillside slopes. *Hard Fibres Series No. 22*, FAO, Rome.

Neal, I. G. (1988) The effect of shelter element configuration and vertical structure of windbreak fences on leeward shelter. MSc Thesis, Silsoe College, Cranfield Institute of Technology.

Njoroge, S. N. J. (1985) Use of mulches in controlling soil erosion. MSc Thesis, Silsoe College, Cranfield Institute of Technology.

Oehler, R. (1986) Jute mats used as basis to stop erosion and regenerate Alpine plant life. *World Farmers Times*, June 1986.

Okwach, G. E., Palis, R. G. and Rose, C. W. (1992) Sediment concentration and characteristics as affected by surface mulch, land slope and erosion mechanisms, in *Erosion, Conservation and Small-Scale Farming* (eds H. Hurni and Kabede Tato), Geographica Bernensia, Bern, pp. 91–105.

Othieno, C. D. (1980) Effects of mulches on soil water content and water status in tea plants in Kenya. *Exper. Agric.*, 16, 295–302.

Othieno, C. O. and Ahn, P. M. (1980) Effects of soil temperature and growth of tea plants in Kenya. *Exper. Agric.*, 16, 287–94.

Othieno, C. O., Stighler, E. J. and Mwampaja, A. R. (1985) On the use of Stighlers Ratio in expressing the thermal efficiency of grass mulches. *Exper. Agric.*, 21, 169–74.

Parkinson, R. (1989) Environmental impact of a large scale expedition. *British Schools Exploring Society (BSES) Arctic Norway Science Report*, 1989.

Pickles, S. M. (1984) The effect of straw mulch on the erosion process. MSc Thesis, Silsoe College, Cranfield Institute of Technology.

Pierson, B. J., Lewis, C. E. and Birklid, A. (1988) Observed differences in determining crop residue cover in the Alaskan subartic. *J. Soil and Water Conserv.*, 43(6), 493–5.

Pla, I., Florentino, A. and Lobo, D. (1987) Soil and water conservation in Venezuela through asphalt mulching, in *Soil Conservation and Productivity* (ed. I. Pla Sentis). Sociedad Venezolana de la Ciencia del Suelo, Maracay, Venezuela, pp. 481–95.

Poesen, J. (1986) Surface sealing as influenced by slope angle and position of simulated stones in the top layer of loose sediments. *Earth Surface Processes and Landforms*, 11, 1–10.

Prove, B. G. and Truong, P. N. V. (1988) Research into soil erosion control practices on sugarcane lands in Queensland (Australia), in *Land Conservation for Future Generations* (ed. S. Rimwanich). Department of Land Development, Ministry of Agriculture and Cooperatives, Bangkok, Thailand, pp. 669–80.

Reynolds, K. C. (1976) Synthetic meshes for soil conservation use on black earths. *Soil Conserv. J. N.S.W.*, 34, 145–60.

Rickard, P. (1985) In-field blow protection systems, in *Arthur Rickwood Annual Review, 1985.* MAFF/ADAS, pp. 12–16.

Rickson, R. J. (1987) Geotextile applications in steepland agriculture, in *Steepland Agriculture in the Humid Tropics* (eds T. H. Tay, A. M. Mokhtaruddin and A. B. Zahari). MARDI Press, Ministry of Agriculture, Selangor, Malaysia, pp. 352–76.

Rickson, R. J. (1988) The use of geotextiles in soil erosion control: comparison of performance on two soils, in *Land Conservation for Future Generations* (ed. S. Rimwanich). Department of Land Development, Ministry of Agriculture and Cooperatives, Bangkok, Thailand, pp. 961–70.

Rickson, R. J. (1990) The role of simulated vegetation in soil erosion control, in *Vegetation and Geomorphology: Processes and Environments* (ed. J. B. Thornes). Wiley, Chichester, pp. 99–112.

Rickson, R. J. (1992a) Control of sediment production using geotextiles: Results of experimental testing using simulated rainfall and runoff, in *The Environment is Our Future*. Proceedings of the XXIII IECA Annual Conference, Reno, Nevada. International Erosion Control Assocn, Steamboat Springs, CO, p. 353.

Rickson, R. J. (1992b) The application of geotextiles in the protection of grassed waterways, in *Erosion, Conservation and Small-Scale Farming* (eds H. Hurni and Kebede Tato). Geographica Bernensia, Bern, pp. 415–21.

Rickson, R. J. and Morgan, R. P. C. (1988) Approaches to modelling the effects of vegetation on soil erosion by water, in *Erosion Assessment and Modelling* (eds R. P. C. Morgan and R. J.

Rickson). Commission of the European Communities, DG VI. EUR 10860 EN, pp. 237–54.

Rickson, R. J. and Vella, P. (in press) Experiments on the effectiveness of natural and synthetic geotextiles for the control of soil erosion. Paper presented to the Congress on Geosintetici per la costruzione in terra – Il controllo dell'erosione, Bologna, October 1992.

Roose, E. J. and Asseline, J. (1978) Mesures des phénomènes d'érosion sous pluies simulées aux cases d'érosion d'Adiopodoumé. Les charges solide et soluble des eaux de ruissellement sur sol nu et diverses cultures d'ananas. Cah ORSTOM. *Séries pédologie*, 16(1), 43–72.

Rose, S. J. C. (1989) The Three Peaks Project: tackling footpath erosion, in *Erosion Knows no Boundaries*, Proceedings XX IECA Annual Conference, Vancouver, Canada, pp. 369–78.

Ruiz, J. F. and Valentin, Ch. (1987) Effects of various types of cover on soil detachment by rainfall, in *Soil Conservation and Productivity* (ed. I. Pla Sentis). Sociedad Venezolana de la Ciencia del Suelo, Maracay, Venezuela, pp. 1071–88.

Sanders, T. G., Abt, S. R. and Clopper, P. E. (1990) A quantified test of erosion control materials, in *Erosion Control: Technology in Transition*, Proceedings XXI IECA Annual Conference, Washington, DC. International Erosion Control Assocn, Steamboat Springs, CO.

Schürholz, M. (1991) Erosion control on cut slopes with the light weight coir fibre at new European high speed railroads, in *Erosion Control: A Global Perspective*, Proceedings of the XXII IECA Conference, Florida, US. International Erosion Control Assocn, Steamboat Springs, CO, pp. 213–16.

Schürholz, H. (1992) Use of woven coir geotextiles in Europe. Paper presented to the 1992 UK Coir Geotextile Seminar. Organised by ITC, UNCTAD/GATT, Coir Board of India and Swedish SIDA, May 1992.

Scruby, M. (1991) The effect of fertiliser, jute netting and stone scatter on grass seed germination along eroded footpaths in the Brecon Beacons. MSc Thesis, Silsoe College, Cranfield Institute of Technology.

Seginer, I., Morin, J. and Sachori, A. (1962) Runoff and erosion studies in mountainous terra-rosa regions in Israel. *Bull. Int. Assoc. Sci. Hydrol.*. 4, 79–82.

Sheldon, J. C. and Bradshaw, A. D. (1977) The development of a hydraulic seeding technique for unstable sand slopes. I. Effects of fertilisers, mulches and stabilisers. *J. Applied Ecol.*, 14, 905–18.

Siddoway, F. H., Chepil, W. S. and Armbrust, D. V. (1965) Effect of kind, amount and placement of residue in wind erosion control. *Trans. Am. Soc. Agric. Engrs.*, 8(3), 327–31.

Singer, M. J. and Blackard, J. (1978) Effect of mulching on sediment in runoff from simulated rainfall. *Soil Sci. Soc. Am. J.*, 42, 481–6.

Singer, M. J., Matsuda, Y. and Blackard, J. (1981) Effect of mulch rate on soil loss by raindrop splash. *Soil Sci. Soc. Am. J.*, 45, 107–10.

Smika, D. E. and Wicks, G. A. (1968) Soil water storage during fallow in the Central Great Plains as influenced by tillage and herbicide treatments. *Soil Sci. Soc. Am. Proc.*, 32(4), 591–5.

Sprague, M. A. and Triplett, G. B. (1986) Tillage management for a permanent agriculture, in *No-Tillage and Surface-Tillage Agriculure* (eds M. A. Sprague and G. B. Triplett). Wiley, Chichester.

Stallings, J. H. (1949) Crop residues conserve soil and water. *J. Soil and Water Conserv.*, 14(1), 103–6, 107.

Swanson, N. P., Dedrick, A. R., Weakley, H. E. and Haise, H. R. (1965) Evaluation of mulches for water erosion control. *Trans. Am. Soc. Agric. Engrs.*, 8, 438–40.

Thomson, J. C. and Ingold, T. S. (1986) Use of jute fabrics in erosion control. Report to the Jute Market Promotion (Western Europe) Project, International Jute Organisation (IJO), International Trade Centre, UNCTAD/GATT. Project No. RAS/77/04.

Thomson, J. and Sembenelli, P. (1987) High altitude erosion control: some experience and trials in the Alpine ski area. *Proceedings XVII IECA Annual Conference on Erosion Control Practice and Research*. Reno, Nevada, IECA, Steamboat Springs CO.

Unger, P. W. and Parker, J. J., Jnr (1968) Residue placement effects on decomposition, evaporation and soil moisture distribution. *Agron. J.*, 60, 469–72.

Valentin, C. and Roose, E. J. (1980) Soil and water conservation problems in pineapple plantations of south Ivory Coast, in *Soil Conservation: Problems and Prospects* (ed. R. P. C. Morgan). Wiley, Chichester, pp. 239–46.

Van Liew, M. W. and Saxton, K. E. (1983) Slope steepness and incorporated residue effects on rill erosion. *Trans. Am. Soc. Agric. Engrs.*, **26**, 1738–43.

Wischmeier, W. H. (1973) Conservation tillage to control water erosion, in *Proc. Nat. Conserv. Tillage Conf.* Ankeny, IA. pp. 133–41.

Wischmeier, W. H. and Smith, D. D. (1978) Predicting rainfall erosion losses. *USDA Agr. Res. Serv. Handbook No. 537*.

Worku Bekele, M. and Thomas, D. B. (1992) The influence of surface residue on soil loss and runoff, in *Erosion, Conservation and Small-Scale Farming* (eds H. Hurni and Kebede Tato), Geographica Bernensia, Bern, pp. 439–52.

Zingg, A. W. (1954) Wind erosion problem in the Great Plains. *Trans. Am. Geophys. Union*, **35**, 252–85.

WATER EROSION CONTROL   5

R. P. C. Morgan and R. J. Rickson

5.1 INTRODUCTION

Soil erosion by water occurs whenever the dislodging forces of raindrop impact and running water exerted on individual particles and aggregates of soil are greater than the forces resisting removal. Erosion is a two-phase process, comprising the detachment of soil materials and their transport downslope or downstream. When the rate of soil particle detachment is low relative to the ability to transport sediment, erosion takes place at the detachment rate: this is the detachment-limited case. When the detachment rate is very high, the erosion rate is controlled by the ability of the erosive agents to entrain and transport soil material: this is the transport-limited case. Knowing whether erosion is detachment- or transport-limited is important for erosion control. First, if erosion is detachment-limited, there is the potential for further erosion to occur until the transport-limited condition is reached. Second, the strategies required to deal with the processes of detachment are different from those for controlling sediment transport. In both cases, the rate of erosion is a function of the energy of the erosive agents, the resistance of the soil, the slope of the land and the protection afforded by the vegetation cover.

As explained in Chapter 2, vegetation affects the efficiency of water erosion processes in a number of ways. It interacts directly with the erosive agents by modifying the energy of the rainfall reaching the soil surface and altering the velocity of flowing water. Through its effect on the hydrological cycle, vegetation is a major control over the generation of runoff. It also acts mechanically to increase the resistance of the soil to erosion through increases in the cohesion of the soil–root matrix. The exact nature of these effects in the short term depends upon the morphology of the individual plants, in terms of canopy, stems, roots and residue cover, and the architecture of the plants in combination, which determines the structure, density and uniformity of the vegetation cover.

In the long term, there is a complex link between vegetation and the environment, in which erosion is frequently an important part. On the one hand, a good vegetation cover can undoubtedly control water erosion, whereas, on the other, severe erosion can either prevent vegetation from growing or seriously affect its composition and structure (Thornes, 1988, 1990). One reason for this is that infiltration capacities within the soil generally decrease with depth so that, as the soil cover is reduced by erosion, less rainfall passes into the soil. Another is that the water-holding capacity of the subsoil is generally less than that of the top soil. In either case, biomass production becomes limited by the available moisture. Erosion also affects the availability of nutrients in the soil for plant growth. Nitrogen can be lost in surface runoff, and phosphorus and organic matter are preferentially removed when adsorbed to the clay particles which are often selectively eroded whilst the coarser material remains behind. Soluble phosphorus is also removed in the surface runoff. However, under most circumstances, water availability operates as a limiting factor to vegetation growth long before any

Slope Stabilization and Erosion Control: A Bioengineering Approach. Edited by R. P. C. Morgan and R. J. Rickson. Published in 1995 by E & FN Spon, 2-6 Boundary Row, London, SE1 8HN. ISBN 0 419 15630 5.

effects of the loss of mineral matter are observed. Figure 5.1 shows some of the key interactions involved in hydrological, erosion and nutrient systems which affect the engineering functions of vegetation. These interactions need to be understood if vegetation is to be managed in a way that maximizes its engineering benefits.

Erosion can take place in a wide range of environments, from sloping land under arable cultivation to road banks, river banks and shorelines. Often there are considerable human consequences both on-site where, for example, the productivity of the soil for future food production may be affected, and off-site, as a result of pollution and sedimentation. Although soil erosion is frequently publicized in the media as an agricultural problem, its occurrence on mining spoils, cut-and-fill slopes, channel and reservoir banks, and in recreational areas makes it an engineering problem too. Whilst both agricultural engineers and civil engineers have

traditionally looked to structural solutions such as terraces, gabions, stone revetments and concrete walls, the last 20 years have seen an increasing emphasis within agriculture on agronomic or biological solutions which rely on the engineering functions of vegetation. With some modifications, these solutions can also be applied to engineering situations, particularly where the use of vegetation is already demanded for landscaping, notably aesthetic, purposes.

5.2   SOME BASIC PRINCIPLES

Numerous plot studies carried out in the field under a wide range of agricultural conditions show that vegetation cover can control soil erosion by water on sloping land. Typical data are shown in Table 5.1 from research carried out at Séfa, Senegal (Roose, 1967). Erosion and runoff are lowest under protected forest cover and highest on bare ground with about three orders

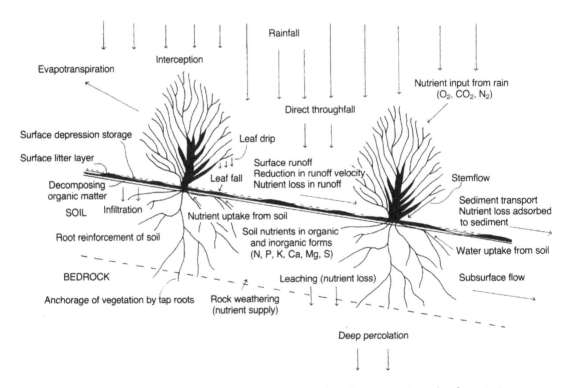

Figure 5.1 Hydrological, erosion and nutrient systems related to the engineering role of vegetation.

Table 5.1 Relationship between mean annual soil erosion on slopes and vegetation cover (after Roose, 1967)

Vegetation cover	Replications	Annual runoff (% of rainfall)		Annual soil loss (t/ha)	
		Mean	Range	Mean	Range
Protected forest	7	0.67	0.10–1.18	0.08	0.02–0.22
Burned forest	9	0.90	0.30–1.52	0.18	0.02–0.51
Groundnut	24	22.8	8.1–42.5	6.89	2.91–16.30
Cotton	3	28.0	0.9–42.7	7.75	0.47–18.52
Sorghum	9	20.6	11.2–35.0	7.82	1.19–22.71
Maize	1	30.9		10.34	
Millet	2	34.7	26.4–39.7	10.34	8.10–12.57
Bare soil	7	40.1	22.3–53.1	25.13	6.93–54.48

of magnitude difference between the erosion rates for the two conditions. Crops, such as groundnuts, cotton, sorghum, maize and millet, result in about 100 times more erosion than the protected forest but only about one-third to one-half of the erosion recorded from bare ground. One reason for these differences in erosion is that the vegetation is interacting with the erosion processes in two ways: one relates to the effect of the overall vegetation cover and the other to its spatial configuration or layout. 'Vegetation cover' is used here as a broad term to encompass the combined effects of canopy, plant stems and roots.

5.2.1 ROLE OF VEGETATION COVER

The data in Table 5.1 show a very strong relationship between the mean annual rate of erosion and the mean annual runoff. This implies that the major role played by vegetation in controlling water erosion is hydrological. As indicated in Chapter 2, this role is exerted largely through the infiltration process. The infiltration rates of vegetated soils are higher than those of bare soils because the growth of the root network and the presence of soil fauna open up the pore system; also the return of organic material to the soil contributes to the stability of the soil aggregates and therefore to the stability of the pores which are less likely to

close as the soil wets up. In addition, vegetation leads to lower antecedent moisture contents because of the removal of water from the soil through evapotranspiration, and to changes in the effective rainfall intensity at the soil surface because of interception of rainfall by the canopy. The combination of the ability to take in water at a higher rate and in greater quantity means that vegetated soils are less likely to generate runoff.

A simple way of evaluating the effect of vegetation on runoff volume is to use the runoff coefficients from Cook's method (United States Soil Conservation Service, 1953) and Hudson's (1981) method of estimating runoff for small catchments, relate them to a percentage vegetation cover and then express the coefficients relative to the coefficient value for bare soil to obtain a set of soil loss ratios (Table 5.2). These ratios represent the soil loss under a given vegetation cover as a proportion of that under bare soil. If, for simplicity, it is assumed that soil loss varies directly with the volume of runoff – in practice it varies with runoff raised by a power of between 0.67 and 1.8 (section 2.3.2) – these ratios show an exponential decrease with increasing percentage cover (Figure 5.2; Rickson and Morgan, 1988). In reality, there will be some degree of variability from this relationship because of vegetation effects other than cover; for example, stem density, and differences in the

Table 5.2 Derivation of soil loss ratios expressing the effect of vegetation cover on the volume of runoff

Cover type	Estimated % cover	Catchment characteristic[a]	Soil loss ratio (CCbare/CCveg)[b]
Bare soil	0	25	1.0
Cultivated land with poor cover	10	20	0.8
Cultivated land with fair cover	25	15	0.5
Cultivated land with good cover, scrub or grass	50	10	0.4
Forest, grass or scrub	90	5	0.2

[a]Catchment characteristic (CC) values based on Cook's Method (United States Soil Conservation Service, 1953) and Hudson's (1981) method.
[b]CC bare = catchment characteristic value for bare soil; CCveg = catchment characteristic value for vegetated soil.

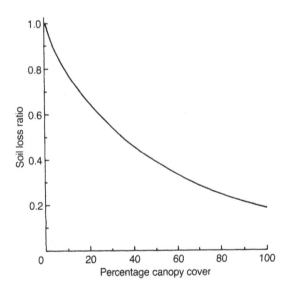

Figure 5.2 Relationship between the soil loss ratio and percentage vegetation cover taking account of runoff volume (after Rickson and Morgan, 1988).

types and species of the plants making up the vegetation community.

A further way in which vegetation cover controls erosion was demonstrated by Hudson (1981) in a mosquito gauze experiment. Two experimental plots, 1.5 m wide and 27.5 m long, were set up in Zimbabwe on a clay loam soil with a 5% slope. They were kept bare of vegetation by hand-weeding. Both plots were exposed to the same natural rainfall but one plot was covered by a double layer of fine-mesh wire gauze, simulating a dense (90–100%) vegetation cover. As can be seen from Table 5.3, the soil loss from the plot covered by the gauze is less than 1/100th of that from the uncovered plot. This experiment indicates the protective effect of a cover close to the soil surface in breaking up the raindrops so that they reach the soil from

Table 5.3 Soil losses recorded in the mosquito–gauze experiment (after Hudson, 1981)

	Soil loss (t/ha)	
Year	Plot covered by gauze	Bare plot
1953/54	nil	146.2
1954/55	2.0	204.5
1955/56	4.5	135.6
1956/57	0.2	132.4
1957/58	0.2	49.5
1958/59	2.5	202.0
1959/60	nil	7.4
1960/61	nil	121.4
1961/62	nil	138.5
1962/63	nil	128.2
Ten-year totals	9.4	1265.7

very low fall heights and, therefore, have very low impact energies. These results were confirmed in a similar study carried out on plots, 5 m wide and 20 m long, on a marine clay soil in Toscana, Italy by Zanchi (1983). Here, the difference in erosion between the two plots was much less with the soil loss from the covered plot being about one-tenth of that from the uncovered plot (Table 5.4). Runoff on the covered plot was about one-third of that on the uncovered plot, showing that, even without a rooting system, a cover close to the soil surface can reduce runoff.

These experiments simulate the protective effect of a low-growing vegetation cover described in Chapter 2. The cover reduces the energy of the rainfall at the soil surface which, in turn, reduces the rate of soil particle detachment by raindrop impact. This means that there is less material available for transport by any runoff that is generated and, also, less infilling of the soil pore spaces by detached fine particles. The soil is thus protected against surface crusting and sealing so that high infiltration rates are maintained.

Whilst the data and experiments cited above demonstrate the effect of vegetation cover, how that effect is brought about is better understood by considering the individual roles played by the plant canopy, stems and roots.

Plant canopy

As shown in Chapter 2, it is important that the vegetation canopy is uniform and either close to or in contact with the soil surface to obtain the maximum protection. In this instance, research with grasses (Lang and McCaffrey, 1984), crop residues (Laflen and Colvin, 1981) and stones (van Asch, 1980) indicates that the soil loss ratio, defined here as relating to soil detachment by raindrop impact, decreases exponentially with increasing percentage canopy cover (Figure 5.3; Rickson and Morgan, 1988). As the height of the cover above the ground increases, the relationship becomes linear and the cover is less effective because of the greater fall height of the

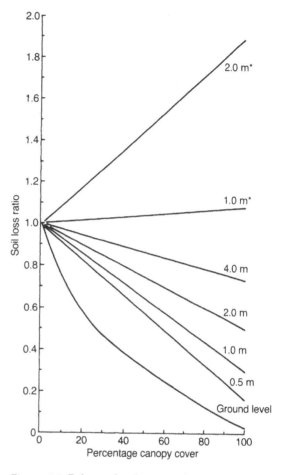

Figure 5.3 Relationship between the soil loss ratio and percentage vegetation cover taking account of the effect of soil detachment by raindrop impact. Asterisks denote relationships allowing for the effect of leaf drainage (after Rickson and Morgan, 1988).

leaf drainage. Where leaf drainage occurs with large diameter drops because of the coalescence of raindrops on the leaves, and the canopy is 1 m or more above the ground, the vegetation no longer affords any protection and soil loss ratios increase with increasing percentage cover. This effect has been modelled theoretically by Styczen and Høgh-Schmidt (1986) from a consideration of the physics of raindrop impact as an inelastic collision between the water drop and the soil surface. It is supported

Table 5.4 Soil losses and runoffs recorded in a net-covering experiment (after Zanchi, 1983)

Year	Soil loss (t/ha)		Runoff (mm)	
	Covered plot	Bare plot	Covered plot	Bare plot
1978	0.5	6.3	45	98
1979	1.3	66.3	88	392
1980	1.2	48.0	67	302
1981	0.8	19.1	82	509
1982	16.7	83.2	292	615
1983	2.7	35.2	111	216
Six-year totals	23.2	258.1	685	2132

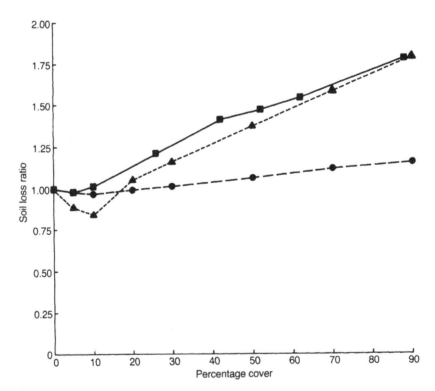

Figure 5.4 Application of the relationship shown in Figure 5.3 to selected crops showing how the soil loss ratio changes with as the vegetation cover increases with plant growth. ●, Brussels sprouts; ▲, potatoes; ■, maize.

by field measurements of soil detachment rates under maize (Morgan, 1985) and laboratory studies under Brussels sprouts (Noble and Morgan, 1983). Figure 5.4 shows a typical relationship of the soil loss ratios for detachment by raindrop impact as a plant grows. Initially the soil loss ratio is high because little of the ground surface is covered. The ratio then reduces exponentially because the cover is close to the soil surface in the early stages of vegetative growth. As the height of the canopy rises, however, with further growth, the ratio increases in value (Morgan *et al.*, 1986).

Plant stems

In addition to the hydrological and protective roles, vegetation can influence water erosion through its hydraulic effect arising from the roughness that vegetation imparts to flowing water. For typical shallow flows on slopes, it has been shown (Morgan, 1980) that the sediment transport capacity of the runoff varies with Manning's n raised by the power of -0.15. If it is assumed that $n = 0.01$ for bare soil, the $n^{-0.15}$ value for this condition is 1.99. If it is further assumed that the bare soil condition is represented by a soil loss ratio of 1, ratios taking account of the effect of vegetation on flow velocity can be obtained by calculating $n^{-0.15}$ values for different values of Manning's n and expressing them as a ratio of 1.99. The results, plotted in Figure 5.5 (Rickson and Morgan,

1988), show that the ratio decreases rapidly as Manning's n increases from 0.01 to 0.05 but that further increases in n have little additional effect.

Plant roots

In addition to the effects on infiltration described above, plant roots have a mechanical effect on the soil. By penetrating the soil mass, they reinforce it, bringing about an increase in cohesion and, hence, in soil shear strength. Also, a fine root mat close to the soil surface may act like a mulch or low-growing vegetation cover and protect the soil from erosion. Dissmeyer and Foster (1985) propose soil loss ratios to take account of the root effect (Figure 5.6). An increase in the percentage area occupied by fine roots produces an exponential decay in the soil loss ratio. As expected, the effect is much greater for rooting systems which spread laterally close to the soil surface than for systems with a strong vertical development characterized by tap roots.

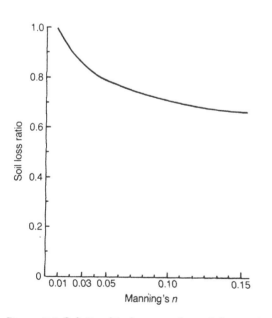

Figure 5.5 Relationship between the soil loss ratio and Manning's n (after Rickson and Morgan, 1988).

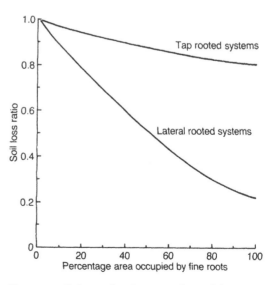

Figure 5.6 Relationship between the soil loss ratio and the percentage area occupied by fine roots (after Dissmeyer and Foster, 1985).

Effects of vegetation growth

A comparison of the slopes of the lines on Figures 5.2, 5.3, 5.5 and 5.6 reveals how the separate effects of vegetation on erosion change as the vegetation develops. A summary of this comparison is given in Table 5.5 which shows the changes in the soil loss ratios that result from given changes in the percentage vegetation cover (Morgan, 1987). In the early stages of vegetation growth (0–20% cover), the most important effect is through the reduction in soil detachment by raindrop impact. This means that it is extremely important to aim for a good ground cover of vegetation to obtain the maximum effect at this stage. If this is not achieved, increasing the vegetation cover from 0 to 20% will enhance soil particle detachment although at a rate which would be offset by reductions in the soil loss ratio due to lower runoff volumes and greater soil cohesion.

As the vegetation cover increases from 20 to 60%, the effects on soil detachment, runoff volume and soil cohesion are roughly equal, assuming that the vegetation is at ground level and provides a uniform cover which means that tussocky and tufted species must be avoided. With increases in vegetation cover above 60%, the most important effect is through increases in soil cohesion.

Although this analysis supports the view, first expressed above, that vegetation plays an important hydrological role, it also implies that this is not the most important mechanism by which erosion is reduced. More important is its protective role in reducing soil detachment, as illustrated by the mosquito-gauze experiments, and, for long-term effectiveness, the reinforcement of soil strength by the root network. The hydraulic effect of vegetation in reducing runoff velocity is always subsumed by these other effects.

5.2.2 VEGETATION LAYOUTS

Vegetation can be organized in a layout that will reduce erosion risk on slopes. A vegetation layout aligned across slope, ideally on or close to the contour, will reduce effective slope length and impede or obstruct overland flow due to increased surface roughness. These effects will reduce the accumulation of runoff volume downslope and reduce the flow velocity which in turn will reduce the kinetic energy and, therefore, the capacity of the flow to detach and transport soil particles. Indeed, the reduction in velocity may be sufficient to prevent potentially erosive velocities from being attained. Any small reduction in the erosive power of the flow will have a dramatic impact on reducing the transporting capacity of that flow, as this varies with the fifth power of the velocity (see Chapter 2). Reduced flow velocity results in localized

Table 5.5 Changes in the soil loss ratio as a function of changes in percentage vegetation cover (after Morgan, 1987)

Change in % cover	Change in Manning's n	Change in soil loss ratio			
		Detachment	Runoff volume	Roughness	Fine roots
0–20	0.01–0.03	0.43 (0.17)	0.36	0.16	0.22
20–40	0.03–0.05	0.19 (0.18)	0.19	0.06	0.19
40–60	0.05–0.07	0.14 (0.17)	0.12	0.03	0.15
60–80	0.07–0.09	0.11 (0.17)	0.11	0.03	0.14
80–100	0.09–0.11	0.09 (0.18)	0.04	0.02	0.08

Figures in parentheses denote increases in the soil loss ratio as a result of leaf drainage. All other values denote decreases.

deposition of transported eroded sediments. In some cases this will reduce local slope steepness, in turn reducing hydraulic gradients. Flow velocities are then further decreased and further deposition occurs. Over time, these processes can result in the formation of a series of benches on the hillside, sometimes referred to as 'erosion-induced' or 'erosion-controlled' terraces.

As runoff velocity is retarded, so more of the flow will infiltrate the soil, especially as vegetation often improves soil permeability through the presence of pipes or holes where roots have decayed, and improved soil structure. Infiltration reduces the volume of potentially erosive surface flow. These mechanisms are described in Chapter 2 with respect to filters where infiltration occurs in an area of ponding upslope of the vegetative barrier.

Across-slope vegetation layouts are best suited to well drained soils with high infiltration capacities and rates. On poorly drained soils there may be a risk of waterlogging and standing water, although this may be advantageous in situations where water is the limiting factor to vegetation growth. In some circumstances, however, water ponded behind vegetation aligned across slope may break through as a concentration of flow, thus increasing the risk of erosion by rilling.

5.3 LEARNING FROM AGRICULTURE

Studies of the way soil erosion can be controlled on agricultural land indicate how the basic principles outlined above can be translated into practice. The examples presented below illustrate procedures designed to maximize vegetation cover and to utilize vegetation layouts.

5.3.1 PRACTICES BASED ON VEGETATION COVER

Generally, agronomic measures of soil erosion control rely on alternating crops that are depleting of soil nutrients and are characterized by high rates of erosion with crops that are not.

Shifting cultivation

Shifting cultivation is a traditional agricultural practice in tropical countries whereby patches of forest are cleared in order to grow a crop for one or two seasons. The land is then abandoned and allowed to revert through natural plant succession over about 15 to 20 years to something approaching its original cover before it is cleared and cropped again. In the meanwhile, other patches of land are cleared and cultivated. The system effectively amounts to rotation of the fields.

The amount of erosion depends upon the way in which the land is cleared – whether traditionally, using a machete or similar implement, or mechanically – and the length of time the same piece of land remains in continuous cultivation (Table 5.6; Kellman, 1969; Lal, 1981).

Crop rotations

In more mechanized farming systems, crop rotation is adopted whereby a succession of crops is grown in the same field; for example, separating years of cereal production with two or more years of grass. The high erosion in the years when the land is under cereals is more than offset by the low erosion rates under grass, so that, on average, the annual erosion rate is kept to an acceptable level. It should be recognized that crop rotations are adopted mainly to maintain soil fertility, control pests and diseases, and eliminate weeds. Though the grasses and natural fallows play an engineering role in providing ground cover and root reinforcement of the soil, they are not selected primarily for these functions. Also, since the functions are performed only in certain years, the approach has limited value in non-agricultural situations where continuous protection is normally required. Two other approaches, however, make use of agronomic principles that can be applied elsewhere.

Table 5.6 Soil erosion rates under shifting cultivation

Rates (t/ha per year) under first-year of maize following different types of land clearance (after Lal, 1981).

Traditional method, incomplete clearing, no tillage	0.01
Manual clearing, no tillage	0.4
Manual clearing, conventional tillage	4.6
Shear blade, no tillage	3.8
Tree-pusher and root rake, no tillage	15.4
Tree-pusher and root rake, conventional tillage	19.6

Rates (g/day) under different covers (after Kellman, 1969).

Primary rain forest	0.20
Softwood tree fallow	0.29
Imperata grassland	0.40
New abaca plantation	0.47
Ten-year old abaca plantation	0.59
New maize swidden (cropping period)	3.03
New hill rice swidden (cropping period)	1.45
Two-year old maize swidden (cropping period)	12.05
Twelve-year old hill rice swidden (cropping period)	119.31

Table 5.7 Effect of planting density and management on runoff and erosion under maize (after Hudson 1957)

	Plot A	*Plot B*
Planting density	25 000 plants/ha	37 000 plants/ha
Fertilizer application	N 20 kg/ha P_2O_5 50 kg/ha	N 100 kg/ha P_2O_5 80 kg/ha
Crop residues	removed	ploughed-in
Crop yield	5 t/ha	10 t/ha
Runoff	250 mm	20 mm
Erosion	12.3 t/ha	0.7 t/ha

Data are for 1954/55 cropping year (rainfall 1130 mm).

High density planting

One of the crops most commonly associated with high erosion rates is maize. This arises partly from it being a row crop, which reduces the overall density of cover given to the soil and allows runoff to become concentrated between the rows. More important, however, is the architecture of the plant, which results in high volumes of leaf drainage and stemflow. As a consequence, rates of soil detachment under a 90% canopy cover of maize can be twice those of a bare soil (Morgan, 1985) and, with canopy covers of 67–78%, between 30 and 49% of the rainfall intercepted by the canopy can be concentrated as stemflow at the base of the plant (Quinn and Laflen, 1983), causing local generation of runoff.

As part of a research programme designed to increase maize yields in central Africa, Hudson (1957) conducted an experiment to compare two management systems. He found that increasing planting density and fertilizer applications resulted not only in higher yields but, also, in less runoff and less erosion (Table 5.7). Higher yields also mean higher overall production of biomass and it is this which plays the important engineering role. The effects of greater leaf drainage and stemflow are more than offset by the effects of the rooting system in promoting higher infiltration and increasing soil cohesion. Higher biomass production also means that more crop residue is produced for ploughing back into the soil. In the long term, this helps to maintain the level of organic matter which further contributes to the resistance of the soil to erosion. After ten years of continuous maize production with this system, the soil was in better physical condition than at the start of the experiment (Hudson, 1981). Although this approach is associated with high cost and a high level of management, it is economic because it is repaid by the higher yield. Even if it cannot be transferred directly to non-agricultural conditions where soil fertility is lower and the management costs cannot be justified nor sustained, the approach emphasizes the importance of maximizing biomass production for maximum engineering effect and how, if it can be maintained over a decade or more, it can become self-sustaining through improvements brought about in soil structure and fertility.

Multiple cropping

Multiple cropping takes the form of either sequential cropping, which is the growing of two or more crops a year in sequence on the same piece of land, or intercropping, which is the growing of two or more crops simultaneously. In some instances, the two forms can be combined whereby two crops are grown but one matures faster. This system is helpful in providing maximum canopy cover as early as possible in the growing season. Figure 5.7 shows how intercropping of maize and cassava at Ibadan, southwest Nigeria, provides better cover than growing cassava on its own. The rainy season in this part of Nigeria begins on about 1 March and ends on 10 November (Walter, 1967) but towards the end of July and early August there is a 'little dry season'. This rainfall regime results in two cropping periods, one from April to July, when rainfall intensities and the risk of soil erosion

are highest, and the other from August to November.

Intercropping of maize and cassava produces 50% canopy cover 50 days after planting in April, compared with 63 days for cassava as a monoculture. This reduces the period at risk and from the end of May results in a two-storey canopy of tall-growing maize and short-growing cassava capable of intercepting 40% of the rainfall compared with only 28% for cassava alone (Aina, Lal and Taylor, 1979). The maize is harvested in July, by which time the cassava has attained 60–70% cover. During the second rainy season, the residue of the maize crop can be used as a mulch to provide ground cover underneath the cassava. Despite this intercropping system, the absence of ground vegetation, particularly in the early part of the wet season whilst the crops are establishing, means that overall erosion rates are still rather high, also confirming that these crops are associated with high rates of soil loss. Annual erosion rates on an alfisol near Ibadan were 110 t/ha for the cassava monoculture and 69 t/ha for the maize–cassava intercrop. Nevertheless, maize–cassava intercropping is adopted quite widely in the humid areas of west Africa. In the semihumid areas, maize can be intercropped with cowpeas, soya bean and phaseolus bean (Okigbo, 1978). The importance of establishing cover as early as possible in the wet season to control splash erosion was also emphasized by Shaxson (1981) in support of early rather than delayed planting in the Indore region of central India.

The beneficial effects of multiple cropping emphasize the importance of using a mixture of vegetation species with different growth rates in order to maximize vegetation cover over as long a period of the year as possible.

Ground cover crops

The establishment of a ground cover is widely used to control erosion in plantations of tree crops such as rubber, oil palm and coffee. Traditionally, these crops were clean-weeded which

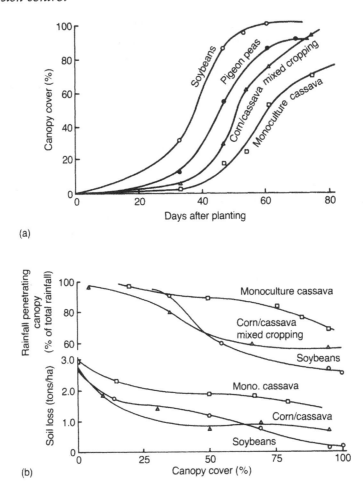

Figure 5.7 Effect of different cropping systems in Nigeria on: (a) rapidity of canopy cover growth after planting; and (b) rainfall interception and soil loss (after Aina, Lal and Taylor, 1979).

resulted in unacceptably high rates of soil loss because of the failure to protect the soil from the damage caused by leaf drips from the canopy. Research on rubber plantations in Malaysia shows that best results are achieved with leguminous creepers rather than grasses. Table 5.8 (Pushparajah, Tan and Soong, 1977) compares the effects of a mixed leguminous cover of *Pueraria phaseoloides* (tropical kudzu), *Centrosema pubescens* and *Calopogonium mucunoides* with a mixed grass cover of *Axonopus compressus* (carpet grass) and *Paspalum conjugatum*. Whilst both covers produce similar quantities of biomass and similar reductions in erosion, the leguminous plants return more nutrients to the soil, particularly nitrogen as a result of nitrogen fixation, and maintain higher infiltration rates. This means that any competition for nutrients between the rubber trees and the cover crop is offset by the greater availability of nutrients for both. The soils under legumes also maintain a higher content of organic material than those under grass.

The maximum engineering effect of the rubber–legume mix is probably achieved after four to five years. After this, the closure of the

Table 5.8 Comparison of the effect of grasses and leguminous creepers as cover crops in young rubber plantations (after Pushparajah, Tan and Soong, 1977)

	Grasses	*Leguminous creepers*
Biomass after 20–24 months	2696 kg DM/ha	5427 kg DM/ha
Litter production after 20–24 months	6140 kg DM/ha	6038 kg DM/ha
Nutrients returned to the soil over 5 years	N 24–65 kg/ha P 8–16 kg/ha K 31–86 kg/ha Mg 9–15 kg/ha	N 226–353 kg/ha P 18–27 kg/ha K 85–131 kg/ha Mg 15–27 kg/ha
Soil carbon content at 0–15 cm depth (years after planting)	2nd 1.82% 3rd 1.64% 4th 1.68% 6th 1.47% 8th 1.31%	2nd 1.75% 3rd 1.73% 4th 1.74% 6th 1.55% 8th 1.41%
Bulk density of soil (0–15 cm depth)	1.11 Mg/m^3	1.04 Mg/m^3
Permeability of soil (0–15 cm depth)	65 cm/h	90 cm/h
Depth of soil deposited on terraces[a]	12.7 cm	11.0 cm

Data on soils are for a heavy clay soil (Munchong Series).
[a] Depth of soil deposited on terraces should also be compared with a value of 19.0 cm recorded for bare soil. Length of period of observation is not stated.

Table 5.9 Effect of cover crops on soil properties and erosion under mature rubber (after Pushparajah, Tan and Soong, 1977)

	Bare soil	*Grass*	*Nephrolepis*
Bulk density (Mg/m^3)	1.36	1.24	1.19
Permeability (cm/h)	7	45	45
Erosion (kg/ha)	132	117	59

Data are for a sandy soil (Serdang Series) over a 15-month period.

rubber tree canopy causes the ground cover to decline because of shade, and the organic content of the soil begins to fall. However, reasonable protection of the soil is maintained because the greater vigour of the rubber trees with legumes results in a higher turnover of leaf litter. Also, ferns, such as *Nephrolepis*, may take over from the legumes and give a dense cover which will maintain erosion at very low levels compared with either a bare soil or a grass cover (Table 5.9).

The procedure in Peninsular Malaysia is to plant the legumes from seeds immediately the land has been cleared of either forest or a previous plantation crop. The seeds are scarified or scraped to aid germination and planted at about 0.5 m spacing. Fertilizer, in the form of rock phosphate, is applied at a rate of 0.5 t/ha. Once the cover crop has emerged, the rubber trees, usually clonal seedlings, are transplanted from the nursery. Weed control is very important until both the rubber trees and cover crop are established. After this, no management of the cover crop is required. The mutual benefit of combining rubber trees and legumes that is observed in Malaysia, in areas where the annual rainfall is between 1800 and 2200 mm with no dry season, may not be repeated elsewhere. In eastern Java, where the rainfall amount is similar but there is a marked dry season, cover crops may reduce the soil moisture in the dry season by up to 50% compared with clean-weeding (Williams and Joseph, 1970).

The combination of ground covers and tree crops not only emphasizes the importance of having a mixed vegetation cover but shows how the engineering effects of each component species change over time in what is effectively a controlled ecological succession. The Malaysian experience also demonstrates that leguminous species can play an important role in the succession and that it is not necessary for engineers to rely solely on grass to obtain the desired effects.

5.3.2 PRACTICES BASED ON VEGETATION LAYOUTS

Grass strips

Grass strips are used in agriculture where they are also referred to as buffer strips or grass filter strips. Ideally grass strips should be aligned on the contour. Species which do not spread and yet have fast and dense growth are often used. The denser and more uniform the vegetative growth, the more effective the strip will be in filtering out sediment already transported in the flow because of the greater roughness imparted to the flow. If the species chosen form a dense low-growing mat, then the strips can be traversed by machinery, without reducing their effectiveness. Any 'clumpiness' of growth may concentrate flows and build up differential drag velocities which lead to localized scour. The strip itself should not compete with the surrounding vegetation for light, nutrients or water. Grass strips can also be used to control the encroachment of weeds on to surrounding areas.

Grass strips can help reduce erosion on slopes where hedgerows have been removed. Ankenbrand and Schwertmann (1989) report on the use of grass strips to reduce slope lengths and thus runoff generation on fields which have been enlarged as part of a land consolidation project in Bavaria, Germany. With the removal of bench terraces and field boundaries, the number of fields in the area was reduced from 1084 to 339, the average field size increased from 0.31 to 1.13 ha and the average field length from 102 to 172 m. To offset the increase in erosion risk resulting from larger and longer fields, grass strips were established on the contour along the field boundaries in the expectation that, over time, they would develop into new 'erosion-induced' terraces.

Grass strips can be simulated with the use of vertical mulching techniques, where crop residues or other mulch material are placed in contour trenches, so intercepting flow and reducing slope lengths (see Chapter 4). The practical disadvantage here is that the strips cannot be traversed by machinery.

The species used in grass strips for erosion control are numerous. An increasingly common one is vetiver grass, particularly *Vetiveria zizanioides*. It is claimed that this species grows almost universally, from sea-level to 2000 m, from 600 mm to 6000 mm annual rainfall, on soils with pH from 4.5 to 10.5, and in temperature ranges from 5 °C to 45 °C (Anon., 1990). Other species include alfalfa, kudzu and various close-growing crops. Abujamin, Abdurachman and Suwardjo (1988) tested a number of grasses including bahia grass (*Paspalum notatum*), broad-leaf paspalum (*Paspalum wettsteinii*), pangola grass (*Digitaria decumbens*), signal grass (*Brachiaria decumbens*) and *Brachiaria brizantha*. These authors stress the importance of species selection based on soil conservation characteristics, i.e. the ability of the vegetation to perform the required engineering role, as well as choosing species which are acceptable to the local farmers.

Abujamin, Abdurachman and Suwardjo (1988) observed that on slopes of 15–22% a bahia (*Paspalum notatum*) grass strip of 1 m width, planted on the contour, reduced soil erosion by more than 95 % compared with the control plot after just one year from time of planting. The effectiveness of these grass strips in retarding sediment movement and promoting sedimentation was so great that the authors go on to claim that the use of grass strips in this way is the first step of cheap bench terrace construction, again by the mechanism of erosion-induced terraces referred to above.

There are some practical difficulties associated with the use of grass strips. Farmers are often concerned by the extent of the land which the strips occupy and which is therefore taken out of crop production. Abujamin, Abdurachman and Suwardjo (1988) show that the land area taken out of production is a function of slope steepness (which affects spacing of the grass strips) and strip width. They found that the strips can take up between 5 and 27 % of the land area (Table 5.10).

In mixed farming communities any loss of productive land can be compensated for by selecting species that can be fed to livestock or which, when dried, can be used for household applications such as thatching. There is often a conflict between selecting a species such as *Vetiveria zizanioides* which does not spread into farmers' fields to compete with the crops but which is unsuitable for grazing, and a species which has other uses but which spreads easily.

Strip cropping

As applied in agriculture, strip cropping refers to strips of different vegetation species aligned across the slope, ideally following the natural contours. Strips with high erosion risk species are alternated in space with crops associated with low erosion rates (e.g. row crops and grass respectively). The principle is that any erosion generated in the erosion-prone strips will be intercepted by the soil-saving strip, and deposited there. The steepest slope on which this technique can be used successfully on its own is about 8.5° (Morgan, 1986).

Table 5.10 Width and percentage of cultivated area for contour grass strips in relation to strip width and slope steepness (after Abujamin, Abdurachman and Suwardjo, 1988)

Slope (%)	0.5 m wide strip		1.0 m wide strip	
	Width	Percentage	Width	Percentage
8	9.4	95	9.4	91
15	5.0	91	4.5	82
18	3.9	89	3.4	77
22	3.2	87	2.7	73

The vegetation in each strip is usually rotated over time, so that fertility is maintained over the rotation as a whole. Crops with high erosion risk (e.g. row crops) are effectively 'subsidized' in future years by soil-saving crops which have low erosion risks and, since they are often legumes, will return nitrogen to the soil. In this way the erosion rate for the whole rotation is kept within the soil loss tolerance limits for the site. Where erosion risk is high, such as on steep slopes and on erodible soils, the grass strips may be permanent.

There are a number of types of strip cropping used for control of water erosion. Contour strip cropping involves the precise alignment of the cropped strips on the natural contour. This is excellent in intercepting runoff, but is unpopular because of the practical disadvantages. On very complex land where the contours form an awkward pattern, the strips will also assume highly irregular shapes which may be difficult to manage. Also the effective cross-slope length is increased by virtue of the complicated strip pattern. This means that energy consumption and time needed to manage these strips is much greater than if the field was conventionally laid out.

Field strip cropping attempts to overcome irregularly shaped areas of land by having the strips normal to the predominant slope direction. Whilst this practice simplifies the field operations, it ignores low spots where concentrations of flow may occur leading to breaching or breakdown of the system. On very complex slopes, it may be difficult to simplify the natural slope shape in a satisfactory way.

Buffer strip cropping is a combination of contour and field strip cropping. Uniform strip widths (e.g. 20 m) are laid out close to the contour as keyline strips following the general lie of the land. Where there is deviation from the natural contour line, buffer strips (often about 2–4 m wide, usually grassed or planted to legumes) are used to fill in the intervening land between adjacent strips (Figure 5.8; Hudson, 1981). Generally, deviation from the contour of up to 10% is permissible without the need for buffer strips, but where greater deviations occur, buffer strips should always be used.

All strip cropping practices have practical disadvantages. Each strip is a relatively small area to manage. Different strips require different treatments so that managing each strip separately takes longer than treating the field as a whole as would be the case in monocropping. Many farmers and contractors are uneasy about running machinery across a slope because of the risk of tipping over, especially on steep slopes. Harvesting is difficult when operating across the slope, especially for root crops, unless specialized and expensive machinery is used. The small management areas are less of a problem on smallholdings, although in these cases, the large proportion of the land given over to non-economic permanent grassland may be an important constraint on adopting the system.

Often grassed waterways are incorporated in strip cropping schemes. They are located in depressions in order to drain excess runoff at a safe velocity downslope.

Alley cropping

Alley cropping is a form of agroforestry, where food crops (often associated with high erosion risks) are grown in alleys aligned across the slope and framed by hedgerows of trees or shrubs (with low erosion risk). Agroforestry is the collective name for land use systems in which woody perennials are grown in association with herbaceous plants and/or livestock and in which there are both mutually beneficial ecological and economic interactions between the tree and non-tree components. Alley cropping is used as a soil conservation technique which aims at maintaining soil fertility as well as controlling soil erosion. Yields of the food crops are maintained by cutting back the hedgerows to reduce shading effects and using the prunings as a mulch to reduce erosion of soil and nutrients. The prunings also decompose to add organic matter and additional nutrients, especially nitrogen, to the soil.

Lal (1988) found that the integration of trees with annual crops helps to reduce runoff velocity as the dense hedgerow forms a barrier to flow. As runoff is retarded, it has more time to infiltrate the soil. Indeed, experimental data support these assumptions as infiltration rates were increased on plots planted to perennial hedges. *Leucaena leucocephala* was more effective at reducing runoff and soil erosion than hedges of *Gliricidia sepium*.

Lal's work reported that a hedge spacing of 2 m was more effective than one of 4 m, reducing soil erosion by 92.3% and 42.3% respectively, compared with conventionally ploughed land. The dramatic differences between the two spacings illustrates the importance of maintaining adequate soil protection over as large an

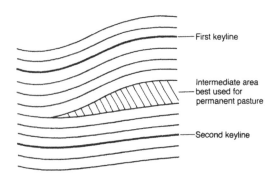

First keyline

Intermediate area best used for permanent pasture

Second keyline

Figure 5.8 Use keyline strips and butter strips in contour strip cropping.

area of the soil as possible. With the wider spacings, additional conservation measures (e.g. minimum tillage and mulching) are required between the hedgerows. Lal argues this is because the contribution of splash and interrill erosion to sediment yields should not be under-estimated, as evidenced by the formation of 'erosion-induced' terraces on the alley cropped plots. Although erosion control was enhanced on the plots with the closely spaced hedges, this practice had negative effects on the yields of the annual arable crops (maize and cowpea) grown between the hedges.

Despite the observed reductions in soil erosion and runoff volume under the alley cropping practice, the effectiveness of the shrub or tree components will not be apparent in the first few years because it takes time for these species to establish.

5.4 DESIGN OF VEGETATIVE SYSTEMS FOR WATER EROSION CONTROL

Some uses of vegetation for controlling erosion by water are well researched. Design procedures, based on an empirical understanding of the processes involved and engineering experience, are well established, e.g. the design of grass-lined waterways (United States Soil Conservation Service, 1954; Temple *et al.*, 1987; Hewlett, Boorman and Bramley, 1987). Design procedures for other uses have not been developed and the way in which vegetation is used relies on judgement, based on an understanding of the underlying principles. Knowledge of the following will help ensure the successful use of vegetation for water erosion control:

1. the engineering role that vegetation is required to perform;
2. the properties that vegetation must possess in order to perform that role;
3. the ability of vegetation species with the above properties to survive in the local environment;
4. the seasonal growth pattern of those species;

5. the type of vegetation structure and plant community that these vegetation species will provide;
6. the way in which the vegetation cover needs to be managed over time both in the early stages, to ensure that plants with different growth rates and levels of competitiveness succeed one another and lead to the proposed plant community, and in the long term, to maintain the plant community in the required state;
7. the extent to which vegetation alone will give adequate control of the erosion or whether, for example on steep slopes, it will need to be supplemented by inert materials or structural works.

At present, this knowledge, brought about by both research and practical experience, is concentrated on three broad environments: slopes, channels and shorelines.

5.5 EROSION CONTROL ON SLOPES

The limited studies on rates of erosion on steep slopes under forests (e.g. Hatch, 1981; Singh and Prasad, 1987) and grasslands (e.g. Veloz and Logan, 1988) demonstrate the effectiveness of vegetation in keeping annual soil loss below 1 t/ha. Provided a satisfactory vegetation cover can be established, there appears to be no reason why vegetation alone cannot control erosion by water on even the steepest slopes. Only where, because of soil or climate, a satisfactory vegetation cover cannot be obtained or where processes other than water erosion also need to be controlled, should it be necessary to supplement the vegetation with additional (e.g. structural) measures.

One reason for the effectiveness of vegetation is that the steepest slopes do not always record the highest erosion rates. Odemerho (1986) found that erosion by water on cut road banks in Nigeria followed a curvilinear relationship with slope steepness (Figure 5.9) with peak rates occurring on slopes of 15–20%. The decline in erosion on slopes greater than 20% may

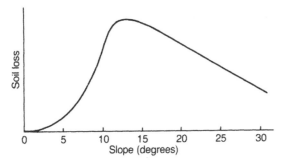

Figure 5.9 Relationship between soil loss and slope steepness (after Odemerho, 1986).

be explained by changes in the hydrological processes. Dunn (1975) suggests that with increasing slope steepness above 20%, the transport capacity of the runoff increases more rapidly than the detachment rate of soil particles. Erosion thus becomes limited by the rate of detachment. Further, the main agent of particle detachment, which is the raindrop impact, declines with the greater slope angle, because the drop impact is spread over a larger surface area and becomes less effective. Under some conditions, the transport capacity of the runoff may actually decrease. Heusch (1970) found that on slopes up to 63% in the Rif Mountains of Morocco surface runoff declined and rates of subsurface flow increased. When these hydrological processes are considered together, they imply that if vegetation can control erosion on slopes up to 20%, it will do so on steeper slopes.

5.5.1 SALIENT PROPERTIES OF VEGETATION

Table 5.11 lists the salient properties of vegetation cover required for controlling soil erosion by water on slopes. Based on the studies by agricultural engineers (section 5.3), in order to achieve the maximum effect the vegetation must provide:

1. a dense uniform cover (at least 70%) close to the ground surface;
2. a dense laterally-spreading root system.

This means that the vegetation must be of low height (< 50 cm) with a single-stemmed or spreading habit (not clumped) and have a shallow (80% of roots within 15 cm of the soil surface), fine or fibrous rooting system. The root network must promote an even pattern of infiltration of water into the soil in order to avoid concentrations of water which might initiate or feed a network of subsurface pipes.

Additional properties of the vegetation which will be important are:

1. rapid growth, both in the first year of revegetation and at the beginning of each growing season;
2. ability to produce the maximum effect at the time of year when the rainfall intensities are highest;
3. resistance to mechanical damage, e.g. by walkers, livestock and vehicles;
4. a high rate of litter production, so as to build-up the organic content of the soil as rapidly as possible and to cover the soil surface with a protective layer of decaying humus.

The ability to fix nitrogen may, as seen above, be advantageous though it is not essential.

In agricultural situations, it will be necessary to consider the economic value of the vegetation (e.g. for fodder, fuel or as a green manure) and how the plant cover will be managed. It should be stressed, however, that an agricultural harvest represents a removal of all or part of the vegetation and therefore an impairment to its engineering role. A high level of maintenance and a fertile top soil are generally prerequisites in agricultural areas with management aimed at establishing a single-species, often single-variety, ecosystem. In contrast, on road banks, where there is no need to harvest the biomass, high-yielding varieties of vegetation are not required; nor are high levels of soil fertility essential. Indeed, the often-prevailing low level of soil fertility is an advantage because it limits the vigour of the vegetation growth and, thereby, reduces the cost of maintenance. With the low soil fertility, however, it is generally not possible to obtain satisfactory vegetation cover

Table 5.11 Salient properties of vegetation for controlling erosion by water (after Coppin and Richards, 1990)

	Ground cover	Height	Stem density	Stem rigidity	Rooting depth	Root density	Root strength	Weight	Growth cycle
Soil detachment	x	x	x						x
Mechanical strength	x	x	x	x		x	x		x
Rainfall interception	x	x							
Runoff	x		x		x				
Infiltration			x		x	x			
Subsurface flow						x			
Evapotranspiration			x			x			x
Soil moisture depletion					x				x
Root reinforcement					x	x	x		x
Anchorage					x	x	x		
Arching/Buttressing Surcharge					x		x		
Wind loading		x	x	x	x		x	x	x
Root wedging					x	x			

with a single species. A mixed vegetation species is preferred to allow for the likely failure of some species to grow or regenerate in the more hostile conditions. A similar focus applies to mining spoils with the additional constraint that the vegetation must be able to survive an often toxic environment. In recreational areas, it will be necessary to combine the ability to survive in the local soil and climatic environment and the need for low maintenance with the ability of the vegetation to withstand mechanical damage and to create interesting wildlife habitats of diverse species.

5.5.2 SELECTION OF VEGETATION SPECIES

Vegetation species must be selected according to their ability to perform relevant engineering roles, particularly those of controlling soil detachment by raindrop impact and increasing soil cohesion. Species selection must take into account:

1. the bioclimate;
2. the soil quality, based on soil fertility and soil water regime, as well as considering any toxicity or salinity;
3. fauna, both species, such as earthworms, which will promote soil mixing and species which will cause damage to seeds and young shoots;
4. the time of year when the risk of erosion is highest;
5. the seasonal pattern of vegetation growth and therefore the ability to maximize the engineering role at the time of greatest need;
6. the long-term successional trends of the vegetation community and how these can be manipulated or controlled by management.

In addition, the effects of the vegetation on other erosion processes must be considered, particularly if these are adverse (Chapter 2). For example, if surface vegetation reduces soil erosion by promoting greater infiltration of water into the soil, will this lead to either increased subsurface erosion by piping or to slope instability through rises in pore-water pressures? The likelihood of adverse effects will vary with local site conditions.

Vegetation species are normally chosen to replicate the type of colonization and ecological succession that would occur naturally (Chapter 3), but at the same time to try and speed up the process. Under natural conditions, a few pioneer species, well-adapted to the local, often adverse, environment colonize an area, stabilize it and, over time, modify the environment so that other vegetation species are able to invade and displace the pioneers. In practice, the need to establish a ground cover rapidly means that grasses generally form the major component of the colonizing species. In order to allow for the risk of failure of individual species, particularly on soils of low fertility, mixtures are often used, containing some six to ten species including a 10–50% content of legumes or herbaceous plants. The mixture may be adapted to include both quick- and slow-growing species with the intention of providing a plant succession of 'nurse' species to give immediate protection followed by other species to give a more diversified ecology. Also, once established, leguminous species generally have a better survival rate and a lower management requirement than grasses. Although there are some cultivars of grasses that grow slowly to only a short height and, therefore, require little mowing or cutting, they usually produce very limited root growth and so are unsuitable for an engineering role (Coppin and Richards, 1990).

Native species are preferred because they help to retain the local ecology and also can be chosen with greater confidence in their ability to grow in the local environment. Exotic species should only be chosen if there are no suitable native species that will perform the required engineering function. Species selection will depend on whether the aim is to:

1. maintain a single species on highly fertile soils and with a high management input, as in agriculture;

2. develop and maintain with a low level of management a particular mixed species ecosystem, as might be the case on land devoted to grazing or in nature conservation areas; or

3. provide the basis for a natural ecosystem to establish which may ultimately evolve to a climax or subclimax vegetation.

In each of these cases, species selection will be dependent on whether the need is to provide cover rapidly at the beginning of the growing season (e.g. to protect against erosive rainfall that comes mainly in the spring and early summer), or to ensure cover at the end of the growing season to counteract erosive rains in late summer or early winter.

5.5.3 DESIGN OF VEGETATION COVERS

In designing an ecosystem to control erosion by water on slopes whereby the vegetation cover will perform an engineering function, it is necessary to take account of the role played by the individual plant species and the time taken for them to develop sufficiently to perform that role. As with cover crops in tree-crop plantations, grass–legume mixtures are preferred because the nitrogen-fixing nodules produced by the legumes benefit the grasses which would otherwise disappear from the plant community through lack of nitrogen. Legumes, however, take two to three years to establish complete cover. Low-growing sod-forming grasses which spread rapidly by stolons above ground or rhizomes below ground are therefore an essential requirement of any species mix. Suitable species include *Cynodon dactylon* (Bermuda grass) and *Digitaria decumbens* (Pangola grass) for warm climates, *Axonopus* spp. (carpet grass) for humid tropical climates, and *Agrostis* spp. (bent grasses) and *Festuca rubra* (red fescue) in temperate climates. Where very rapid vegetation growth is required, barley, oats and *Lolium* spp. (rye grasses) will need to be included. Typical leguminous species in temperate areas include *Trifolium* spp. (clovers), *Coronilla varia* (crown vetch) and *Lespedeza*

Table 5.12 Recommended grass-based seed mixtures for erosion control in the UK (after Coppin and Richards, 1990)

	Standard mixture (%)	*Other recommended mixtures*					
		Loamy soils free-draining (%)	Clayey soils prone to waterlogging (%)	Sandy soils prone to drought (%)	Saline, clayey (%)	Acid conditions (%)	Alkaline conditions (%)
Lolium perenne	45						
Phleum pratense			10				
Poa pratensis	15	15	15	15	20		20
Poa trivialis			10				
Poa compressa		10	5	20		30	10
Festuca longifolia		10		10			
Festuca fallax		10					
Festuca rubra	15	40	40	40	35		55
Festuca ovina						30	
Agrostis canina		10	5	10	10		
Agrostis stolinifera			10		10		10
Deschampsia flexuosa						25	
Cynosurus cristatus	15						
Pucinella distans					35		
Trifolium repens	10	5	5	5	2.5	2.5	
Lotus corniculatus					2.5	2.5	

Note: the above mixtures are given as a guide and should not be taken as prescribed mixtures for particular circumstances.

spp. A range of grass mixtures is commercially available, each designed for specific site conditions. Examples of mixtures for the UK are listed in Table 5.12 (Thompson, 1986; Coppin and Richards, 1990).

Table 5.13 lists species of grasses and legumes recommended for surface protection of slopes in the UK (Coppin and Richards, 1990). Table 5.14 gives similar information for the USA (Troeh, Hobbs and Donahue, 1980), taking account of climate, acidity (important for mining spoils) and salinity (important for arid and semi-arid areas).

Road banks

Studies on road banks in western Oregon (Dyrness, 1975) with 1:1 (45°) backslopes formed in soils derived from tuffs and breccias showed that seeding in the late summer or early autumn, along with an application of ammonium phosphate at 0.45 t/ha and a straw mulch of 4 t/ha,

resulted in limited ground cover (15–45%) during the first winter but satisfactory cover (>70%) by the end of the following summer. Annual ryegrass (*Lolium multiflorum*) germinated and established faster than any other species and dominated the ground cover at the end of the first winter. Although the legumes were visible in the first spring, they had virtually disappeared by the end of the first summer, being unable to compete with the vigorous growth of bent, fescue and perennial rye grasses. Possibly the original fertilizer application was too high and gave too great a stimulus to grass production. With the loss of the legumes, lack of nitrogen reduced the soil fertility and, after three years, the ground cover started to decline (Figure 5.10). After eight years, the cover was reduced to 10% but the soil was protected by a sizeable amount of dead grass litter. Refertilization of the plots, however, quickly revived vegetation growth which developed to 90% cover in one year. Erosion

Table 5.13 Plant species recommended for protection of slopes against water erosion in the UK (after Coppin and Richards, 1990)

Plant species	Characteristics
Grasses	
Agrostis capillaris	Wide soil tolerance, rhizomatous; cultivars available for tolerance of heavy metal contamination
Agrostis castellana	Wide soil tolerance, spreads by stolons; prefers damp soils, tolerates occasional flooding and salt
Arrhenatherum elatius	Wide soil tolerance, natural colonizer of embankments and cuttings; tall habit
Elymus (Agropyron repens)	Suitable for deeper soils, not tolerant of extremes; rhizomatous; strongly competitive
Festuca rubra	Very wide soil tolerance, rhizomatous; cultivars available for tolerance of heavy metals and salt
Festuca longifolia	Drought tolerant; wear tolerant; not rhizomatous
Lolium perenne	Quick-growing; wear tolerant; suitable for fertile soils
Lolium multiflorum	Very quick-growing but does not persist long
Poa annua	Annual, quick-growing on bare ground; wear tolerant
Poa compressa	Rhizomes; tolerant of infertile soils
Poa pratensis	Wide soil tolerance, strong rhizomes; wear tolerant
Legumes	
Coronilla varia	Wide tolerance, dense growth, slow to establish especially in the north
Lotus corniculatus	Wide soil tolerance, salt-tolerant
Lupinus polyphyllus	Wide soil tolerance, subject to winter-kill
Medicago sativa	Suitable for neutral–alkaline soils; drought-tolerant
Onobrychis vicifolia	Suitable for neutral–alkaline soils with some fertility; drought-tolerant
Trifolium repens	Requires moderate fertility
Trifolium hybridum	Tolerates waterlogging

over the first five years of measurement resulted in a lowering of the ground surface by 3.5–9.7 mm, most of it during the first year when vegetation cover was still low. In contrast, erosion on a site where the vegetation cover was allowed to develop naturally was about 25 mm over the same period, by which time the cover had developed to only 10%. This study thus shows the importance of speeding-up the rate of vegetation growth compared with the natural succession. Whilst this can be achieved by the addition of fertilizers, fertilizer application must be controlled in order to limit growth of highly competitive species which might otherwise dominate the ecological succession and force it to take a different course from that originally designed.

Similar studies in Georgia (Richardson, Diseker and Sheridan, 1970) show that the ecological succession can be controlled to allow legumes gradually to take over from the grasses and produce a sustainable vegetation cover which will reduce annual erosion from over 200 t/ha to less than 10 t/ha (Table 5.15).

Perennial legumes are widely used for erosion control on road banks in Pennsylvania (Duell, 1989), being ideally adapted to low maintenance. *Coronilla varia* (crown vetch) is the most common of these, especially on steep slopes that will not be mown. Although it takes about three years to become established, it will then crowd out the grasses and weeds and provide more than 50% of the plant community until finally invaded by brush after 20 to 30

Table 5.14 Plant species used successfully for surface erosion control on construction sites and mine spoils in the USA (after Troeh, Hobbs and Donahue, 1980)

Species	Climate[a]				Soil				
					pH[b]			salinity[c]	
	wm	cl	cd	dr	lw	md	hg	md	hg
GRASSES									
Agropyron desertorum	x	x		x					x
Agropyron intermedium		x		x					x
Agropyron smithii	x			x					x
Alopecurus arundinaceus			x	x					
Astragalus cicer		x							
Bromus inermis		x		x				x	
Cynodon dactylon	x								x
Dactylis glomerata		x		x			x	x	
Deschampsia caespitosa			x						
Elymus junceus		x		x					x
Eragrostis curvula	x				x				
Eragrostis ferruginea	x				x				
Eragrostis trichodis	x					x			
Festuca arundinacea	*x*	*x*				*x*			*x*
Festuca rubra			x						
Lolium perenne	x	x		x				x	
Miscanthus sinensis	x	x					x		
Panicum amarulum	x	x					x		
Panicum clandestinum	x	x			x				
Panicum virgatum	x	x			x				x
Paspalum notatum	x								
Phalaris arundinacea	x	x					x	x	
Poa pratensis			x						
Sporobolis airoides	x			x					x
Trifolium hybridum			x						
LEGUMES									
Atriplex canescens	x			x					x
Coronilla varia	x	x				x			
Lathyrus silvestris	x	x				x			
Lespedeza striata	x								
Lespedeza vergata	x								
Lotus corniculatus	x	x				x		x	
Medicago sativa		x				x		x	
Pueraria lobata	x	x					x		
Sericea lespedeza	x	x				x			
Trifolium hybridum		x				x			

[a] wm = suitable for warm climates; cl = suitable for cool climates; cd = suitable for cold climates, e.g. central Alaska; dr = drought-resistant.
[b] pH values; lw < 4.0; md = 4.0–5.5; hg = > 5.5.
[c] Salinity levels: md = 4–8 mmhos/cm; hg = 8–12 mmhos/cm.

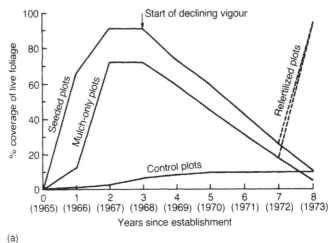

(a)

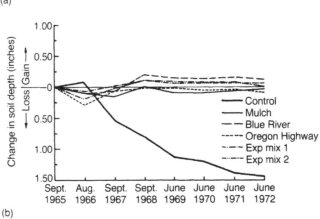

(b)

Figure 5.10 Trends in (a) vegetation cover and (b) soil loss on road banks with 1:1 slopes cut in tuffs and breccias, western Oregon, USA (after Dyrness, 1975).

Table 5.15 Mean annual rainfall, runoff and soil loss from roadbanks on Cecil subsoil in Georgia, USA, from 1965 to 1967 (after Richardson, Diseker and Sheridan, 1970)

Cover	Slope	Rainfall (mm)	Runoff (mm)	Soil loss (t/ha)
None	1:1.4	1344	293	338
None	1:1.25	1344	282	278
Crown vetch and Abruzzi rye	1:2.5	1353	210	10
Sericea lespedeza and lovegrass	1:3.3	1353	109	3
Kentucky fescue	1:1	1353	116	6
Pensacola Bahia grass and Bermuda grass	1:1.1	1353	141	5

years. The flat pea (*Lathyrus sylvestris* L.) is equally vigorous and has been used successfully in Pennsylvania and New York. Less competitive with grasses are birdsfoot trefoil (*Lotus corniculatus* L.) and the perennial pea (*Lathyrus latifolia* L.). Suitable tall grasses that can be mixed with these less vigorous legumes are weeping lovegrass (*Eragrostis curvula*) and switchgrass (*Phalaris arundinacea*).

Grasses are also used to control erosion on road banks of 30° to 65° covered with shallow weathered debris in the Outer Himalayas of eastern Nepal (Howell *et al.*, 1991). Mean annual rainfall in the area ranges from about 1500 to 3500 mm, depending upon altitude and localized rain shadows. Most of the rain is concentrated into the four-month period from June to September and falls in storms lasting 2–3 h separated by dry spells of similar length. Peak intensities greater than 100 mm/h are common and produce considerable runoff and erosion. *Cynodon dactylon* (locally known as 'dhubo') and *Pennisetum clandestinum* (kikuyu grass) are the recommended grass species.

Although shallow-rooted grasses can help reduce surface erosion on these steep slopes, they can also have the adverse effect of enhancing shallow slope failures. There appear to be two main mechanisms involved: first, the increase in infiltration brought about by the vegetation results in higher soil moisture; and second, the weight of the grass and the root mat adds a surcharge to the slope. In order to reduce the risk of shallow slides, it is recommended that, on slopes greater than 35°, deeper rooted grasses, such as *Saccharum spontaneum* (khans), *Neyraudia arundinacea* (sito) and *Pennisetum purpureum* (napier) be used to try to anchor the soil and root mat to the underlying rock and weathered debris. These species are not always successful, however, because their clumpy nature causes concentrations of runoff and the roots do not adhere well to the coarse debris, often pulling-out when the material starts to move, before their tensile strength has been fully mobilized. Thus, although an almost complete cover of grass can be obtained within

one growing season, it is often necessary to supplement the vegetative measures with structural controls.

In drier parts of the Himalayas, there is a risk of many of the plant species cited above failing if there is no rain soon after treatment. In the North West Frontier Province of Pakistan, the Pakistan Forest Institute has successfully used *Saccharum* hedges under these conditions (Shah, undated). They are planted as tufts in 30 × 30 cm trenches dug on the contour. The space between the *Saccharum* hedges is planted with local tree species at 1.5 × 1.5 m spacing.

A key issue in the revegetation of road banks is the extent to which top soil should be used. Bradshaw and Chadwick (1980) stress the following disadvantages of top soil: its inherent fertility results in vigorous vegetation growth which then has to be managed by frequent mowing; it often contains a seed bank of weeds, such as docks and thistles, which are extremely competitive and also not very beautiful; it is generally more erodible than the substrate and more prone to compaction by raindrop impact; and it is more expensive to use because of the need to spread it over the slope. As an alternative, it may be better to prepare the substrate by chisel ploughing, apply fertilizer and seed with a mixture of grasses, legumes, shrubs and trees to give a varied and more interesting ecology. Where access is difficult, hydroseeding may be used; where erosion is a problem, geotextiles may help stabilization and vegetation growth. In all cases, the requirements of creating an ecological habitat must be balanced by the need for the vegetation to perform its appropriate engineering function and the need for low-cost long-term maintenance.

Recreational areas

Rose (1989) describes the use of vegetation in the Three Peaks area of the Yorkshire Dales National Park to control erosion along footpaths. The work involves two projects, one to revegetate heavily damaged ground where a footpath has been diverted to relieve pressure, and the other to reinforce vegetation on paths where there is still residual vegetation cover ranging from 20 to 70%. In the first project initial success has been achieved with a specially prepared mix of native grasses comprising 70% *Deschampsia flexuosa* (wavy hair grass), 14% *Nardus stricta* (mat grass), 14% *Agrostis tenuis* (bent), 1% *Potentilla erecta* (tormentil) and 1% *Campulana rotundifolia* (harebell). Reseeding was accompanied by the application of 350 kg/ha of slow-release nitrogen fertilizer. After 18 months, the vegetation cover on the paths ranged from 53 to 95%. Surprisingly, control areas where no reseeding occurred yielded a similar cover indicating the possibility of establishing vegetation from the self-sown seed bank within the soil. In the second project, similar application rates of slow-release nitrogen fertilizer to the existing vegetation produced increases in plant cover of up to 20%.

There is no mention in the Three Peaks study of the importance of selecting species for erosion control in recreational areas that are resistant to damage from trampling. Nevertheless, *Nardus stricta*, *Agrostis* spp., and *Deschampsia* spp., are recognized as being relatively resistant to trampling (Coleman, 1981). In contrast, most heathland species and bracken (*Pteridium aquilinum*) are highly susceptible to damage.

According to Coppin and Richards (1990), the following characteristics of vegetation make it resistant to wear:

1. short or prostrate growth form;
2. flexible rather than rigid stems and leaves;
3. basal or underground growth points;
4. ability to spread by stolons or rhizomes as well as seed;
5. deciduous habit;
6. rapid rate of growth;
7. long growing period;
8. ability to withstand burial by soil or rock;
9. ability to withstand exposure of the roots.

These authors list the following species as meeting these requirements: *Festuca longifolia* (hard fescue), *Lolium perenne* (perennial rye grass), *Poa annua* (annual meadow grass) and *Poa pratensis* (smooth meadow grass). *Poa pratensis* and *Poa annua* were also identified by Liddle (1974), along with *Festuca rubra*, as being resistant to damage. A *Festuca idahoensis–Poa pratensis* meadow was found by Weaver and Dale (1978) in the northern Rocky Mountains to be more resistant than shrubby vegetation to damage by hikers and horses.

Mining spoil

Mining spoils are another example where vegetation has been used successfully to control surface water erosion. Haigh (1979) compares unvegetated and naturally vegetated slopes of about 20° on surface coal mine dumps at Waunafon, near Blaenavon, south Wales. The mine was opened in 1942 and closed in 1947. Erosion was monitored between 1972 and 1977, at which time a large part of the site remained unvegetated and was finely dissected by gullies. The naturally vegetated sites had about 60–80% vegetation cover dominated by grasses, mainly *Nardus stricta*, *Festuca ovina*, *Festuca rubra* and *Agrostis tenuis*. Average annual ground loss on the vegetated slopes varied between 2.1 and 2.3 mm whereas that on the bare slopes was between 3.6 and 5.9 mm.

Studies on reclaimed surface-mine spoils at two coal mining sites in Wyoming (Lusby and Toy, 1976) emphasize the importance of the dual role of vegetation in surface protection and root reinforcement of the soil. Both sites show differences in erosion recorded during rainfall simulation experiments between natural and rehabilitated slopes. In both cases, higher erosion occurs where the vegetation cover and the root density are less (Table 5.16). These studies also show that rehabilitation work does not always produce better results than allowing land to revegetate naturally. In this case, the rehabilitated sites have been reformed into steeper slopes and have more clay in the soil, both of which lead to greater runoff.

Table 5.16 Site characteristics and erosion recorded from rainfall simulation experiments on mine spoils in Wyoming (after Lusby and Toy, 1976)

	Dave Johnston Mine		Big Horn Mine	
	Natural site	Rehabilitated site	Natural site	Rehabilitated site
Average slope (%)	18	23	15	21
Clay in topsoil (%)	20	48	25	35
Bare ground (%)	30	13	25	48
Root density in top 10 cm of soil (g/g of soil)	37	14	62	35
Runoff (mm)	21.6	31.2	3.3	20.8
Soil loss (t/ha)	1.1	4.7	0.2	9.5

Runoff and soil loss are from a simulated rainstorm of 38.1 mm over 45 min.

5.5.4 DESIGN OF VEGETATIVE BARRIERS

Two design parameters are important for vegetative barriers such as grass strips; namely the distance apart or spacing and the barrier width.

Spacing

The spacing or distance apart of vegetative barriers is based on engineering experience rather than calculated from a specific design formula. Guidelines can be developed from the experience of agricultural engineers with contour grass strips. Spacing is normally expressed in terms of a vertical interval, i.e. the vertical height difference between two consecutive grass strips on a slope. From trigonometry, a vertical interval of 3 m would produce strips about 60 m apart on a 5% slope, about 100 m apart on a 2% slope and about 7 m apart on a 57% slope.

For vetiver grass strips a vertical interval of 1–2 m is normally recommended (Anon., 1990), although a smaller value may be used for gentle slopes (less than 15%) and a greater value on steeper slopes. Abujamin, Abdurachman and Suwardjo (1988) adopt the same vertical interval for grass strips in Indonesia (i.e. 0.75 m) as that used for bench terrace construction. Hurni (1986) recommends 1 m wide strips with a vertical interval of 1 m on slopes of 15% or less; this gives a spacing of 33 m on a 3% slope and a 7 m spacing on a 15% slope. For slopes greater than 15%, a vertical interval of 2.5 times the soil depth is recommended. Another approach, particularly relevant in arid and semi-arid conditions, would be to space the strips according to the soil moisture recharge. In this case, the spacing would be such as to reduce the runoff generated between the strips to that which can be retained on the upslope side of each strip and there infiltrate the soil. As an example, in the Doon Valley, India, spacing is based on a 3:1 ratio between the area contributing runoff and the area of infiltration behind a bench terrace. A typical system on a 2.5% slope has an 18 m long contributing area and a 6 m long infiltration area (Figure 5.11; Singh, 1990), giving an overall spacing of 24 m.

Recommendations on spacing can be tested by using empirically or theoretically derived critical slope lengths. The LS factor of the Universal Soil Loss Equation (Wischmeier and Smith, 1978) can be used as an approximation of the critical slope conditions which will keep annual erosion rates to within a set soil loss tolerance. Thus:

$$LS = \frac{A}{R.K.C.P} \qquad (5.1)$$

where A is the soil loss tolerance rate, R is the rainfall erosivity factor, K is the soil erodibility factor, C is the cropping practice factor and P is the conservation practice factor. Once the value

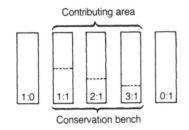

Conservation bench

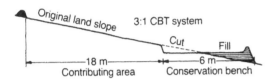

Figure 5.11 Plot layout and cross-section on a conservation bench terrace (CBT) system for water harvesting (after Singh, 1990).

For example, the critical slope length on a sandy soil on a 3° slope and a peak rainfall excess $(R - i)$ with a 10-year return period of 0.2 mm/s can be calculated by assuming a value of $n = 0.01$ for shallow overland flow over bare soil and $v_{cr} = 0.75$ m/s (Table 5.17). Equation 5.2 becomes:

$$L_c = \frac{0.75^{5/2} \times 0.01^{3/2}}{0.0002 \times 0.1094 \times 0.9886},$$

$$L_c = 22.5 \text{ m},$$

which implies that placing grass strips on the contour every 22.5 m down the slope will prevent runoff from attaining erosive velocities for storms with rainfall intensities equal to or less than that of the design storm.

of LS has been obtained, Figure 5.12 can be used to determine the critical slope length for the local slope steepness. This critical slope length can be taken as the maximum spacing between strips along the slope at which soil loss will be maintained below the chosen tolerance level. Examples of the Universal Soil Loss Equation used in this way are found in Hudson (1981) and Gray and Leiser (1982).

A theoretical approach to determine critical slope lengths can be based on the velocity of overland flow, the roughness imparted to the flow (expressed as the Manning's n value), the rainfall, soil infiltration characteristics and the steepness of the slope. Thus:

$$L_c = \frac{v_{cr}^{5/2}\, n^{3/2}}{(R - i)\sin^{3/4} S \cos S} \tag{5.2}$$

where L_c represents the theoretical critical slope length at which flow becomes erosive, v_{cr} is the maximum permissible velocity of overland flow (i.e. non-erosive), n is the Manning's roughness coefficient, R is a typical rainfall intensity, i is the infiltration capacity of the soil and S is the slope.

Barrier width

The width of each strip will affect the potential runoff that can be generated over that width and the extent to which it can reduce the velocity of runoff from above and filter out any sediment being carried by the flow. Strip width is usually based on the steepness of the slope, which affects the energy of any runoff that is generated. Recommended strip widths are given in Table 5.18. They can also be tested for given soil and rainfall conditions by applying equations 5.1 and 5.2.

Wattling

Wattling is a technique in which, instead of grass, barriers are formed by packing live bundles of freshly cut, leafy brush wood such as willow (*Salix* spp.) into cigar-shaped bundles or fascines and placing them in small trenches dug on the contour (Figure 5.13). The bundles are then securely staked and hopefully will begin to root and sprout. The stakes used in the installation of wattles are usually inert, but live stakes, which will also sprout and root in time, can be

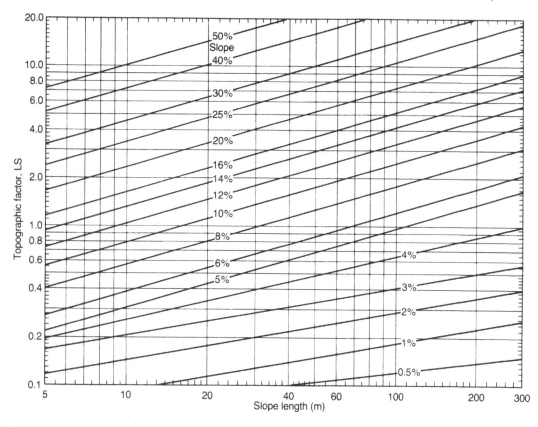

Figure 5.12 Nomograph for determining the value of the LS factor in the Universal Soil Loss Equation (after Wischmeier and Smith, 1978).

Table 5.17 Maximum safe velocities (m/s) in channels (after Hudson, 1981)

Material	Maximum velocity on cover expected after two seasons		
	Bare	Medium grass cover	Very good grass cover
Very light silty sand	0.3	0.75	1.5
Light loose sand	0.5	0.9	1.5
Coarse sand	0.75	1.25	1.7
Sandy soils	0.75	1.5	2.0
Firm clay loam	1.0	1.7	2.3
Stiff clay or stiff gravel	1.5	1.8	2.5
Coarse gravel	1.5	1.8	n/a
Shale, hardpan, soft rock	1.8	2.1	n/a
Hard cemented conglomerate	2.5	n/a	n/a

n/a = condition unlikely to exist because grass will not grow to give an appropriate cover.
Intermediate values may be selected.

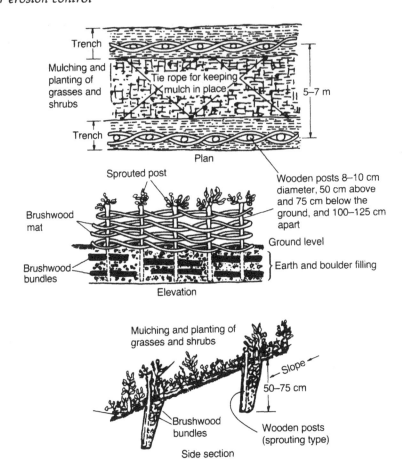

Figure 5.13 Details of contour wattling used to stabilize the surface of landslide scars in the Outer Himalayas (after Sastry, Mathur and Tejwani, 1981).

Table 5.18 Recommended strip widths for contour strip-cropping (after FAO, 1965)

Slope (%)	Strip width (m)
2–5	30
6–9	25
10–14	20
15–20	15

used on relatively loose and moist sites (Gray and Leiser, 1982).

Development of the dense root mat will ensure increased cohesion, high rates of evapotranspiration, improved infiltration of surface water and anchoring effects. The obstruction to flow will reduce runoff velocity, filter out any sediment within the flow and encourage deposition within and upslope of the barrier, thereby creating a set of benches over time. Soil trapped behind the wattle bundles will enhance vegetation establishment. When the surface vegetation is established this will provide further soil protection through canopy effects. Wattling is used for slope stabilization and erosion control in a variety of applications from stabilization of cut-and-fill slopes on road banks to the regeneration of gullied areas. A comprehensive guide to its use can be found in Gray and Leiser (1982).

Fascines are used in eastern Nepal (Howell

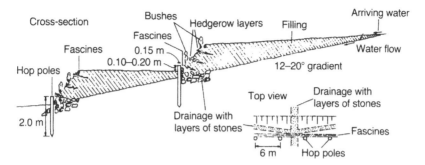

Figure 5.14 Details of terrace construction in Bavaria using fascines (after Ankenbrand and Schwertmann, 1989).

et al., 1991) to control water erosion on slopes up to 45° covered with unconsolidated but compacted debris. Species used are *Vitex negundo* (simali) and *Lantana camara* (phul kanda). The technique is not recommended on slopes prone to slumping. On steeper slopes (>45°) where fascines are not appropriate, it is recommended that wooden cuttings are used to create palisades across the slope. Again *Vitex negundo* and *Lantana camara* are the species used. On slopes less than 35°, a vertical interval of 4 m is used; this reduces to 2 m on slopes of 35–60°.

Fascines were used in the construction of new terraces on a land consolidation scheme in Bavaria (Ankenbrand and Schwertmann, 1989). The terraces were built up from soil in several layers between which fascines made from a local willow (*Salix purpurea*) were placed (Figure 5.14). Bushes (e.g. *Ligustrum vulgare, Prunus spinosa, Cornus sanguinea, Rosa canina, Viburnum lantana* and *Euonymus europaeus*) were planted to provide additional protection and anchorage. Where the terrace banks were very high (>1.5 m), wooden poles were driven into the soil and joined by fascines to give extra support.

Experiments were carried out with wooden dams placed across the slope to control water erosion on 54–68% slopes under hill rice and garlic in Luzon, Philippines (Cuevas and Diez, 1988). However, the dams, made of small tree branches, were not totally effective in trapping all the eroded sediment. Their efficiency also declined greatly after the first year following installation, probably due to decomposition and decay of the wood.

Contour brush layering

With contour brush layering, live branches of green, leafy and shrubby material are inserted on the contour, into the slope (Figure 5.15). The tips of the branches should protrude slightly beyond the face of the slope, although this can lead to the formation of mini terraces above the brush layer and scour immediately beneath it. To overcome this hazard, the brush tips can be placed flush with the slope face, still allowing for some filtering of the runoff running over the brush layer.

The distance between successive layers is dependent on the erosion risk on the slope. Vertical intervals of 1–3 m are normally used with a closer spacing at the bottom of the slope, where the erosion potential is greatest due to the accumulation of runoff down the slope. The vegetative material will root into the slope, so stabilizing it. Contour brush layering can be incorporated during construction of embankments or used as a remedial treatment for eroding slopes (Gray and Leiser, 1982).

Directional layouts

It has been assumed in all the examples of strip systems described above that the layouts are

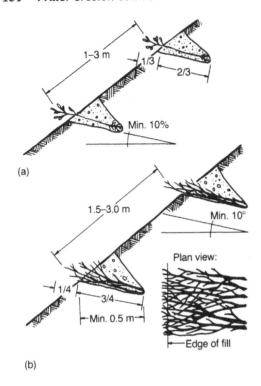

Min. 10%

(a)

1–3 m

1/3

2/3

1.5–3.0 m

Min. 10°

Plan view:

1/4

3/4

Min. 0.5 m

Edge of fill

(b)

Figure 5.15 Contour brush layering. (a) Hedge layering with live twigs; (b) brush layering and cuttings.

aligned across the slope, either on the contour or with only slight deviation from it. Howell *et al.* (1991) discuss the value of alternative alignments, however, when using grass strips for controlling surface erosion on road banks in eastern Nepal. On steep slopes (15–70°), with soils of low cohesion, horizontal or contour alignments are not entirely effective. Either the rainfall is so heavy that runoff flows through any weakness in the barrier, creating rills, or water ponds up behind the barrier, saturates the soil, reduces soil strength and causes shallow mudflows.

The prime consideration on these slopes is to reduce the risk of shallow soil failure. This means that water must be removed rapidly from the slope instead of allowing it to infiltrate the soil. Surface drainage of the slopes can be enhanced if the vegetation is planted in lines running straight down the slope, so that chan-

nels are formed between. This approach, however, has enormous dangers because the runoff water will quickly become erosive and rates of soil loss will remain high. It is only acceptable where the consequences of mass soil failure are greater than those of high rates of surface erosion.

A compromise approach (Figure 5.16) is possible whereby the grass strips are aligned diagonally across the slope. The angle of alignment and, therefore, the grade of the channel formed between the strips can be adjusted according to the relative needs of controlling soil failure versus surface erosion. On gentle slopes with reasonably well-drained soils, grades of 0.5° or 1° may be appropriate but, on the steep slopes in Nepal, grades of 45° are being used. A variant on the diagonal alignment uses a chevron layout whereby the planted lines of grass grade into stone-lined channels which take the excess runoff downslope to road drains at the base.

Typical grass strip spacings used in Nepal are:

1. for horizontal planting, 2 m vertical intervals on slopes less than 35°, 1 m on slopes of 35–55° and 0.5 m on slopes steeper than 55°;
2. for downslope planting, 50 cm between the rows;
3. for diagonal planting, 50–75 cm between the rows.

5.5.5 COMPOSITE SOLUTIONS

In the above examples, vegetation is used on its own as a method of controlling erosion by water on slopes. It can also be used alongside traditional engineering measures employed to deal with specific processes. One example is the way vegetation has been used to control surface erosion on the steep slopes of an abandoned limestone quarry near Dehra Dun, India, in the foothills of the Himalayas. The vegetation is integrated with mechanical measures such as temporary check dams of brushwood, loose stones and gabions which are used to stabilize the steep channels with average slopes of 38%.

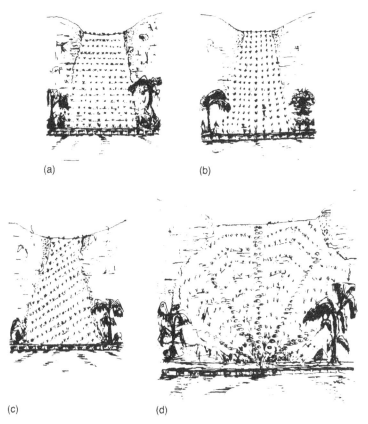

(a)   (b)

(c)   (d)

Figure 5.16 Examples of grass strip alignments proposed for bioengineering work on road banks in eastern Nepal: (a) contour; (b) downslope; (c) diagonal; (d) chevron (after Howell *et al.*, 1991).

A planting programme was adopted comprising quick-growing tree species, stumps and cuttings of shrubs, and rooted slips of grasses. The tree species included *Salix tetrasperma* (jalmala), *Bauhinia retusa*, *Leucaena leucocephala*, *Lannea grandis* (jhingora) and *Erythrina suberosa* (pangara or mandar). The main shrubs were *Vitex negundo* (samalu) and *Ipomoea carnea* (beshram). The chief grasses were *Chrysopogon fulvus* (gorda) and *Eulaliopsis binata* (bhabar). The rehabilitation programme started in 1984. By 1987 the vegetation cover had increased from 10% to 50% and the flow in the main drainage channel had become clear and perennial, whereas previously it was intermittent, with no flow between October and the onset of the summer monsoon, and turbid (Katiyar, Sastry and Adhikari, 1987). By preventing the deposition of material on to the neighbouring road, an annual saving of 100 000 IndRs has accrued to the Public Works Department.

Table 5.19 gives a list of recommended tree, shrub and grass species that, based on experience of similar projects on surface erosion control on landslide scars and abandoned quarries (Sastry, Mathur and Tejwani, 1981; Gupta and Arora, 1983), can be recommended for the Himalayan foothills.

Under the high rainfall volumes and intensities described above in eastern Nepal, grass covers can be difficult to establish on the steep road banks involved because the seeds can be

Table 5.19 Recommended species for bioengineering work in the Lower Himalayas (after Sastry, Mathur and Tejwani, 1981; Gupta and Arora, 1983)

Recommended species	Common name
Grasses	
Eulaliopsis binata	Bhabar
Chrysopogon fulvus	Gorda
Dactylus glomerata	
Lolium perenne	
Eragrostis curvala	
Cenchrus spp.	
Arundo donax	Narkato
Pennisetum purpureum	Napier
Shrubs	
Vitex negundo	Samalu
Ipomoea carnea	Beshram
Woodfordia fruticosa	
Wendlandia exserta	
Boehemeria rugulosa	
Moringa petrygosperma	
Melia azadirachta	
Populus ciliata	
Trees	
Salix tetrasperma	Jalmala
Bauhinia retusa	
Leucaena leucocephala	
Lannea grandis	Jhingora
Erythrina suberosa	Pangara or mandar
Alnus nepalensis	

washed away before they germinate. In order to prevent this, a jute geotextile is laid on the surface. This helps to reduce surface erosion and provides a suitable microclimate for seed germination and plant growth (Chapter 4).

5.6 EROSION CONTROL IN CHANNELS

Vegetation can be used to control surface erosion within river channels, canals and waterways. This is done primarily in three ways:

1. Vegetation imparts a roughness to water flows in channels, causing flow retardance, resulting in reduced flow velocity (Chow, 1981) and thus lower flow energy for detachment and transport of sediment. The roughness imparted by the vegetation is often termed 'hydraulic resistance'. This impedance to flow is exerted by many different forms of vegetation, from, at the micro scale, individual plant stems to macro-scale grade stabilization structures made of vegetative materials and used to control channel erosion and gully development. The principles of hydraulic resistance, flow disturbance and reduction in velocity are the same, however, at all scales.

2. Shear stress within the water flow is imparted to the vegetation rather than to the channel floor and sides.

3. The roots of the vegetation bind the soil mass, giving mechanical protection as well as soil/root cohesion, so that higher flow energy is needed to detach soil particles from the channel bed.

The interaction between vegetation and flowing water is affected by many factors. The degree of impedance to flow is related to the height of the vegetative obstruction relative to the flow depth. A simple regression of vegetation height against mean flow velocity gives a strong r value of 0.94 (Watts and Watts, 1990), indicating a close association between the two. The critical height for maximum roughness effect is determined by the flow conditions. When the depth of flow, relative to the vegetation height is shallow, the vegetation stands rigid and imparts a high degree of roughness associated with internal distortion of the flow by the individual plant stems. Deeper flows may cause the vegetation to bend and even lie down. This brings about a decline in the level of roughness because the retardance is due mainly to skin resistance rather than interference with the flow.

The type of vegetation is also important in affecting the way vegetation interferes with and modifies flow conditions. The following physical parameters will influence the interaction between vegetation and flow:

1. **Inherent characteristics of the vegetation:** the relevant characteristics are the size, shape and surface texture of the plant stems and

leaves. Plants with leafy stems will reduce flow velocity to a greater extent than species devoid of leafy, fleshy growth.

2. **The distribution of the vegetation**: this can be considered at two scales: first, the distribution of stems within a plant stand and their areal and numerical density; second, the frequency and pattern of plant stands along the channel.

3. **The behaviour of vegetation in water flows**: the flexibility or stiffness of both individual stems and the whole plant stand affects the deflection and frequency of vibration of the vegetation within the water flow and, therefore, the degree of roughness imparted to the flow. The bending resistance of the stems also determines how the permeability of the stand changes in response to changing flow velocity. The dimensions of the plant stand affect the level of form resistance offered to the flow.

Each of these parameters will vary with vegetation type and the way in which the vegetation has been established, either naturally or by reseeding or planting. Vegetation is the only component of channel flow roughness which varies seasonally (Watts and Watts, 1990). A full bibliography of vegetation and its channel hydraulic resistance is given by Dawson and Charlton (1987).

Vegetation is used in channels in two main ways: first, as a lining to the sides and base, and second, to form barriers placed across the path of the flow.

5.6.1 VEGETATION AS A CHANNEL LINING

Channels are susceptible to erosion when subjected to surface water flows. Vegetation can be used as a lining or protective buffer between potentially erosive water flows and the channel floor, operating on the same principles which apply to the protection of the soil surface from erosive sheet or overland flow (sections 2.3.1 and 5.2).

Velocity profiles (Figure 5.17) can be used to illustrate how little of the flow contacts the bed or banks of the channel once vegetation is established (Watts and Watts, 1990). However, it must be remembered that these profiles are limited to two dimensions (channel width and depth at a point, i.e. cross-section), with little consideration of how up- and downstream conditions affect the flow velocity at any given point along the channel.

Any reduction in flow velocity drastically reduces the transporting capacity of the flow, and deposition of suspended sediment may result. This accretion of sediment may have to be removed as part of the maintenance programme of the channel, or the channel capacity will be filled with sediment rather than water. The other implication of reduced flow velocity is that there is an increase in water depth and the level of the water table adjacent to the channel as the flow is restricted (Dawson and Robinson, 1984). This increase in water depth was observed by Watts and Watts (1990), even when flow discharges were reduced. As a result, vegetation may increase the risk of overbank flow and flooding, especially in the summer when vegetation is most luxuriant.

Species requirements

Ideally, the vegetation lining should take the form of a uniform sward and have maximum percentage cover so that the unvegetated area exposed to the erosive flow is minimal. Species with clumpy growth habits should be avoided because of the danger of flow concentrations which can give rise to high local flow drag and localized erosion. Care must also be taken to avoid invasive species which become difficult to eradicate and shade out all other species, including even those of engineering value. One example is Japanese knotweed (*Reynoutria japonica*) which has invaded many river banks and disturbed land sites in the UK over the last decade and which spreads very rapidly by flooding or by earth-moving operations. On no account should this or similar species be chosen as part of an ecological succession.

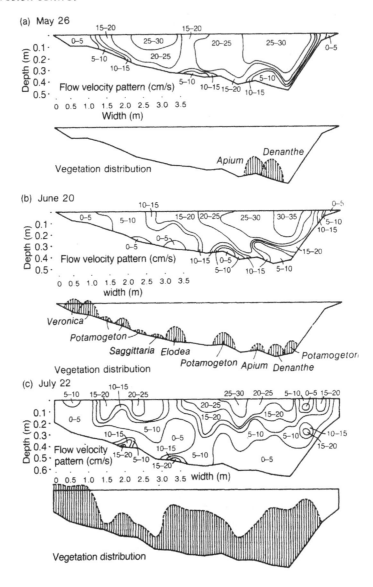

Figure 5.17 Velocity profiles for flows in a vegetated channel on the River Yare, Norfolk (after Watts and Watts, 1990).

Species chosen to line channels are usually fast-establishing but slow-growing, so minimizing maintenance requirements as well as avoiding filling up the storage capacity of the channel with vegetation. It is important that species selection takes account of the hydraulic requirements of any channel (Ash and Woodcock,

1988). Using vegetation to protect a channel from erosion does not require such dense plant growth as to impair its drainage function. Species selection and subsequent management often represent a compromise between reducing flow velocity enough to minimize erosion yet maintaining an adequate throughput of water.

Species selection is very important as different species can aggravate or dampen turbulence of flow through their surface biomass and bind the soil by their root structure to different extents. The effectiveness of vegetation lining in controlling erosion is related to the length or height of the vegetation elements, the robustness and flexibility of the stems and leaves, the growth habit, the density and uniformity of the vegetation cover and the structure of the biomass above and below ground. In addition to varying with individual species, these properties depend on the age of the vegetation and the way it is managed (Hewlett, Boorman and Bramley, 1987).

Management of the vegetation is crucial as unmanaged vegetation in channels may interfere with angling, navigation and recreation as well as impair the passage of water through the channel. Changes in the magnitude and direction of currents within the channel flow can cause localized erosion depending upon the location, extent and density of vegetation. Remedial measures will need to be taken to deal with this.

Design principles

Kouwen, Unny and Hill (1969) derived a quasi-theoretical equation for flow and vegetation conditions in a channel:

$$v/u^* = C_1 + C_2 \ln (A/A_v). \qquad (5.3)$$

where v is the mean flow velocity: u^* is the shear velocity, defined as $(gR_1S_1)^{0.5}$ where g is the acceleration due to gravity, R_1 is the hydraulic radius and S_1 is the energy gradient (hydraulic slope); C_1 is a parameter based on vegetation density; C_2 is dependent on the stiffness of the vegetation; A is the area of the channel cross-section; and A_v is the area of the vegetated part of the cross-section.

Roughness of the channel lining will affect the flow characteristics within the channel. Ree and Palmer (1949) quantified empirically the changes in the roughness imparted to the flow as the vegetation grew. This imparted roughness is

not usually measured directly but, instead, is described by the effect it has on the flow. The most commonly used descriptor is Manning's roughness coefficient 'n' which is based on the relationships between flow velocity, hydraulic radius and channel slope. As discussed in Chapter 2, when flow depth is relatively shallow compared with the vegetation height, the vegetation stands rigid and imparts a high degree of roughness, with Manning's n values around 0.25–0.30. As the flow depth increases the plant stems begin to oscillate, imparting more disturbance to the flow, as evidenced by n values increasing to about 0.4 (Figure 2.12). When the vegetation is submerged by deeper flows, the Manning's n value reduces dramatically by as much as an order of magnitude. This is because the vegetation is laid flat by the flow and, as indicated earlier, retardance is due to the effect of skin resistance, rather than interference by individual leaves and stems.

Kouwen and Li (1980) were able to characterize the critical shear velocity (u_{crit}^*) at which vegetation becomes prone by an index of stiffness, MEI, defined as the flexural rigidity of the vegetation elements per unit area. In this index, M is the number of roughness elements per square metre, E is the modulus of elasticity of the vegetative material (N m) and I is the second moment of the cross-sectional area of the stems (m^4). Multiplying these components of the index together gives a value of MEI in N.m^2. Where the vegetation bends to flatten downstream, the critical shear velocity (m/s) for becoming prone is defined by:

$$u_{\text{crit}}^* = 0.028 + 6.33(MEI)^2. \qquad (5.4)$$

Where the vegetation breaks on becoming flattened, the critical shear velocity is lower and defined by:

$$u_{\text{crit}}^* = 0.23 (MEI)^{0.106}. \qquad (5.5)$$

Kouwen and Li (1980) give some typical values of MEI for grasses and these are listed in Table 5.20. The authors claim that the differences in values between grasses are much greater than the variability for a particular grass and that the

Table 5.20 Values of the stiffness index (MEI; N m²) for selected grasses (after Kouwen and Li, 1980)

Grass type	MEI value
Alfalfa, green and uncut	2.9–6.2
Bermuda grass, green and long	1.5–47.4
Bermuda grass, green and short	0.03–0.6
Buffalo grass, green and uncut	0.03–0.7
Blue grama, green and uncut	4.2–6.0
Weeping love grass, green and long	3.1–15.4
Kentucky grass, green and short	0.01–0.2
Common lespedeza, green and short	0.005
Common lespedeza, green and long	0.02–3.0
Serica lespedeza, green and short	0.015
Serica lespedeza, green and long	6.3–15.9
Kikuyu grass, green and long	35–57
Kikuyu grass, green and short	0.14–0.21
African star grass, green and long	3.8–5.7
Bluegrass, green and long	6.3–15.7
Rhodes grass, green and long	96–212

The values given are typical ranges. Some extreme values quoted by the authors have been omitted.

values are not significantly affected by soil and climatic conditions.

Values of Manning's *n* vary with different vegetation species. Flexible, trailing stems such as *Potamogeton perfoliatus* allow more flow through than a more dense, rigid clump of *Oenanthe* for example (Watts and Watts, 1990). Different grass covers can be classified by their length into retardance categories, where, as shown in Chapter 2, curves can be plotted to relate the *n* value to a discharge intensity parameter (Figure 2.11). The latter is based on the flow discharge per unit width and is represented by the product of velocity and channel hydraulic radius (United States Soil Conservation Service, 1954; Coppin and Richards, 1990).

Manning's *n* is a dynamic term, changing as the vegetation goes through growth and recession cycles. Watts and Watts (1990) observed changes in Manning's *n* for aquatic vegetation on the River Yare, Norfolk, with values ranging from 0.02 to 0.15 during the year. Powell (1978) reports summer Manning's *n* values of 0.25 on a Lincolnshire river, reducing to a winter range between 0.04 and 0.021. Dawson (1978) found

that the Manning's *n* value for *Ranunculus calcareus* in a chalk stream increased with biomass from 0.05 for a biomass value of 1 g DM/m to 0.40 at a biomass of 350 g DM/m.

Whilst vegetation height appears to be strongly correlated with reduction in flow velocity, Watts and Watts (1990) found a surprisingly poor relationship between roughness estimates and the height of vegetation. One criticism of Manning's *n* is that it is based on the channel hydraulic radius. In reality, once vegetation is established in the channel, the vegetation itself creates a different effective wetted perimeter which may be smoother and smaller than that of an unvegetated channel. This effective wetted perimeter is also variable with the flexibility of the vegetation because deflection of the vegetation downstream alters its effective height. Resistance estimates such as the Manning's *n* value attach a friction value to the channel hydraulic radius which may not be the most appropriate term to consider in such vegetated channels (Watts and Watts, 1990). It has been suggested that more accurate roughness parameters should be based on vegetation height relative to water depth to achieve a 'relative roughness estimate'. This approach considers the uniformity of the plant type and the 'flexural rigidity' of individual species.

Kouwen and Li (1980) adopt this approach and present the following equation for estimating Manning's *n* as a function of the deflected roughness height (*k*; m):

$$n = \frac{y^{1/6}}{(8g)^{0.5}[a + b \log (y/k)]}, \quad (5.6)$$

where *y* is the depth of flow (m), *g* is the gravity term (m/s²) and values of *a* and *b* depend upon the ratio of shear velocity to critical shear velocity. For $u^*/u^*_{crit} \leqslant 1.0$, $a = 0.15$ and $b = 1.85$; for $1.0 < u^*/u^*_{crit} \leqslant 1.5$, $a = 0.20$ and $b = 2.70$; for $1.5 < u^*/u^*_{crit} \leqslant 2.5$, $a = 0.28$ and $b = 3.08$; and for $u^*/u^*_{crit} > 2.5$, $a = 0.29$ and $b = 3.50$. The value *k* can be determined as a function of the stiffness (*MEI*) index, defined above, from:

$$k = 0.14\,h \left[\frac{\left(\frac{MEI}{\gamma y S}\right)^{0.25}}{h} \right]^{1.59}, \qquad (5.7)$$

where h is the undeflected roughness height of the vegetation (m) and γ is the unit weight of water.

The above discussion has concentrated on Manning's n because this is the roughness parameter most frequently used by engineers in practice. As a result, values have been determined for a wide range of vegetation types and soil conditions (Tables 2.9 and 2.10). Roughness has also been quantified using Chezy's 'C' coefficient and the Darcy–Weisbach friction factor 'f' (Watts and Watts, 1990) but values of these are less widely available (Thornes, 1980). Equations 2.16, 2.17 and 2.18 allow conversions from one roughness term to another.

The implication in the design of vegetated watercourses, such as grassed waterways, is that where vegetation is used to line the channel, a higher velocity of flow is permissible without causing erosion of the channel (Ree and Palmer, 1949). Design is therefore based on the concept of a maximum permissible mean velocity of flow. This is also known as the maximum safe, allowable or non-eroding velocity. The permissible velocity within a channel is related to the soil type and vegetation (Temple *et al.*, 1987), and whether the channel has any additional protection from geotextiles or other inert armouring (Hewlett *et al.*, 1985). Ree and Palmer (1949) established maximum permissible mean velocities for different soil and vegetation conditions using test channels and flumes. Gregory and McCarty (1986) obtained the following relationship between the maximum allowable velocity (v_t) in a channel and the percentage vegetation cover (F) and channel slope (S):

$$v_t = v_b \left(\frac{(1 + 1.5\,F)^{4/3}}{(1 - 0.219\,F)^{2/3}} \right) \left(\frac{2.5}{S} \right)^{1/6}, \qquad (5.8)$$

where v_b is the maximum allowable velocity for the bare soil. Table 5.21 summarizes the information from these studies.

Channels with permanent flows

Species chosen to protect channels with continuous flow will have to be able to withstand submergence to different degrees depending on their position with respect to the water level (Figure 5.18). Vegetation will only survive in the aquatic plant zone if there is sufficient light and the flow velocities are low. The plants in this zone will obstruct the flow if their growth is too great and normally regular clearance of the vegetation is carried out in order to maintain flow. Protection of the marginal zone requires plants that can survive submerged to depths of at least 0.5 m. These are normally reeds. Most grasses cannot be used because their roots cannot withstand prolonged submergence. Reeds will be effective provided the velocity remains below 1 m/s but will need to be combined with other measures, such as geotextile meshes and rip-rap, for faster flows (Coppin and Richards, 1990). These additional measures will almost certainly be required if the banks are subject to substantial wave action, for example from boat wash.

The seasonally flooded zone can be protected with fast-growing shrubs and trees (e.g. willow and alder species) as well as grass. In addition to providing the foliage cover to impart roughness to the flow, these plants, particularly the trees, will add to the strength of the bank through reinforcement of the soil by the roots. Although the trees must be planted close enough to this part of the bank for the roots to extend into this zone and the reinforcement to be effective, care must be taken to prevent large trunks from projecting into the flow where they can be pushed over or pulled out during floods or where they will reduce channel capacity. Again, geotextiles and structural measures may be required alongside the vegetation in channels subjected to fast flows or wave action.

Grasses, shrubs and trees will also occupy the dry zone. A dense tree growth should be avoided here, however, so as not to shade the vegetation lower down the bank. On the other hand, some shade is desirable to prevent the bank vegetation from becoming too luxuriant.

Table 5.21 Maximum allowable velocities (m/s) in vegetated channels for different soil and slope conditions (after Temple, 1980; Gregory and McCarty, 1986)

Cover expected after two seasons	*Percentage slope in channel*		
	0–5	*5–10*	*>10*
Easily eroded soils (sands, sandy loams, silt loams, silts, loamy sands)			
Very good cover (100%) e.g. creeping grasses such as Bermuda grass	1.8	1.5	1.2
Good cover (88%) e.g. sod-forming grasses such as Blue grama, Buffalo grass, Kentucky blue grass, smooth brome, tall fescue	1.5	1.2	0.9
Moderate cover (29%) e.g. bunch grasses, legumes such as Kudzu, lespedeza, weeping lovegrass, alfalfa	0.8	n/r	n/r
Erosion resistant soils (clay loams, clays)			
Very good cover (100%)	2.4	2.1	1.8
Good cover (88%)	2.1	1.8	1.5
Moderate cover (29%)	1.1	n/r	n/r

n/r = not recommended.

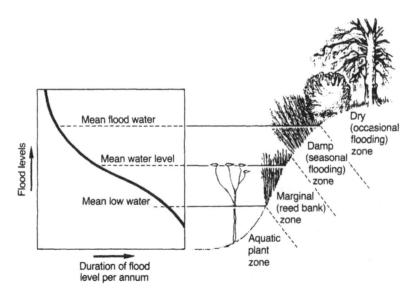

Figure 5.18 Vegetation zones on a river bank.

The experience of New Zealand workers in using vegetation for river bank protection shows that willows, poplars and alders are the most suitable species (Hathaway, 1986a) with willows being considered the most important. Shrub willows (osiers and sallows) are preferred because they require less management, in the form of layering, to prevent them becoming too heavy and creating bank instability. Their multiple-stemmed growth habit imparts greater roughness to stream flow than is obtained from single-stemmed tree willows. The latter are also more brittle so that branches break off more readily and can choke the channels further downstream. Bitter osiers (*Salix purpurea* and *Salix elaeagnos*) are often used because they are less susceptible to damage by livestock, possums and rabbits.

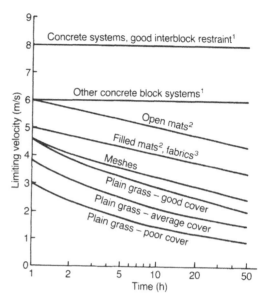

Figure 5.19 Recommended limiting flow velocities for control of water erosion in channels using grass and grass reinforced with geotextiles and inert structures (after Hewlett, Boorman and Bramley, 1987). 1 = minimum superficial mass of 135 kg/m^2; 2 = minimum nominal thickness of 20 mm; 3 = installed with 20 mm of soil surface or in conjuction with a surface mesh. All reinforced grass values assume well-established good grass cover.

Channels with discontinuous flow

Engineers often construct channels to deal with discontinuous flows. Such channels include drainage ditches, auxiliary spillways, crests of flood banks, grassed waterways and dam spillways. In certain conditions, all may be susceptible to surface erosion and slumping. Species chosen to reduce erosion risk in these channels will have to withstand periodic submergence, followed by periods of relatively dry conditions. The ability of the vegetation to regenerate after periods of inundation at varying velocities is thus extremely important. Research into the recovery time of different species has identified the time that must elapse between periods of immersion (Hewlett, Boorman and Bramley, 1987). This recovery time may influence the choice of species used in any given application.

Duration of flow in intermittent channels has a bearing on permissible velocity. Graphs have been derived linking erosion resistance, as related to the limiting velocity, and estimated flow duration (Figure 5.19). Hewlett, Boorman and Bramley (1987) show that a well-chosen and established grass cover can withstand a limiting velocity of 2 m/s for over 10 h. Under these conditions some superficial scour may take

place but the scars are often quickly healed. When the velocity is increased to 3–4 m/s, however, failure in the form of loss of vegetation cover and uncontrolled erosion of the channel would occur after several hours. At a velocity of 5 m/s failure will occur within 2 h. Long-term stability is only possible if the flow velocity is limited to 1 m/s. The critical velocity at which failure occurs is reduced if the quality of the vegetation cover is detrimentally affected when subjected to flows over time.

Grassed waterways are usually used to convey, without erosion, surplus water generated in terrace channels. They can also be used as a form of gully erosion prevention and, after filling-in of an eroded gully, as a reclamation technique. Here, the gully would be tilled and reshaped, and the waterway designed to be longer in length and therefore gentler in gradient than the original gully (Heede, 1976). This technique would be supplemented by an adequate drainage system, such as a network of tile drains to encourage infiltration and subsurface drainage, rather than surface flows. Heede (1968) argues, that this approach is more successful than one using check dams (section 5.6.2) in the first few years of construction. This is because the flat cross-sections of waterways present low erosion risk and favour the establishment of vegetation. Check dams, however, do not modify gradients until sediment has accumulated behind them. The success of the waterway approach is dependent on uncontrollable factors such as rainfall and temperature, which will directly influence vegetation growth but will only indirectly affect the performance of a series of check dams. Design procedures for grass waterways are presented in Hudson (1981) and Morgan (1986).

Reference is often made in the literature to the selection of the following species for the lining of grassed waterways: alfalfa/grass mix, lespedeza, Bermuda grass (*Cynodon dactylon*), orchard grass (*Dactylis glomerata*), centipede grass (*Eremochloa ophiuriodes*), Sudan grass (*Sorghum vulgare sudanense*), Dallis grass (*Paspalum dilatatum*), crabgrass (*Digitaria sanguinalis*), kudzu (*Pueraria thunbergiana*), brome grass (*Bromus inermis*) and redtop (*Agrostis alba*). The following are recommended for sections of waterways on gully fill: smooth brome (*Bromus inermis* Leyss.), intermediate wheatgrass (*Agropyron intermedium*), and yellow sweet clover (*Melilotus officinalis*). Ideally, perennial species should be chosen for use in all waterways, although a more rapid cover may be attained with annuals and biennials such as ryegrass (*Lolium* spp.) and sweet clover.

The time at which the seeding of waterways takes place is critical, particularly as the period of rainfall early in the growing season is often the time when erosion risk is also highest. Where it is difficult to obtain rapid vegetation growth at this time, erosion can be reduced with the use of erosion mats or geotextiles (Chap-

ter 4) or by minimizing soil compaction by machinery. Heede (1968) mentions the use of horses rather than vehicles to prepare the channel for seed drilling in order to reduce compaction associated with wheelings and thus avoid undesirable concentrations of runoff.

Any establishing vegetation will require dressings of fertilizers to encourage healthy and sustainable growth. Management of the established grass is equally important as the degree to which the vegetation is cut will affect its height and effective percentage cover and, therefore, its ability to curb erosive flow velocities (Ree and Palmer, 1949).

Dam spillways

Reinforced grass is often used to withstand the erosion hazard presented by low frequency but high intensity, short-duration flow events, such as experienced on dam spillways (Hewlett, Boorman and Bramley, 1987). This technique provides an alternative to the more expensive and aesthetically unattractive option of hardlining or heavy armouring of the channel. The vegetation reinforcement is in the form of geotextiles or cellular concrete products, through which the grass can grow, thus forming a composite protective layer. Hewlett, Boorman and Bramley (1987) indicate that grass reinforced with a geotextile can resist significantly higher, short-duration flow velocities compared with where a geotextile has not been used. They also note that the time before failure occurs is extended when some reinforcement of the vegetation is carried out. For example, a good plain grass cover will withstand flow velocities of 4.5, 3.2 and 2.8 m/s for durations of 1, 5 and 10 h respectively; with a combination of grass and 20 mm thick open mat geotextiles, these velocities are increased to 6.0, 5.5 and 5.0 m/s (Figure 5.19).

Gullies

Gullies are large erosion features which are characterized by discontinuous and often unpredictable flow. Although barriers are used to stabilize the gully sides and floor (section 5.6.2), this is mainly to provide a suitable environment within which vegetation can establish. The most effective vegetative cover in gully channels is a dense sward, with a deep and dense root network and low plant height. Long, flexible plants tend to lie down when subjected to flow, providing a smooth interface between the flow and the channel bed. Whilst the long stems may protect the channel at this point, they do not, when prone, impart sufficient roughness to reduce flow velocity substantially. This means that further downstream, where perhaps the vegetation may not be as luxuriant, velocities remain high and the sides and the floor of the gully are susceptible to the highly erosive flow. A uniform dense sward is required to maintain an even pattern of infiltration of water into the substrate so as to avoid the concentrations of water which might feed subsurface pipes or tunnels.

Special care is required when selecting vegetation for gully control. Certain species, such as trees when fully established, will restrict the flow within the gully, reduce channel capacity and may divert the flow out of the gully altogether causing it to cut a fresh gully elsewhere. Heede (1960) illustrates problems with the establishment of willows for gully stabilization. He found that plants and branches floated downstream, where they took root and grew into dense stands, in places where obstruction to the waterflow was undesirable. The choking of the channel in this way led to undercutting of the channel banks and widening of the channel bottom. This experience is rather contradictory to the recommendations of Hathaway (1986b) who considers *Salix* species particularly valuable for gully control in New Zealand because of their ability to form a continuous mat of fibrous roots across the gully floor, giving protection against scour. Poplars will perform the same function but take longer to establish. *Alnus* species, particularly *Alnus incana* which has a better shallow rooting system than other species, are used in wet areas, but they have the disadvantage of being palatable to possums.

5.6.2 CHANNEL BARRIERS

Objectives

Barriers are constructed in channels with the primary aim of reducing erosion of the bed. Once this has been achieved, stability of the side walls or banks is more likely as the toe of the banks is at rest. Barriers restrict the cross-sectional area of the channel and obstruct the flow, reducing its velocity and, hence, its erosivity. The magnitude of the restriction is an important factor in the success or failure of the barrier in terms of erosion control (Heede, 1960). Barriers can be constructed either wholly or partially from vegetative materials (Gray and Leiser, 1982) and range from simple hedge weirs to more sophisticated brushwood dams.

Any retardation of flow velocity will encourage infiltration of runoff, so reducing the volume of water in the channel itself, as well as bringing about a significant reduction in the ability of the flow to transport already eroded particles. Thus sedimentation will occur upslope of the vegetative barrier, so reducing local slope gradients and modifying the overall long profile of the channel. This process is particularly relevant for gully erosion reclamation, where the modification of the usually convex long profile reduces the hydraulic gradient of the channel and hence the risk of further gully channel development. Fang *et al.* (1985) describe reclamation work in the $0.87 \, km^2$ Yangjiagou gully in Qingyang County, Gansu, China, where 75 willow check dams were constructed at 20–30 m intervals along the gully bed, which was also planted with 31 000 trees, mainly poplar and willow, to fix the channel and prevent scour. The land formed by siltation behind the dams is used for crop production and generally gives yields that are five times those on unterraced farmland on adjacent hillsides and 2.5 times those on terraced fields. The higher yields are explained by the properties of the reclaimed soil which holds 86% more moisture and has higher levels of nitrogen and organic matter.

Since gullies are deep channels, sedimentation

does not generally result in over-topping of the channel and increased flood risk. Instead, sediment within the channel may help reduce peak flow discharge as the loose, unconsolidated sediment represents extra channel storage capacity, rather like an aquifer within the channel. This additional storage of water helps to raise the water table on land adjacent to the gully. As illustrated by the Chinese example above, sediments deposited behind the barrier are both fertile and have high water-holding capacities. They therefore provide a suitable site for rapid vegetation establishment and development. Once this vegetation is established within the channel, it too exerts a hydraulic resistance to the flow thereby reducing flow velocities even further (section 5.6.1).

The permeability of vegetative or 'quasi-vegetative' barriers (i.e. composite barriers, constructed partly from vegetative material and partly from inert material) is such as to impede runoff, whilst allowing some flow to permeate through the barrier. This leads to a dissipation of high hydrostatic forces, and thus the barrier itself does not need to withstand the full force of the flow. It is important that the vegetation chosen is deep-rooting, so that it anchors properly to the bed, and not too dense in its growth habit when established, so that it produces the required level of porosity and does not inhibit the flow too much. No guidelines have been published on the most effective level of porosity but design procedures have been established for different types of barriers which, from engineering experience, are known to work. These are discussed below with reference to gullies.

Gully stabilization structures

The most widespread use of barriers to control erosion is in gullies. One of the key principles of gully erosion control is to stabilize the channel so that a suitable environment is created in which vegetation can quickly establish and become self-regenerating in time. Heede (1976) argues that well-established vegetation perpe-

tuates itself and can therefore be regarded as a permanent method of gully erosion control. Erosion features such as gullies are notoriously difficult to control because of their erratic behaviour, unpredictability of flows and their exceptionally high sediment concentrations and rates of sediment loss. These conditions make it virtually impossible to establish any sort of vegetation unless gully conditions are altered first. A comprehensive review of the processes of gully formation and development is given in Heede (1976).

Channel stabilization is often achieved with 'grade stabilization structures' or check dams. Check dams are usually placed at critical points along the length of the gully. These include 'knick points', where there is some sudden change in gradient of the gully long profile, or at headcuts, where the gully is actively extending up the slope, often into uneroded areas. These critical locations are easily identified by the recent exposure of soil and removal of any vegetative cover. They can often be protected by plantings of trees and shrubs around the wings of the check dam (Heede, 1960).

In addition to placing check dams in these critical positions, dams must be constructed to a design spacing that is related to the height of the dam and the gully gradient. In turn, selection of dam height is dependent on the objective. If the aim is to maximize the amount of sediment deposited, high widely-spaced dams would be most appropriate. If, however, the objective is to reduce overall gully gradient, small closely-spaced dams are more effective (Heede, 1976). The most efficient and economical, spacing, as defined by Heede (1976), is to place each check dam at the upstream edge of the final sediments trapped by the next dam downstream (the 'head to toe' rule of Heede, 1960). This can only be estimated at the time of construction, with the spacing modified when the real deposition pattern confirms or disputes these estimates. Overdesign results in unjustifiable expenditure, whilst underdesign can cause damage to all other installations upstream and downstream.

Heede and Mufich (1973) developed an equation to simplify the calculation of check dam spacing:

$$X = \frac{H_E}{K \tan S \cos S} \qquad (5.9)$$

where X is the spacing (m); H_E is the effective dam height (m) as measured from the gully bottom to the spillway crest; S is the slope of the gully floor; and K is a constant ($K = 0.3$ when $\tan S \leqslant 0.20$; and $K = 0.5$ when $\tan S > 0.20$). Figure 5.20 shows this relationship as a function of effective dam height and gully gradient. Figure 5.21 shows the number of dams required along a 600 m gully of variable gradient. The number of dams increases with increasing gully

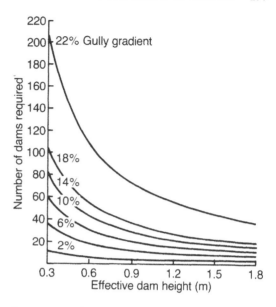

Figure 5.21 Number of check dams required in a 600-m long gully as a function of gradient and dam height (after Heede and Mufich, 1973).

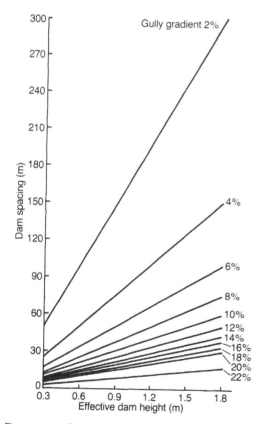

Figure 5.20 Spacing of check dams in gullies as a function of dam height and slope steepness (after Heede, 1976).

slope and decreases with increasing effective dam height. Further design specifications can be found in Heede and Mufich (1973). Unfortunately, this approach takes no account of expected peak flows, but there are often insufficient data in gullied areas on which to estimate these.

Brushwood or fascine check dams (Figure 5.22) are suitable for relatively small gullies. They comprise wooden posts (about 5–10 cm in diameter) which are hammered into holes (about 50 cm apart) dug across the gully floor, to a depth of at least 35 cm. The gully bottom should be as level as possible at this point, to avoid concentration of flow in any low spot, or undermining of the check dam itself. Once the posts are in place, their height above ground should be 50 cm at most and they should lean slightly upstream. Freshly cut branches of poplar, willow or alder are woven between the poles, allowing for a slightly lower level of brush material in the centre of the dam to

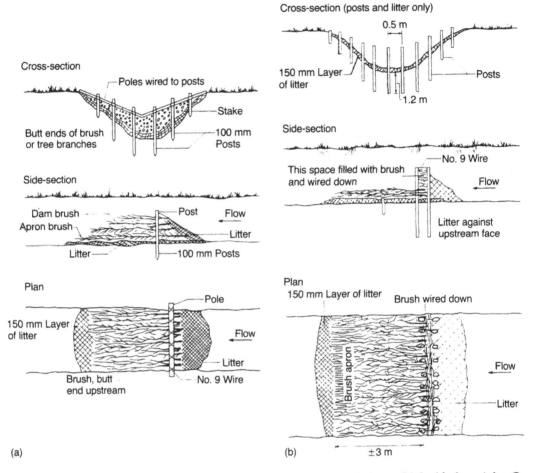

Figure 5.22 Examples of brushwood dams for gully stabilization: (a) single-fence; (b) double-fence (after Gray and Leiser, 1982).

simulate a spillway which will concentrate flows away from the gully sides. The brush should extend downstream of the dam to form an apron and protect the structure from being undermined. The length of the apron should be about 1.5 times the effective height of the structure on slopes less than 8.5° and 1.75 to 2 times the height on steeper slopes.

Plant species which easily take root in the gully should be selected and pushed into the banks to allow the generation and anchorage of a rooting network to stabilize both the gully sides and the check dam itself. Often, fresh branches of freely rooting species are then inserted upslope of the completed dam. To ensure that the 'head-to-toe' principle referred to above is maintained, the next check dam downstream will have the tops of its poles in line with the bottom of the upper check dam. Brushwood check dams can be constructed next to each other at points where maximum erosion control is required.

Grassed waterways

Where grassed waterways have been designed for steep (i.e. >11°) or erodible slopes then small, low dams of brush material can be used at regular intervals along the length of the waterway. The design spacing would be based on the principles discussed above, related to height of the dam and steepness of the slope. Often, live brushwood posts are used, with brush material woven between the posts. These dams act in the same way as check dams within gullies; they retard runoff velocity, encourage runoff to infiltrate the grass sward and filter out any sediment within the flow. The structures are less expensive and sophisticated than the concrete drop structures which are also used to dissipate flow energies in waterways.

Channel training

Another use of vegetative barriers where concrete, and sometimes gabion, structures are normally employed is in channel training. This technique is used to control bank erosion in wide gently sloping channels where the river is often braided or meanders between channel bars, undercutting on the outside of successive bends. To protect bridges, settlements and adjacent farmland, deflecting type spurs or groynes are constructed pointing into the channel to divert flow away from the banks and encourage sedimentation in the side channels. If the structures are built outwards from both banks, they can help confine flows to the centre of the channel.

Hathaway (1986a) describes a system whereby willows or poplars are laid in a trench, which has been bulldozed in the river bed and extends into the channel itself. The trees are weighted with bolsters and connected with wire cable to each other as well as to concrete piles or trees on the bank. The trench should be aligned downstream and, in some cases, will need to be almost parallel with the flow. The land between successive tree lines may be planted with willow stakes. An alternative is to use willow poles to construct a trestle which can be set in the trench and anchored with wire in the same way. Since both these structures will be subject to considerable damage in high flows, continuous monitoring and maintenance are required if they are to be successful. Their advantage is their low cost for what, assuming that the river ultimately follows a new course, is a temporary requirement.

5.7 SHORELINE PROTECTION

There is much visual evidence to show that the shorelines of lakes and reservoirs are subject to erosion by waves generated either by wind or from boat wash. However, few data are available on the rates of erosion involved. Except for considerable work on the reclamation of salt marshes, little experience exists with designing vegetation-based systems for erosion control.

5.7.1 BASIC PRINCIPLES

Although there are some similarities between the processes on a shoreline and those in rivers, there is also one important difference. The movement of water on the shoreline is oscillatory whereas that in a river can be considered unidirectional. Our knowledge of the mechanics of erosion by oscillatory flow comes from studies by engineers and geomorphologists working in coastal environments. The oscillatory action of a wave generates a velocity field for which, based on the Airy wave theory, the bottom orbital velocity (v_m; cm/s) can be calculated from (Clark, 1979):

$$v_m = \pi h / [T \sin h(2\pi d/\lambda)], \quad (5.10)$$

where h is the wave height (cm), T is the wave period (s), d is the water depth (cm) and λ is the open water wave length (cm) (Figure 5.23). Clifton (1976) has calculated the threshold bottom orbital velocities for the initiation of sand particle movement (Figure 5.24).

Equation 5.10 can be used as long as the wave action is oscillatory. It is therefore restricted to seaward of the point where water depth is equal to one-half of the wave length. Inshore of this, the circular orbits are flattened into ellipses and the action of the waves depends upon the slope of the beach (β). According to Galvin (1968), if

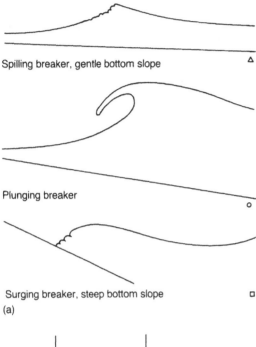

Spilling breaker, gentle bottom slope

Plunging breaker

Surging breaker, steep bottom slope

(a)

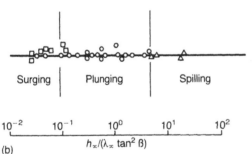

Surging | Plunging | Spilling

10^{-2} 10^{-1} 10^{0} 10^{1} 10^{2}

$h_x/(\lambda_x \tan^2 \beta)$

(b)

Figure 5.23 (a) Main types of breaking waves on shorelines related to (b) their typical wave heights (h_∞), lengths (λ_∞) and beach slopes (β) (after Galvin, 1968; Clark, 1979).

$$h_\infty/(\lambda_\infty \tan^2 \beta) > 4.8, \qquad (5.11)$$

the waves form spilling breakers which generally are not erosive and may even transport material upslope where it is deposited. Values < 4.8 denote a more erosive condition. The subscript ∞ refers to conditions in open water before the wave breaks.

Erosion of the shoreline of Loch Lomond, Scotland, was of particular concern in the 1970s,

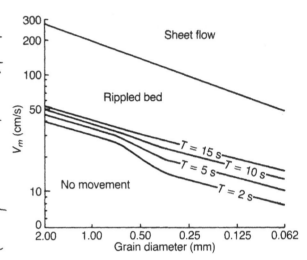

Figure 5.24 Threshold orbital velocities for initiation of movement of quartz grains in water as a function of wave period (T) and bed form (after Clifton, 1976).

as evidenced by the exposure of tree roots and visible loss of land resulting, for example, in undermining of the access road to Portnellan and the loss of the Cashel Camp Site, north of Balmaha (Scott-Park, 1979). Figure 5.25 shows the characteristics of deep water waves on Loch Lomond, Scotland (Smith, 1979), based on equation 5.10. For a depth of 3.25 m, bottom velocities of about 30–32 cm/s are predicted which, for wave periods of about 2 s, are sufficient to move particles smaller than 1 mm in diameter. From equation 5.11, it can be predicted that for the typical wave heights and lengths on the Loch, there will be a risk of erosion as long as the slope of the beach exceeds about 5.5°. Many of the beach slopes in the north of the Loch exceed this value with typical values about 5.8° (Smith, 1979). Typical fluctuations in water level in the Loch show a range of about 2 m during the year with low levels in summer and the highest in the period November to January (Poodle, 1979).

With the high orbital velocities quoted above acting on steep beach slopes there is a need to protect not only against wave action but also against bank instability.

5.7.2 VEGETATION ZONES

A similar zonation of vegetation is found along a shoreline to that along a river bank (Figure 5.26), reflecting the gradual succession of vegetation in a hydrosere, i.e. that initiated in a freshwater environment. Since plants have difficulty surviving in a wave-breaking zone, the vegetation immediately below the mean low water level is sparse and restricted to sheltered areas where algae, mosses and some floating-leaf plants, such as pond weeds, are to be found. In the wave run-up zone, sedges and reeds can survive, the species depending upon the frequency of inundation and the climate. In temperate areas, where the inundation is more than 90–100 days, species of *Juncus* predominate but where inundation is from 10–20 to 90–100 days, sedges (*Carex* spp.) and some herbaceous vegetation develop. The uppermost zone, where the frequency of inundation is less than 10 days per year, is characterized by a mixed plant community of woodland, shrubs and grasses which can withstand very short periods of flooding. Alders and willows are common species in this zone. In tropical climates, the common reed (*Phragmites communis* agg.) and the narrow-leafed cattail (*Typha angustifolia*) dominate the reeds, giving way to almost pure stands of palms as the period of inundation becomes less. Sedges and terrestrial herbs are generally few compared to temperate areas and are replaced by epiphytic orchids and ferns.

A major factor affecting the development of shoreline vegetation is that the level of wave action varies both seasonally and, in reservoirs, with periodic drawdown of the water. Under these conditions two main erosion processes can operate, both contributing to bank instability. First, wave erosion will steepen and, sometimes, undercut the toe of the bank. This may initiate bank collapse and increase the slope of the lower reaches of any streams draining into the lake sufficiently to cause gully erosion. Second, mass soil failure can occur when a rapid lowering of the water level takes away the support to a

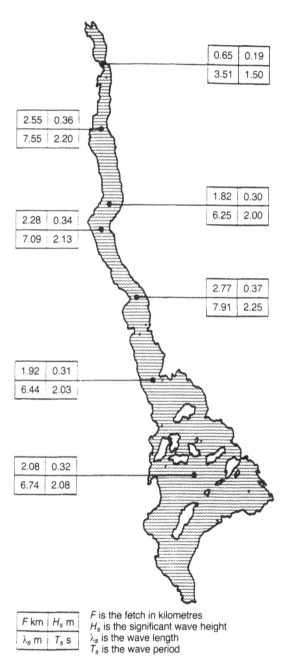

0.65	0.19
3.51	1.50

2.55	0.36
7.55	2.20

1.82	0.30
6.25	2.00

2.28	0.34
7.09	2.13

2.77	0.37
7.91	2.25

1.92	0.31
6.44	2.03

2.08	0.32
6.74	2.08

F km	H_s m
λ_s m	T_s s

F is the fetch in kilometres
H_s is the significant wave height
λ_s is the wave length
T_s is the wave period

Figure 5.25 Characteristics of waves on the shoreline of Loch Lomond, Scotland (Source: Smith, 1979).

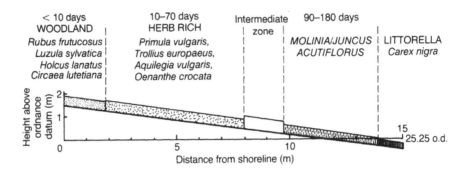

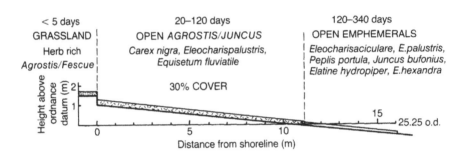

Figure 5.26 Vegetation zones on a shoreline in relation to the annual duration of inundation (days): examples from Loch Lomond (a) Clairinsh; (b) Ring Point (after Idle, 1979).

bank in which high pore-water pressures have accumulated.

5.7.3 COMPOSITE SOLUTIONS

Studies on Loch Lomond found that vegetation alone was not sufficient to protect the shoreline against both wave action and bank collapse. Vegetation had to be combined with structural solutions or geotextiles. A range of techniques proposed by the Countryside Commission for Scotland (1985) and being tested on Loch Lomond is illustrated in Figure 5.27.

The main role of the vegetation is to absorb the wave energy. Although some species will also provide a limited level of surface protection, plants are generally less effective in this role against scour from oscillatory flow motions on shorelines than they are against flows in rivers and on hillslopes. The use of willows,

cutting them back to produce dense and vigorous growth, was found not to provide sufficient protection on its own against wave erosion on Loch Leven, Scotland (Aldridge, 1979). Trials with herbaceous plants on the shoreline and wetland species within the water on the Bluemont Lakes, Fargo, North Dakota, were also unsuccessful. None of the wetland plants survived and, whilst many of the herbaceous plants established themselves successfully, they were ineffective in controlling wave erosion (Burley, 1989).

Two main groups of vegetation are employed in composite techniques. Reeds can be used to create fringes in the seasonally inundated zones which are sheltered and where waves are less than 0.3 m high. Trees can be planted in the uppermost zone to stabilize the bank by providing anchorage through the spread of their root systems. Willows and alders will survive

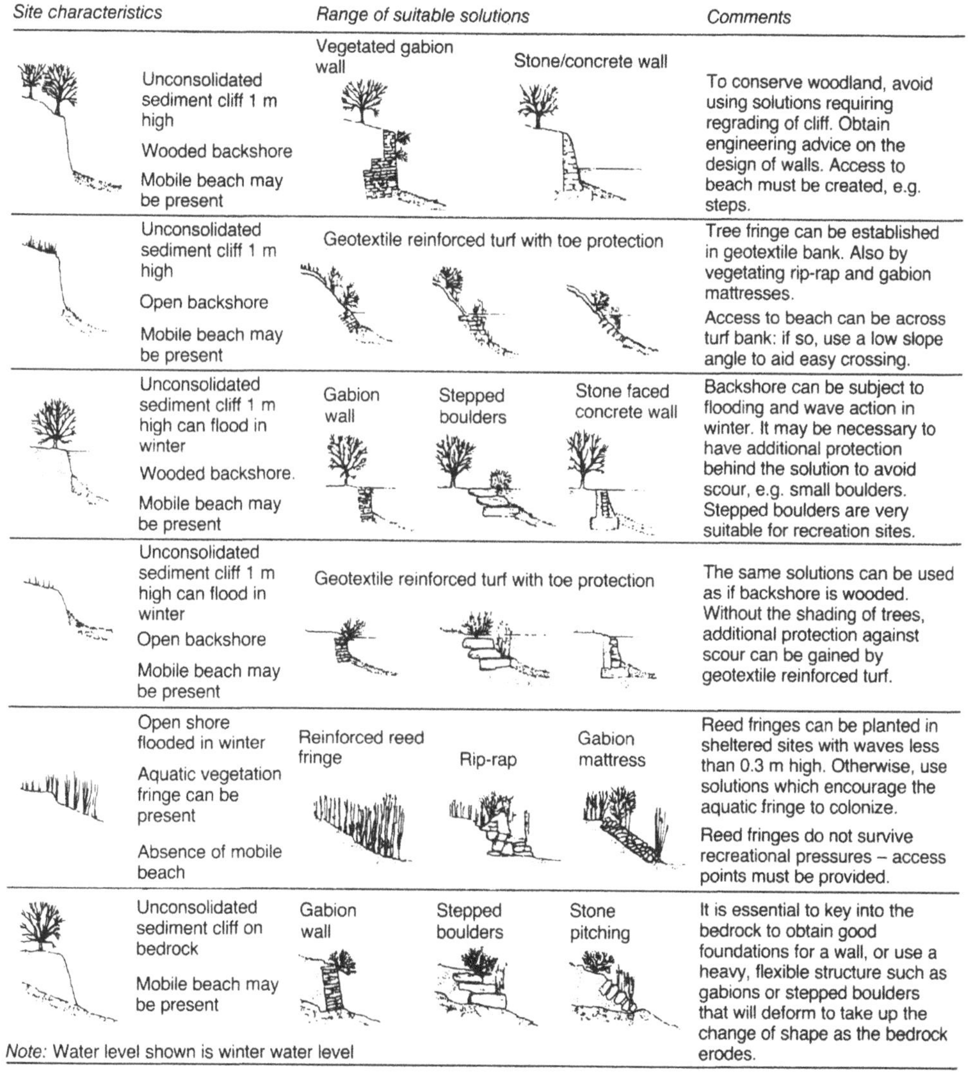

Site characteristics	Range of suitable solutions	Comments
Unconsolidated sediment cliff 1 m high / Wooded backshore / Mobile beach may be present	Vegetated gabion wall — Stone/concrete wall	To conserve woodland, avoid using solutions requiring regrading of cliff. Obtain engineering advice on the design of walls. Access to beach must be created, e.g. steps.
Unconsolidated sediment cliff 1 m high / Open backshore / Mobile beach may be present	Geotextile reinforced turf with toe protection	Tree fringe can be established in geotextile bank. Also by vegetating rip-rap and gabion mattresses. Access to beach can be across turf bank: if so, use a low slope angle to aid easy crossing.
Unconsolidated sediment cliff 1 m high can flood in winter / Wooded backshore. / Mobile beach may be present	Gabion wall — Stepped boulders — Stone faced concrete wall	Backshore can be subject to flooding and wave action in winter. It may be necessary to have additional protection behind the solution to avoid scour, e.g. small boulders. Stepped boulders are very suitable for recreation sites.
Unconsolidated sediment cliff 1 m high can flood in winter / Open backshore / Mobile beach may be present	Geotextile reinforced turf with toe protection	The same solutions can be used as if backshore is wooded. Without the shading of trees, additional protection against scour can be gained by geotextile reinforced turf.
Open shore flooded in winter / Aquatic vegetation fringe can be present / Absence of mobile beach	Reinforced reed fringe — Rip-rap — Gabion mattress	Reed fringes can be planted in sheltered sites with waves less than 0.3 m high. Otherwise, use solutions which encourage the aquatic fringe to colonize. Reed fringes do not survive recreational pressures – access points must be provided.
Unconsolidated sediment cliff on bedrock / Mobile beach may be present	Gabion wall — Stepped boulders — Stone pitching	It is essential to key into the bedrock to obtain good foundations for a wall, or use a heavy, flexible structure such as gabions or stepped boulders that will deform to take up the change of shape as the bedrock erodes.

Note: Water level shown is winter water level

Figure 5.27 Bank stabilization techniques recommended for control of shoreline erosion (after Countryside Commission for Scotland, 1985).

where flooding occurs for less than 100 days per year. Their canopy will, however, provide shade and thereby inhibit plant growth in the lower zones. Very little information is available on the use of herbaceous vegetation in designing plant communities on shorelines (Coppin and Richards, 1990).

A vegetative or bioengineered system has been used successfully, based on the above vegetation groups, to control wave erosion on islands within a lake near Cambridge, England (Anon., 1989). Protection of the islands was important because they supported an electricity transmission line. Between 1 November 1987 and 19 January 1989, the banks were observed to retreat by 1.5 m. The shoreline was reprofiled by hand. Live willow poles were driven into the soil and then osier wands (*Salix viminalis*), 4 m long, were woven between the poles to form a fence or 'spiling'. The area behind the fence was backfilled with top soil to form the bank into which willow setts were inserted. Rhizomes and shoot cuttings of reeds were planted below the spiling through a jute geotextile which had been pegged in place to give temporary protection against erosion. Further protection was provided by a floating log boom and, to keep out grazing waterfowl, a wooden fence. By early 1991 the willows and osiers were performing their function well but the growth of the reed-bed had been hampered by the drawdown of the lake to an unexpectedly low level to provide irrigation water over two dry summers.

The drawdown of reservoirs poses particular problems for vegetative approaches to erosion control because few plants are capable of colonizing an environment subject to alternating periods of 'effective drought' and submergence. Little and Jones (1979) tested 16 possible species for their adaptive response to these conditions and their suitability for erosion control. Plants with tall erect but flexible stems, such as *Phalaris arundinacea* and *Phleum pratense*, proved the best at erosion control because of their ability to attenuate the wave action. These species have the additional advantage of being able to survive in a substrate with limited inherent stability. Erect but more brittle plants produce a passive reduction in wave height but are unlikely to withstand substantial wave action. Semi-erect, prostrate or turf-forming plants do not reduce wave energy levels. Where seasonal die-back, burning or grazing reduce the amount of foliage, the root mass becomes important in protecting and reinforcing the soil. *Phalaris arundinacea* provides a suitable root mat but *Phleum pratense* lacks a dense lateral root network and has a clumpy growth habit which could therefore allow erosion to take place, especially where wave action is concentrated between the clumps.

In old reservoirs, such as Lake Vyrnwy, in Wales, natural colonization of *Salix* spp., provides some protection against wave erosion, as well as providing a more attractive shoreline than a large area of bare mud. Many *Salix* species are very tolerant of flooding for long

periods. How far down the shoreline they will survive, however, depends upon the drawdown regime which, in turn, varies with the objective of the reservoir. Where the reservoir is used for storage, the water levels are generally maintained at their highest levels in the winter. These can be more extensively planted than regulatory reservoirs which have highest water levels in the summer (Bradshaw and Chadwick, 1980).

5.7.4 SALT MARSHES AND MUD-FLATS

As indicated above, greater experience exists in using vegetation to stabilize mud-flats particularly where salt marshes develop on them in sheltered tidal estuaries. In these situations, a marked vertical zonation of vegetation occurs reflecting the ecological succession on a halosere, i.e. that initiated in a salt-water environment. The normal pioneer species are found on the lower slopes, rarely uncovered at low tide, and include green algae, such as *Enteromorpha*, and vascular plants, such as saltworts (*Salicornia* spp.). Above these there is often a bare zone before reaching the 'low marsh' at about mean high water level where the land is inundated about 360 times a year (Chapman, 1974). Typical plant species on the low marsh are either halophytic grasses, such as *Spartina* spp., or alkali grasses, such as *Puccinellia maritima* (common marsh grass), depending on the level of salinity. Higher up still, where inundation is only by the spring tides, a mixed vegetation develops including sea-blite (*Suaeda maritima*), sea plantain (*Plantago maritima*), sea rush (*Juncus maritimus*) and sea pink (*Armeria maritima*). At the highest levels, salt-tolerant sod-forming grasses occur, mainly species of *Festuca* or *Agrostis*.

Where wave attack occurs on salt marshes, it may be necessary to protect them against erosion since they perform a vital function in dissipating wave energy before it can impact on the cliff line behind the marsh. Under these conditions, promoting growth of the salt marsh may be a cheaper way of stabilizing the coast than constructing a sea wall on the toe of the cliff. This may be effected by planting appropriate vegetation species, selected from the natural plant succession found on salt marshes according to the criteria set out in Figure 5.28 (Boorman, 1977).

Studies in the Oosterschelde estuary of The Netherlands, where about 75% of the marsh margins are subject to erosion from wave action on the 0.25–1.25 m high salt marsh cliffs, show the importance of roots in contributing to the strength of the soil (van Eerdt, 1986). The tensile strength of the soft clay material is about 2.5 kPa but this is increased to about 5–6 kPa in a linear relationship with increasing density of the fine root network (Figure 5.29). The root density explains 62% of the variation in tensile strength of the material. The dominant plant species involved is *Spartina anglica* (common cord grass) which has an extensive system of robust horizontal rhizomes and a root network in which more than 85% of the roots are less than 1 mm in diameter. The roots are anchored in the soil by root hairs and branching laterals which means that they may be partly pulled out without failing or losing complete contact with the soil. When the roots break, a strong soil–root bond is still maintained. The increase in strength brought about by the roots applies only when the roots are in tension. Under compression, the roots were found not to increase the strength of the material but merely to bend along with it. Despite this demonstration that *Spartina anglica* can contribute substantial increases in soil strength, it should be remembered that this plant is an invasive species and, in the UK, it should not be planted without reference to the Nature Conservancy Council.

Spartina anglica is the fertile product of a sterile hybrid formed between *Spartina maritima* and *Spartina alterniflora*. An earlier hybrid between *Spartina maritima* and *Spartina alterniflora* produced *Spartina townsendii* which

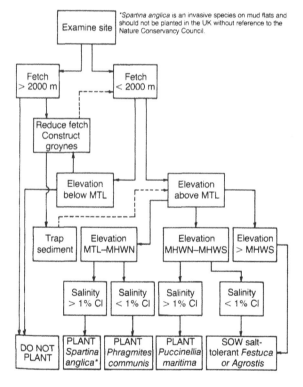

Figure 5.28 Criteria for selection of vegetation species for stabilization of salt marsh in the UK. MTL = mean tide level; MHWN = mean high water neap tides; MHWS = mean high water spring tides; Fetch = maximum extent of open water beyond marsh in metres; salinity in % concentration of chloride ion, sea water being approximately 3.5% (after Boorman, 1977).

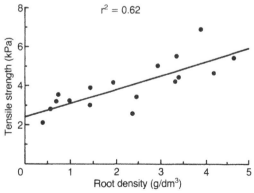

Figure 5.29 Tensile strengths of vegetation on salt marsh in the Oosterschelde, The Netherlands (after van Eerdt, 1986).

was first reported on Southampton Water in the 1870s and quickly spread along much of the south coast of England, particularly in Poole Harbour where it was a menace to navigation and the survival of other plant species. According to Tansley (1939), 'no other species of salt-marsh plant, in north-western Europe at least, has anything like so rapid and so great an influence in gaining land from the sea.' The species proved difficult to eradicate and it was not until the 1950s that it started to die back naturally (Goodman, 1960) following the build-up of the marsh above the spring high tide level. At first this resulted in the loss of vegetation cover and some damage to the salt marshes as they returned to bare mud. Later, however, recolonization occurred with *Salicornia* spp. and *Suaeda maritima*.

Experiments made in tidal areas in North Carolina, Virginia and Maryland (Hamer, 1990) show that plantings of *Spartina patens* (salt meadow cordgrass) and *Spartina alterniflora* (smooth cordgrass) can provide full protection against erosion from the second or third year after planting. Both species spread rapidly by extensive underground rhizomes. It was found that the plantings were more successful if containerized plants were used instead of bare-root materials. Planting was carried out at 45 cm spacings to a depth of 15 cm, and 28 g of slow-release fertilizer (19% nitrogen, 9% phosphate and 23% potash) was placed in each planting hole. Planting at shallower depths resulted in vegetation being washed away by waves on the incoming tides. The *Spartina alterniflora* plants should be salt-conditioned before planting, otherwise their survival can be drastically reduced. Planting was carried out between late March and the end of June to avoid months of high storm frequencies and to take advantage of the full growing season. A further application of fertilizer (10% nitrogen, 10% phosphate, 10% potash) over the whole planting area was necessary in mid-summer of the following year. It was also found that the positioning of the species was critical. Best results were achieved by planting the *Spartina alterniflora* at the mean

high tide level and continuing towards the water and planting the *Spartina patens* between the mean high tide level and the toe of the slope.

REFERENCES

Abujamin, S., Abdurachman, A. and Suwardjo. (1988) Contour grass strip as a low cost conservation practice, in *Soil Erosion and Its Counter Measures*, (ed. S. Jantawat). Soil and Water Conservation Society of Thailand, Bangkok. pp. 112–18.

Aina, P. O., Lal, R. and Taylor, G. S. (1979) Effects of vegetal cover on soil erosion on an alfisol, in *Soil Physical Properties and Crop Production in the Tropics* (eds R. Lal and R. J. Greenland). Wiley, Chichester, pp. 501–8.

Aldridge, T. (1979) Processes of decision-making and cost considerations, in *Shore Erosion around Loch Lomond*. Countryside Commission for Scotland, Battleby, pp. 61–71.

Ankenbrand, E. C. and Schwertmann, U. (1989) The land consolidation project of Freinhausen, Bavaria, in *Soil Erosion Protection Measures in Europe* (eds U. Schwertmann, R. J. Rickson and K. Auerswald). Soil Technology Series 1, Catena Verlag, Cremlingen-Destedt, pp. 167–74.

Anon. (1989) Lakeside erosion: the soft option. *Landscape News*, February 1989, p. 5.

Anon. (1990) How far apart should hedges be planted? *Vetiver Newsletter*, 4, November 1990, 5–6.

Ash, J. R. V. and Woodcock, E. P. (1988) The operational use of river corridor surveys in river management. *J. Inst. Water and Environmental Management*, 2(4), 423–8.

Boorman, L. A. (1977) Sand dunes, in *The Coastline* (ed. R. S. K. Barnes). Wiley, London, pp. 161–97.

Bradshaw, A. D. and Chadwick, M. J. (1980) *The Restoration of Land*. Blackwell, Oxford.

Burley, J. B. (1989) Bluemont Lakes shoreline revegetation study, in *Erosion Knows No Boundaries*. International Erosion Control Association, Steamboat Springs, CO, pp. 287–95.

Chapman, V. J. (1974) *Salt Marshes and Salt Deserts of the World*. Cramer, Lehre.

Chow, V. T. (1981) *Open Channel Hydraulics*. McGraw Hill, New York.

Clark, M. W. (1979) Marine processes, in *Process in Geomorphology* (eds C. Embleton and J. Thornes). Edward Arnold, London, pp. 352–77.

Clifton, H. E. (1976) Wave-formed sedimentary

structures: a conceptual model, in *Beach and Near-shore Sedimentation* (eds R. A. Davis and R. L. Ethington). Special Publication No. 24, Society for Economic Paleontology and Mineralogy, Tulsa, pp. 126–48.

Coleman, R. (1981) Footpath erosion in the English Lake District. *Applied Geography*, 1, 121–31.

Coppin, N. J. and Richards, I. G. (1990) *Use of Vegetation in Civil Engineering*. CIRIA/Butterworths, London.

Countryside Commission for Scotland (1985) *Plants and Planting Methods*. Lochshore Management Information Sheet No. 6.

Cuevas, V. C. and Diez, F. C. (1988) Soil nutrient dynamics and their effects on the productivity of a compost–fertilized, shifting cultivation farming system with wood dams constructed as erosion control measures, in *Land Conservation for Future Generations* (ed. S. Rimwanich). Dept. Land Development, Ministry of Agriculture and Cooperatives, Bangkhen, Bangkok, pp. 1113–26.

Dawson, F. H. (1978) Seasonal effects of aquatic plant growth on the flow of water in a small stream. *Proc. EWRS, 5th Symposium on aquatic Weeds*, Amsterdam, pp. 71–8.

Dawson, F. H. and Charlton, F. G. (1987) Bibliography on the hydraulic resistance or roughness of vegetated watercourses. *Freshwater Biological Association, Occasional Publication No. 25*.

Dawson, F. H. and Robinson, W. N. (1984) Submerged macrophytes and the hydraulic roughness of a lowland chalk stream, in *Proceedings of the International Association for Theoretical and Applied Limnology Congress in France, 1983*. E. Schweitzerbartsche, Verlagsbuchlandlung, Stuttgart, pp. 1944–8.

Dissmeyer, G. E. and Foster, G. R. (1985) Modifying the Universal Soil Loss Equation for forest land, in *Soil Erosion and Conservation* (eds S. A. El-Swaify, W. C. Moldenhauer and A. Lo). Soil Conservation Society of America, Ankeny, IA, pp. 480–95.

Duell, R. W. (1989) Appropriate vegetative covers for erosion control, in *Erosion Knows No Boundaries*. International Erosion Control Association, Steamboat Springs, CO, pp. 279–84.

Dunn, C. S. (1975) Control of erosion on highways. Paper presented at PTRC Summer Meeting, University of Warwick.

Dyrness, C. T. (1975) Grass–legume mixtures for erosion control along forest roads in western Oregon. *J. Soil and Water Conservation*, 30, 169–73.

FAO, (1965) *Soil Erosion by Water*. FAO, Rome.

Fang, H., Hua, S., Meng, Q. and Mou, J. (1985) *Evaluating benefits of reservoir-dam system in Yellow River Valley concerning erosion-preventing, produce-increasing and silt-reducing*. Beijing.

Galvin, C. J. (1968) Breaker-type classification on three laboratory beaches. *J. Geophysical Research*, 73, 3651–9.

Goodman, P. J. (1960) Investigations into 'die-back' in *Spartina townsendii* agg. II. The morphological structure and composition of the Lymington sward. *J. Ecology*, 48, 711–24.

Gray, D. H. and Leiser, A. T. (1982) *Biotechnical Slope Protection and Erosion Control*. Van Nostrand Reinhold, New York.

Gregory, J. M. and McCarty, T. R. (1986) Maximum allowable velocity predictions for vegetated waterways. *Trans. Am. Soc. Agric. Engrs*, 29, 748–55.

Gupta, R. K. and Arora, Y. K. (1983) Technology for the rejuvenation of degraded lands in the Himalaya, in *Proceedings, Seminar on Environmental Regeneration in Himalaya: Concepts and Strategies*. Central Himalayan Environment Association, Naintal, pp. 169–86.

Haigh, M. J. (1979) Ground retreat and slope evolution on regraded surface-mine dumps, Waunafon, Gwent. *Earth Surface Processes*, 6, 183–9.

Hamer, D. W. (1990) Shoreline stabilization with cordgrass, in *Erosion Control: Technology in Transition*. International Erosion Control Association, Steamboat Springs, CO, pp. 43–5.

Hatch, T. (1981) Preliminary results of soil erosion and conservation trials under pepper (*Piper nigrum*) in Sarawak, Malaysia, in *Soil Conservation: Problems and Prospects* (ed. R. P. C. Morgan). Wiley, Chichester, pp. 255–62.

Hathaway, R. L. (1986a) Plant materials for river control and bank protection, in *Plant Materials Handbook for Soil Conservation. Vol. 1: Principles and Practices* (eds C. W. S. Van Kraayenoord and R. L. Hathaway). Soil Conservation Centre, Aokautere, Ministry of Works and Development, Palmerston North, pp. 57–67.

Hathaway, R. L. (1986b) Plant materials for gully control, in *Plant Materials Handbook for Soil Conservation. Vol. 1: Principles and Practices* (eds C. W. S. Kraayenoord and R. L. Hathaway). Soil Conservation Centre, Aokautere, Ministry of Works and Development, Palmerston North, pp. 49–56.

Heede, B. H. (1960) A study of early gully control structures in the Colorado Front Range. *USDA Forest Serv. Paper No. 55*, Rocky Mountain Forest and Range Experimental Station, Fort Collins, Colorado.

Heede, B. H. (1968) Conversion of gullies to vegetation lined waterways on mountain slopes. *USDA Forest Serv. Res. Paper No. RM-40*, Rocky Mountain Forest and Range Experimental Station, Fort Collins, Colorado.

Heede, B. H. (1976) Gully development and control: the status of our knowledge. *USDA Forest Serv. Res. Paper No. RM-169*, Rocky Mountain Forest and Range Experimental Station, Fort Collins, Colorado.

Heede, B. H. and Mufich, J. G. (1973) Functional relationships and a computer program for structural gully control. *J. Env. Management*, 1, 321–44.

Heusch, B. (1970) L'érosion du Pré-Rif. Une étude quantitative de l'érosion hydraulique dans les collines marneuses du Pré-Rif occidental. *Annales de Recherches Forestières du Maroc*, 12, 9–176.

Hewlett, H. W. M., Boorman, L. A., Bramley, M. E. and Whitehead, E. (1985) Reinforcement of steep grassed waterways. *CIRIA Technical Note No. 120*, CIRIA, Westminster, London.

Hewlett, H. W. M., Boorman, L. A. and Bramley, M. E. (1987) Design of reinforced grass waterways. *CIRIA Report 116*, CIRIA, London.

Howell, J. H., Clark, J. E., Lawrance, C. J. and Sunwar, I. (1991) *Vegetation Structures for Stabilising Highway Slopes. A Manual for Nepal.* UK/Nepal Eastern Region Interim Project, Kathmandu.

Hudson, N. W. (1957) Erosion control research. *Rhodesia Agricultural J.* 54(4), 297–323.

Hudson, N. (1981) *Soil Conservation.* Batsford, London.

Hurni, H. (1986) *Soil Conservation in Ethiopia. Guidelines for Development Agents.* Community Forests and Soil Conservation Development Department, Ministry of Agriculture, Ethiopia.

Idle, E. T. (1979) Ecological considerations relating to changes in shoreline, in *Shore Erosion around Loch Lomond*. Countryside Commission for Scotland, Battleby, pp. 27–31.

Katiyar, V. S., Sastry, G. and Adhikari, R. N. (1987) Mechanical measures for mine spoils and landslides. *Indian J. Soil Conservation*, 15(3), 108–16.

Kellman, M. C. (1969) Some environmental components of shifting cultivation in upland Mindanao. *J. Tropical Geography*, 28, 40–56.

Kouwen, N. and Li, R. M. (1980) Biomechanics of vegetative channel linings. *J. Hydraulics Division ASCE*, 106, 1085–103.

Kouwen, N., Unny, T. E. and Hill, H. M. (1969) Flow retardance in vegetated channels. *J. Irrigation and Drainage Division ASCE*, 95, 329–42.

Laflen, J. M. and Colvin, T. S. (1981) Effect of crop residue on soil loss from continuous row cropping. *Trans., Am. Soc. Agric. Engnrs*, 24, 605–9.

Lal, R. (1981) Deforestation of tropical rainforest and hydrological problems, in *Tropical Agricultural Hydrology* (eds R. Lal and E. W. Russell). Wiley, Chichester, pp. 131–40.

Lal, R. (1988) Soil erosion control with alley cropping, in *Land Conservation for Future Generations* (ed. S. Rimwanich). Department of Land Development, Ministry of Agriculture and Cooperatives, Bangkhen, Bangkok, pp. 237–46.

Lang, R. D. and McCaffrey, L. A. H. (1984) Ground cover: its effects on soil loss from grazed runoff plots, Gunnedah. *J. Soil Conservation Service of New South Wales*, 40, 56–61.

Liddle, M. J. (1974) Outdoor recreation: a selective summary of pertinent ecological research. Paper presented to Symposium on Applied Environmental Science, Annual Meeting of the Institute of British Geographers, Norwich.

Little, M. G. and Jones, H. R. (1979) The uses of herbaceous vegetation in the drawdown zone of reservoir margins. *Water Resources Centre Technical Report No. TR 105*.

Lusby, G. C. and Toy, T. J. (1976) An evaluation of surface-mine spoils area restoration in Wyoming using rainfall simulation. *Earth Surface Processes*, 1, 375–86.

Morgan, R. P. C. (1985) Effect of corn and soybean canopy on soil detachment by rainfall. *Trans. Am Soc. Agric. Engrs*, 28, 1135–40.

Morgan, R. P. C. (1986) *Soil Erosion and Conservation.* Longman, Harlow.

Morgan, R. P. C. (1987) Evaluating the role of vegetation in soil erosion control with implications for steepland agriculture in the tropics, in *Steepland Agriculture in the Humid Tropics* (eds T. H. Tay, A. M. Mokhtaruddin and A. B. Zahari) Malaysian Agricultural Research and Development Institute/ Malaysian Society of Soil Science, Kuala Lumpur, pp. 401–23.

Morgan, R. P. C., Finney, H. J., Lavee, H., Merritt, E. and Noble, C. A. (1986) Plant cover effects on hillslope runoff and erosion: evidence from two laboratory experiments, in *Hillslope Processes* (ed. A. D. Abrahams). Allen and Unwin, Winchester, MA, pp. 77–96.

Noble, C. A. and Morgan, R. P. C. (1983) Rainfall interception and splash detachment with a Brussels sprouts plant: a laboratory simulation. *Earth Surface Processes and Landforms*, 8, 569–77.

Odemerho, F. O. (1986) Variation in erosion-slope relationship on cut-slopes along a tropical highway. *Singapore J. of Tropical Geography*, 7(2), 98–107.

Okigbo, B. N. (1978) *Cropping Systems and Related Research in Africa*. Occasional Publication Series, OT. Association for the Advancement of Agriculture in Africa, Addis Ababa, pp. 1–81.

Poodle, T. (1979) Fluctuations in loch levels and the factors affecting them, in *Shore Erosion around Loch Lomond*. Countryside Commission for Scotland, Battleby, pp. 15–26.

Powell, K. E. C. (1978) Weed growth – a factor in channel roughness, in *Hydrometry – Principles and Practices* (ed. R. W. Herschy). Wiley, Chichester, pp. 327–52.

Pushparajah, E., Tan, K. H. and Soong, N. K. (1977) Influence of covers and fertilizers and management on soil, in *Soils under Hevea in Peninsular Malaysia and Their Management*. Rubber Research Institute of Malaysia, Kuala Lumpur, pp. 75–93.

Quinn, N. W. and Laflen, J. M. (1983) Characteristics of raindrop throughfall under corn canopy. *Trans., Am. Soc. Agric. Engrs*, 26, 1445–50.

Ree, W. O. and Palmer, V. J. (1949) Flow of water in channels protected by vegetative linings. *USDA Technical Bulletin No. 967*.

Richardson, E. C., Diseker, E. G. and Sheridan, J. M. (1970) Practices for erosion control on roadside areas in Georgia. *Highway Research Record No. 335*, Highways Research Board USA, pp. 35–44.

Rickson, R. J. and Morgan, R. P. C. (1988) Approaches to modelling the effects of vegetation on soil erosion by water, in *Erosion Assessment and Modelling* (eds R. P. C. Morgan and R. J. Rickson). Commission of the European Communities Report No. EUR 10860 EN, pp. 237–53.

Roose, E. (1967) Dix années de mesure de l'érosion et du ruissellement au Sénégal. *L'Agronomie Tropicale*, 22(2), 123–52.

Rose, S. J. C. (1989) The Three Peaks Project: tackling footpath erosion, in *Erosion Knows No Boundaries*. International Erosion Control Association, Steamboat Springs, CO, pp. 371–8.

Sastry, G., Mathur, H. N. and Tejwani, K. G. (1981) Landslide control in north-western outer Himalayas: a case study. *Central Soil and Water Conservation Research and Training Institute Bulletin No. R8/D6*, Dehra Dun.

Scott-Park, J. H. (1979) Shore erosion as a problem of land management, in *Shore Erosion around Loch Lomond*. Countryside Commission for Scotland, Battleby, pp. 47–53.

Shah, B. H. (no date) *Bio-engineering Techniques for Critical Slope Stabilization*. Pakistan Forest Institute, Peshawar.

Shaxson, T. F. (1981) Reconciling social and technical needs in conservation work on village farmlands, in *Soil Conservation: Problems and Prospects* (ed R. P. C. Morgan). Wiley, Chichester, pp. 385–97.

Singh, A. and Prasad, R. N. (1987) Research experiences in soil and water conservation on steep land in north-east India, in *Steepland Agriculture in the Humid Tropics* (eds T. H. Tay, A. M. Mokhtaruddin and A. B. Zahari). Malaysian Agricultural Research and Development Institute/Malaysian Society of Soil Science, Kuala Lumpur, pp. 387–400.

Singh, G. (1990) Rain water harvesting and recycling for sustainable agricultural production, in *Proceedings, International Symposium of Water Erosion, Sedimentation and Resource Conservation, 9–13 October 1990*. Central Soil and Water Conservation Research and Training Institute, Dehra Dun, pp. 157–68.

Smith, I. R. (1979) The causes of loch shore erosion: an introduction, in *Shore Erosion around Loch Lomond*. Countryside Commission for Scotland, Battleby, pp. 7–14.

Styczen, M. and Høgh-Schmidt, K. (1986) A new description of the relation between drop sizes, vegetation and splash erosion, in *Partikulært Bundet Stoftransport i Vand og Jorderosjon* (ed. B. Hasholt). Nordisk Hydrologisk Program, NHP Report No. 14, pp. 255–71.

Tansley, A. G. (1939) *The British Islands and Their Vegetation*. Cambridge University Press, Cambridge.

Temple, D. M. (1980) Tractive force design of vegetated channels. *Trans. Am. Soc. Agric. Engnrs.*, 23, 884–90.

Temple, D. M., Robinson, K. M., Ahring, R. M. and Davis, A. G. (1987) Stability design of grass lined open channels. *USDA Agricultural Handbook No. 667*, USDA National Technical Information Service, Springfield, VA.

Thompson, J. R. (1986) Roadsides: a resource and a challenge, in *Ecology and Design in Landscape* (eds A. D. Bradshaw, D. A. Goode, and E. Thorp). Blackwell, Oxford, pp. 325–40.

Thornes, J. B. (1980) Erosional processes of running water and their spatial and temporal controls: a theoretical viewpoint, in *Soil Erosion* (eds M. J. Kirkby and R. P. C. Morgan). Wiley, Chichester, pp. 129–82.

Thornes, J. B. (1988) Competitive vegetation–erosion model for Mediterranean conditions, in *Erosion Assessment and Modelling* (eds R. P. C. Morgan and R. J. Rickson). Commission of the European Communities Report EUR 10860 EN, pp. 255–82.

Thornes, J. B. (1990) The interaction of erosional and vegetational dynamics in land degradation: spatial outcomes, in *Vegetation and Erosion* (ed. J. B. Thornes). Wiley, Chichester, pp. 41–53.

Troeh, F. R., Hobbs, J. A. and Donahue, R. L. (1980) *Soil and Water Conservation for Productivity and Environmental Protection*. Prentice-Hall, Englewood Cliffs, NJ.

United States Soil Conservation Service (1953) *Engineering Handbook for Farm Planners. Upper Mississippi Valley Region III*. United States Department of Agriculture, Washington, DC.

United States Soil Conservation Service (1954) *Handbook of Channel Design for Soil and Water Conservation*. Publication SCS-TP61, USDA Washington, DC.

Van Asch, Th. W. J. (1980) Water erosion on slopes and landsliding in a Mediterranean landscape. *Utrechtse Geografische Studies No. 20*.

Van Eerdt, M.M. (1986) The influence of basic soil and vegetation parameters on salt marsh cliff strength, in *International Geomorphology 1986 Part I* (ed. V. Gardiner). Wiley, Chichester, pp. 1073–86.

Veloz, R. A. and Logan, T. J. (1988) Steepland erosion research in the Dominican Republic, in *Land Conservation for Future Generations* (ed. S. Rimwanich). Department of Land Development, Bangkok, pp. 991–1000.

Walter, M. W. (1967) The length of the rainy season in Nigeria. *Nigerian Geographical J.*, **10**, 123–8.

Watts, J. F. and Watts, G. D. (1990) Seasonal change in aquatic vegetation and its effect on river channel form, in *Vegetation and Erosion* (ed. J. B. Thornes). Wiley, Chichester, pp. 257–67.

Weaver, T. and Dale, D. (1978) Trampling effects of hikers, motor cycles and horses in meadows and forests. *J. Applied Ecol.*, **15**, 451–7.

Williams, C. N. and Joseph, K. T. (1970) *Climate, Soil and Crop Production in the Humid Tropics*. Oxford University Press, Oxford.

Wischmeier, W. H. and Smith, D. D. (1978) Predicting rainfall erosion losses. *USDA Agr. Res. Serv. Handbook No. 537*.

Zanchi, C. (1983) Influenze dell'azione battente della pioggia e del ruscellamento nel processo erosive e variazioni dell'erodibilità del suolo nei diversi periodi stagionali. *Annali Istituto Sperimentale Studio e Difesa Suolo*, **14**, 347–58.

WIND EROSION CONTROL 6

R. P. C. Morgan

6.1 INTRODUCTION

The effectiveness of vegetation in controlling wind erosion can be demonstrated by the disastrous consequences of its removal. Examples of catastrophes abound. Arguably the best known is the Dust Bowl of the 1930s when the extension of agriculture into the Great Plains of the USA resulted in the loss of 350×10^6 t of top soil and the transformation of 2×10^6 ha of grazing and cropland into sand dunes. One of the best-documented examples of the effect of wind erosion on a nation, however, is provided by the history of Iceland.

It is estimated that when, as a result of land shortages and political unrest in Norway, Iceland was settled in AD 874 about 65% of the country was vegetated. Birch (*Betula pubescens*) woodlands occupied between 25% (Sigurðsson, 1977) and 40% (Bjarnason, 1974) of the country, and willow (*Salix* spp.) and dwarf shrubs covered large areas with sedges in the wetter parts. Today only about 25% of the country is vegetated and birch woodlands occupy a mere 1% of the land (Figure 6.1; Thorsteinsson and Blöndal, 1986). Settlers cleared the woodlands for farmsteads, hayfields, timber and fuel. Large amounts of timber were used to make charcoal to whet scythes for cutting hay, and for making iron which, until the 15th century, was entirely home-produced from iron deposits within peat (Thórarinsson, 1974). Although the demand for charcoal reduced after 1870 with the development of a scythe that did not require whetting and with the import of iron, the vegetation cover further declined as a result of increasing

grazing pressure. In 1800, it is estimated that there were 304 000 sheep in the country. Sheep numbers increased to 490 000 in 1855, 583 000 in 1924 and 699 000 in 1934. They fell to 450 000 in 1948, partly as a result of destruction of grazing land by volcanic ash from the 1947 eruption of Hekla, but increased to 834 000 by 1960 (Pálsson and Stefánsson, 1968). Today sheep numbers have decreased to about 680 000. Overgrazing and erosion interact to bring about changes in the botanical composition and percentage cover of the rangeland.

Overgrazing by sheep causes the proportion of grasses and forbs to decline in favour of rushes, sedges and mosses (Figure 6.2; Thorsteinsson, 1980). At the same time, increasing bare ground results in higher erosion and a further decline in grasses and also in woody shrubs since these cannot withstand abrasion and burial by moving sand (Arnalds, O., 1984). Figure 6.3 shows the effect of different grazing levels on the above-ground biomass, density of cover and the depth and extent of the rooting system (Thorgeirsson, 1990). Overgrazing results in less surface cover so that more of the soil is directly exposed to wind. It also results in poorer root development reducing the root-reinforcement effect and causing the soil to be more erodible.

The soils of Iceland contain layers of volcanic ash which can be dated to specific eruptions of which between 30 and 40 have been recorded since settlement began. By knowing the age of each layer and the thickness of the soil layers between, it is possible to work out the rates of soil formation. Since most of the soils comprise

Slope Stabilization and Erosion Control: A Bioengineering Approach. Edited by R. P. C. Morgan and R. J. Rickson. Published in 1995 by E & FN Spon, 2-6 Boundary Row, London, SE1 8HN. ISBN 0 419 15630 5.

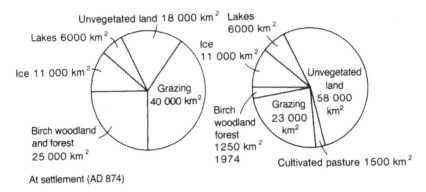

Figure 6.1 Change in land cover of Iceland between time of settlement in 874 and 1974 (from Thorsteinsson and Blöndal, 1986).

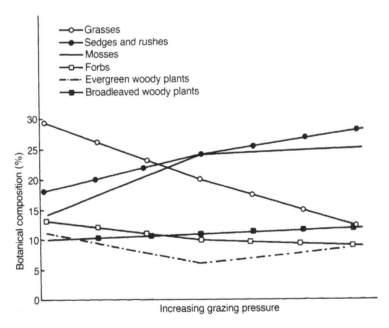

Figure 6.2 Effect of grazing on the botanical composition of Icelandic rangelands (after Thorsteinsson, 1980).

deposits of wind-blown material, the thickness of the soil layers can be used as evidence of rates of wind erosion (Thórarinsson, 1961). These figures show a tenfold increase in wind erosion since settlement with rates of soil accumulation of 0.1 mm/year before settlement, 0.3 mm/year between late 14th and mid 19th centuries, and 1.1 mm/year from late 19th century to 1950

(Thórarinsson, 1968; Larsen and Thórarinsson, 1977; Arnalds, O., 1984) (Figure 6.4; Jóhannesson and Einarsson, 1990). These data cannot be used as absolute erosion rates, however, because much of the evidence comes from soil sections in the vertical erosion fronts or edges (rofabórð; plural rofabarði) of vegetated remnants of land in wind-eroded areas where the

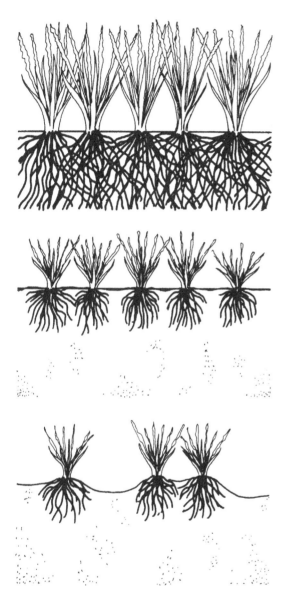

Figure 6.3 Effect of grazing on density of ground cover and root development on Icelandic rangelands: (a) no grazing; (b) moderate grazing; (c) overgrazing (after Thorgeirsson, 1990).

surrounding land has been removed down to the underlying bedrock (Figure 6.5; Arnalds, O., 1989). Since the vegetation has trapped wind blown material, the soil layers are thicker than

they would otherwise be on more open ground. Also, this effect is greater in more recent times. Prior to settlement, there were few rofabarði but as wind erosion has increased, so has their number, which means that apparent rates of soil accumulation have been increasingly exaggerated. Nevertheless, the pattern of increasing wind erosion remains valid, being supported by evidence from pollen analysis of peat layers (Einarsson, 1963) and written historical records.

Historical references support the view that the vegetation cover was reasonably well maintained from the time of settlement to the 12th century when Ari Thorgilsson (*c.* 1100) wrote that 'the country was covered with woods from the beach to the mountains'. Vegetation cover declined as the climate deteriorated around 1300. The winter of 1290 was known as the 'Great Misery' because of heavy snow and severe loss of livestock for lack of hay and shelter. Despite the failure of the hay crop in several summers during the 12th and 13th centuries, however, erosion does not seem to have increased at this time. The 16th century was relatively mild (Figure 6.6; Arnalds, A., 1988a) and vegetation growth must have been good until about 1600 when another climatic deterioration began (Friðriksson, 1972). The increase in erosion from this time is supported by the frequent references to sand drift by Arni Magnússon and Pall Vídalín in their farm survey carried out between 1702 and 1712 (Magnússon and Vídalín, 1913–1917). Hannes Finnsson (1796), the Bishop of Skálholt, wrote of the great famine at the end of the 18th century when a quarter of the population died and the land was worn out by 'sandstorms … disastrous flooding, rivers bursting their banks, landslides, erosion by gales and so forth' so that 'not so many sheep can be supported on the land as formerly'.

The greatest destruction by erosion occurred in the period 1880–1890 during the little Ice Age. It would appear that the pressure exerted by man on the landscape was so strong by the end of the 19th century that the land was unable to withstand what was the end of arguably the

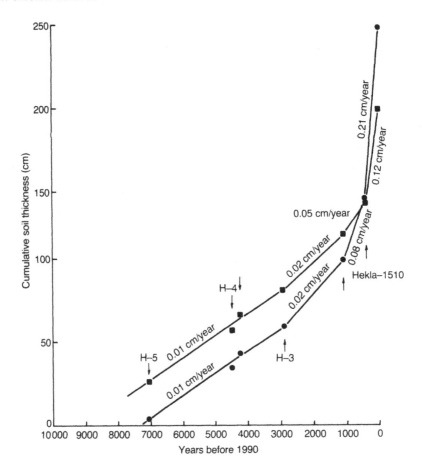

Figure 6.4 Changes in rate of soil erosion over time in the Hekla region of Iceland. Sample sites: ■, Flatahraun; ●, Foss. Values are based on the thicknesses of soil formed between layers of volcanic ash which can be dated to 1510 (Hekla eruption), 2800 BP (H-3), 4000 BP (H-4) and 7000 BP (H-5). The soil surface layer at the time of settlement is also used as a marker (after Larsen and Thórarinsson, 1977; Arnalds, O., 1984; Jóhanesson and Einarsson, 1990).

coldest period in Iceland since settlement began. The year 1882 was known in the south of Iceland as the 'year of the drift sand'. Reports of the time state that

> a severe northerly storm lasted for a long while and literally ground the dry land which was already exhausted by excessive grazing, tore the roofs off the farmhouses which were thatched with green turf, and demolished the walls which were at the time mostly made of sod. . . . For two days not even the bravest men had the courage to leave their homes;

such was the force of the sandstorm and stone shots.

> (Sveinsson, 1953)

Thóroddsen (1914) described the phenomenon of 'mistur', the yellowish-brown cloud of dust that covers the sky during periods when dry northern and northeasterly winds dominate the weather giving very similar conditions to those observed in the United States during the Dust Bowl period.

There is thus strong circumstantial evidence to show that the effect of removal of vegetation

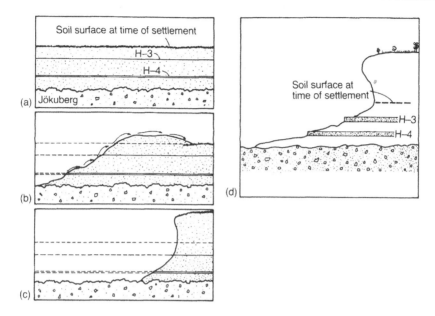

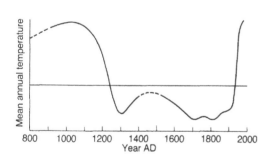

Figure 6.5 Stages in the formation of erosion fronts (rofabarði) and exposure of volcanic ash layers: (a) deposition of ash layers and formation of soil between them; (b) development of sand dune by wind erosion; (c) continued erosion of surrounding land to form the erosion front; (d) typical erosion front in the Hekla region showing position of the soil surface at time of settlement and the volcanic ash layers H-3 and H-4, dated to 2800 BP and 4000 BP respectively (after Arnalds, O. 1989).

Figure 6.6 Diagrammatic representation of mean annual temperature in Iceland since time of settlement. The fluctuation is about ±1 °C (after Arnalds, A., 1988a).

on wind erosion is also dependent upon climate. Whenever the climate deteriorates, plant growth and regeneration are slowed and the land is less resilient to disturbance. Despite this interaction, the extent of erosion today is clearly dependent on the vegetation cover. A comparison of

Figures 6.7 and 6.8 (Arnalds, A., 1987) shows that the areas most affected by erosion generally have less than 25% vegetation cover and that areas with more than 75% cover are not subject to erosion.

Clearly, removal of vegetation can cause wind erosion of such severity as to result in poverty on a national scale. Problems also arise, however, from events of significant but lesser magnitude. Erosion can lead to the long-term loss of soil productivity due to reductions in the nutrient and water-holding status of the soil. Deposition of wind-blown material can bury fields and settlements and clog ditches and reservoirs. Sediment carried in the wind can do considerable damage to vegetation and also to buildings, cars and machinery by abrasion. Air pollution by dust increases the frequency of respiratory ailments and skin disorders. It also reduces visibility with consequent danger to road, rail and air traffic.

Since removal of vegetation enhances the

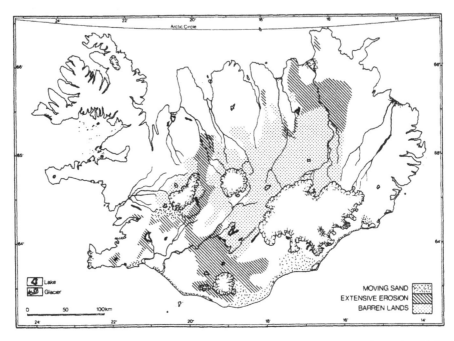

Figure 6.7 Distribution of areas of serious soil erosion, moving sand and barren land in Iceland (after Arnalds, A. 1987).

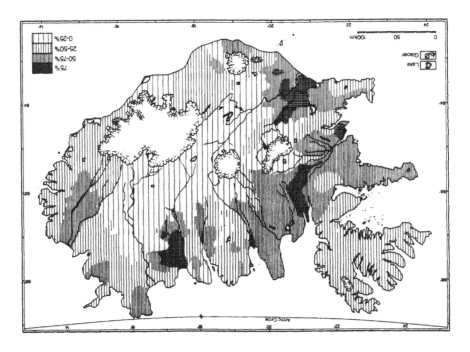

Figure 6.8 Average percentage vegetation cover in Iceland (after Arnalds, A. 1987).

problem of wind erosion, restoration of that cover should be an effective method of prevention or control. This implies that vegetation performs an important engineering role in protecting the soil. This chapter attempts to elucidate that role by analysing how vegetation affects the processes of wind erosion. From this understanding it is possible to determine the salient properties of vegetation which contribute most to the protection of the soil and to analyse the circumstances under which vegetation can, sometimes, fail to protect and even make the situation worse. This knowledge can then be used to build up the bioengineering technology for designing effective strategies for erosion control.

6.2 VEGETATION AND WIND EROSION CONTROL

Although there is a long history of the use of vegetation to control wind erosion, for example by planting shelterbelts, our understanding of the mechanisms by which control is effected is still very limited. Vegetation is not allowed for in the process-based models of wind erosion currently being developed (Nickling, 1988). Its effect has been quantified empirically, however, within the USDA Wind Erosion Equation (Woodruff and Siddoway, 1965) for flattened wheat straw based on the weight of cover per unit area (Figure 6.9). The relationship can be applied to other vegetation covers by converting their biomass into flattened wheat straw equivalents using the equation developed by Lyles and Allison (1981):

$$SG = 0.162\,Rw/\mathrm{ds} + 8.708\,[\,Rw/(\mathrm{ds}\cdot\gamma)\,]^{0.5} - 271, \quad (6.1)$$

where SG = weight of flattened wheat straw equivalent (kg/ha); Rw = weight of standing residue of the crop (kg/ha); ds = average stalk diameter (cm); and γ = average specific weight of the stalks (Mg/m^3).

This approach does not explain how the vegetation cover operates and, except for weight and surface cover, does not indicate what properties of the vegetation are important.

Vegetation can be expected to influence wind erosion in several ways. First, the foliage of the vegetation affects the air flow, usually resulting in a reduction in velocity (Chapter 2). Second, the foliage traps moving sediment. Third, the vegetation cover on the ground surface protects the soil. Fourth, the root system increases the resistance of the soil. Fifth, through the take-up of water for transpiration, the vegetation controls the soil moisture.

6.3 VEGETATION AND SHEAR VELOCITY

It was shown in Chapter 2 that vegetation reduces the shear velocity of the wind by exerting a drag on the air flow. The roughness imparted to the flow by this drag is expressed through an increase in the value of z_0 or the roughness length which is a measure of the effectiveness of the vegetation in absorbing momentum. In addition, the vegetation cover raises the height of the mean aerodynamic surface by a distance d, known as the zero-plane displacement, which is a measure of the mean height at which momentum is absorbed (Thom, 1975). The effect of vegetation in this way on shear velocity can be described by the bulk drag coefficient (CD), defined by equation 2.45.

Though wind speed decreases rapidly from the top of the canopy downwards, it, except in light winds, rarely falls to zero by 70% of the canopy height, the elevation normally taken as the approximate position of the zero-plane displacement. Velocity remains reasonably constant between 70 and 10% of the canopy height and then declines again to zero close to the ground (Landsberg and James, 1971; Figure 2.16). Thus, whilst shear forces are generally negligible throughout most of the vegetation layer below the plane of zero displacement, there is still some shear in the lower part of the vegetation layer which is exerted on the soil surface. The shear velocity in this layer is described by equation 2.54 and is dependent upon the drag exerted by the foliage elements, as

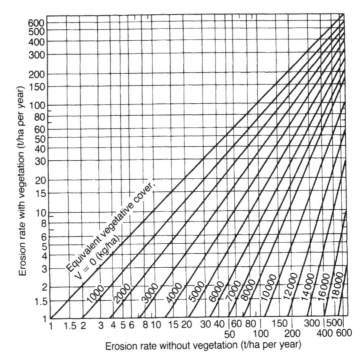

Figure 6.9 Conversion of rates of wind erosion without a vegetation cover to rates with a vegetation cover expressed as flattened wheat straw equivalents (after Woodruff and Siddoway, 1965). Note: the reference standard is 25.4-cm-long dry wheat stalks lying flat on the surface in rows perpendicular to the wind direction with 25.4-cm row spacing and stalks oriented parallel to the wind direction.

quantified by the value of the drag coefficient, Cd. The properties of the vegetation which determine the value of Cd are the biomass, projected area of the foliage facing the wind, leaf area density, leaf orientation and leaf shape (Morgan, Finney and Williams, 1988). Since vegetation is not a rigid material, the values of Cd are not constant as would be the case for solid objects, but vary with both wind velocity and the degree of wind turbulence (Morgan and Finney, 1987). Table 2.12 gives typical values of Cd for a range of agricultural crops.

6.4 EFFECT OF VEGETATION ON SEDIMENT REMOVAL

6.4.1 SEDIMENT ENTRAINMENT

Virtually no studies have been carried out on the effect of vegetation on sediment entrainment.

All winds capable of initiating soil movement are turbulent and have high Reynolds numbers, indicating that inertial forces are more important than viscous ones. With high Reynolds numbers the aerodynamic lift force to raise soil particles directly into the air flow requires that $u^* > [(\sigma - \rho_\alpha)/\rho_\alpha gd]^{0.5}$, where σ is the density of the grains. In reality, however, particles are entrained at u^* values of about one-tenth of this value, implying that there must be another contributing mechanism. This is generally thought to involve particles gaining a horizontal momentum as a result of the drag or shear force of the wind.

Shear induces a difference in pressure between the upwind and downwind sides of a projecting soil particle as a result of acceleration of flow over the obstacle. This induces low pressure above the particle relative to the slowly moving air in the voids beneath the particle, creating a lift force. Shear is therefore a major factor in the detachment of individual soil particles from the soil mass and their entrainment in the air flow. The rate of soil detachment by wind (DTW) can therefore be expressed by the following relationship:

$$DTW = a(u_g^* - u_{gcrit}^*)^b, \qquad (6.2)$$

where u_g^* = the grain shear velocity; u_{gcrit}^* = the critical grain shear velocity for soil particle entrainment; a = a coefficient dependent upon surface geometry, particle size, particle sorting and soil moisture; b = a coefficient also dependent upon soil characteristics but generally assigned a value of 2.0.

The grain shear velocity is that part of the total shear velocity which is dissipated on the soil particles and not on the microtopographic roughness of the surface, the vegetation or other moving grains. On a plane surface, the grain shear velocity can be considered equal to the total shear velocity, as calculated from equations 2.43 or 2.54. At present there are no standard ways of characterizing the way total shear velocity is split, so soil particle entrainment is normally considered in relation to total shear velocity. The rate of soil detachment or dislodgement rate of soil particles has been shown in wind tunnel experiments (Sørensen, 1985; Willetts and Rice, 1988) to follow the relationship:

$$DTW = a(u^* - u_t^*)^{2.8}. \qquad (6.3)$$

Bagnold (1941) evaluated the threshold shear velocity (u_t^*) for particle movement by balancing the drag force on the particle against the movement of the grain about its axes of support. He obtained the expression:

$$u_t^* = A[(\sigma - \rho_\alpha/\rho_\alpha)gd]^{0.5}, \qquad (6.4)$$

where A can be defined as the value of dimensionless shear stress at the velocity threshold:

$$A = \rho_\alpha u_t^{*2}/\rho_s gd, \qquad (6.5)$$

where ρ_s = the immersed sediment density $(\sigma - \rho_\alpha/\rho_\alpha)$.

For particle sizes $(d) \geqslant 0.2$ mm, $A = 0.01$ (Savat, 1982), which is lower than its equivalent for water erosion, calculated by Shields (1936) as equal to 0.05. For particles <0.2 mm in size, A increases in value so that when $d < 0.08$ mm, the value of u_t^* also starts to increase. This is normally explained in terms of the need to overcome the cohesive bonding of the finer particles to the soil mass. In reality, the value of A varies with the density ratio (ρ_α/ρ_s). Iverson (1985) gives values of 0.1 for large particles in air, a value also obtained by Bagnold (1941), rising to 0.2 in water when the density ratio approaches 1.0.

Equation 6.4 applies strictly to a bed of uniform grain size. For particles of mixed grain size, the threshold shear velocity varies with the proportion of erodible to non-erodible grains. Although non-erodible grains are normally defined as those dry stable aggregates and primary particles larger than 0.84 mm (Chepil, 1950), in reality the upper limit to the size of particles that can be moved is dependent on and increases with the wind speed. The lower limit is influenced by the cohesive forces mentioned above and by the protection afforded to small particles from surrounding coarser material.

Threshold velocity is also dependent upon the moisture content of the soil. Generally, the threshold velocity increases with greater soil moisture but a high moisture content may not always be sufficient to stop erosion or reduce it to low levels. De Ploey (1977) has observed that considerable movement of dune sands can occur during or just after heavy rainstorms, presumably because detachment of non-cohesive sand particles by the raindrop impact has created a loose and easily moved surface. Whether vegetation will enhance this process through detachment by leaf drainage has not been studied. Vegetation will definitely affect the moisture status of the soil depending on the relative importance of transpiration, which will tend to reduce the moisture content, and shade, which will reduce evaporation from the surface soil.

This equivocal role of vegetation is difficult to quantify and predict but its effects mean that no single threshold velocity exists for any given soil for field conditions.

Once a particle has been entrained in the air flow, it rises to a maximum height dependent upon the ratio of lift to drag forces. The ratio is about 0.75 at the point of particle entrainment close to the bed but decreases rapidly so that at a height above the bed equal to several grain diameters, lift becomes negligible. Particles obtain horizontal momentum from the aerodynamic drag exerted on them and are thus moved along in the flow until gravity forces cause them to descend. When the particle hits the ground, its momentum either causes it to rebound into the flow or is transferred to another particle which is then ejected into the flow. The movement of grains through a series of bouncing trajectories is known as saltation. This process accounts for between 55 and 70% of soil particle movement. Saltating particles rarely rise more than 1 m above the ground and most form a saltating layer no greater than 30–40 cm depth. Individual particle jump lengths are 30–40 cm but may be as much as 4 m. Height and length of movement vary with particle size, shear velocity of the wind and roughness of the ground surface. On falling to the ground, they have an impact angle of between 10° and 16°.

Once soil particles are in motion, their impact on the ground surface reduces the threshold velocity required to detach and entrain new soil material, at least as far as non-cohesive grains are concerned. As a result of this effect of bombardment and abrasion by moving soil material, the so-called impact threshold velocity is only about 70% of the fluid threshold velocity in value (Bagnold, 1941). Since, with a saltating soil surface, part of the shear stress is carried by the grains as the grain-borne shear stress (Bagnold, 1956), the momentum transferred to the soil by the fluid shear stress is reduced and wind speeds close to the soil surface are increased. Sørensen (1985) has modelled this effect in the wind tunnel to show that the wind

velocity profile in the presence of a saltating layer can be expressed by:

$$u(z) = 2.5\,u^*\ln(z/z_0) + \{\Phi T/(\rho_a k u^*)\}$$
$$\int_z^{z^*} \overline{\Delta x(z)/z}\ dz, \qquad (6.6)$$

where ΦT = the total grain dislodgement rate; z^* = the height where the concentration of sand grains is zero; $\Delta x(z)$ = the average horizontal increase in grain velocity.

The first term is a continuation into the saltating grain layer of the logarithmic velocity profile found in the grain-free air above, and the second term corrects for the effect of the grains. The addition to shear velocity as a result of the moving grains may be about 14% close to the ground, so that:

$$u^*_{sed} = u^* + 0.14\,u^*, \qquad (6.7)$$

where u^*_{sed} refers to the shear velocity with saltating grains, but this addition falls to about 2% at a height of 40 mm (Sørensen, 1985). Application of equation 6.7 to field data collected on sand dunes at Hantsholm, Denmark (Rasmussen, Sørensen and Willets, 1985) was not conclusive. Friction velocities were calculated from wind speeds measured with hot-wire anemometers which had been calibrated only for sediment-free air so the influence of moving sand grains on the measurements is not known. Furthermore the effect of variations in land slope on wind velocity profiles may be more important than the effects of saltating grains.

Wasson and Nanninga (1986) propose two approaches for considering the effect of vegetation. One is to reduce wind velocity as a function of the vegetation cover; the other is to increase the threshold velocity required for erosion. In reality, both effects need to be accounted for, the first as a function of the above-ground components of the vegetation and the second as a function of the root system. As seen in Chapter 2, the effect of vegetation on shear velocity is complex. Implicitly, under many circumstances a plant cover will reduce sediment entrainment, but there will be situations when the likelihood of entrainment is enhanced, particularly with low-growing bladed-leaved vegetation where the cover is not uniform. Since spatial variations in flow over rough surfaces can produce localized pockets of low pressure in which soil grains can be seen to oscillate wildly before being picked up and visually sucked into the flow, a clumpy vegetation of tufted grasses or scattered bushes and shrubs would be expected to have a similar effect and increase the rate of sediment entrainment. Blow-outs may be initiated in this way within gaps in the vegetation cover.

Unfortunately no studies exist of the interaction between vegetation and an airflow with saltating particles. Three situations may be envisaged, however. First, the vegetation may reduce the shear velocity below the threshold level for particle entrainment, and saltation will cease. Second, the vegetation may reduce the shear velocity but not sufficiently to overcome the higher velocities associated with the saltating layer; in this case, saltation will continue but be reduced. Third, the vegetation may increase the shear velocity and thereby enhance the effect of the saltating layer so that sediment entrainment increases. Research is required to investigate these possibilities.

6.4.2 SEDIMENT TRANSPORT

Many formulae have been developed to predict the rate of sediment transport by wind as a function of wind velocity above a threshold value. Bagnold (1941) calculated sediment discharge (Qs; t/m h) from the equation:

$$Qs = B(v - v_t)^3 \qquad (6.8)$$

and

$$B = [0.174/\log{(z/z_0)}]^3 C$$
$$[(d\rho_a)/(Dg)]^{0.5}, \qquad (6.9)$$

where v = wind velocity (m/s); v_t = threshold wind velocity (m/s); C = a sorting factor; d = average grain diameter of the material; D = a standard grain diameter of 0.25 mm.

This formula has been applied successfully by Finkel (1959) in studies of barchans in Peru,

Tsoar (1974) in Israel, and Wasson and Nanninga (1986) to vegetated dunes in Australia. It is the most widely used of all transport equations in wind erosion research and is one of the few validated by field measurements. Similar formulae have been developed by Chepil (1945), Zingg (1953) and Lettau and Lettau (1978).

The limitation of the equation is that it assumes that once the threshold velocity has been exceeded, sediment transport is at transport capacity. Recent work, reviewed by Nickling (1988), shows that where the supply of sediment is limited (for example, by trapping of sediment in surface irregularities produced by tillage or by frictional resistance through interlocking of angular or platy soil particles), the power exponent for velocity is lower than 3.0. Equation 6.8 is therefore best considered as an equation for transport capacity whilst, in reality, the actual transport rate depends upon the supply of particles available for transport. This, in turn, depends on the dislodgement or detachment rate and the transport length of individual soil particles.

For saltating particles, transport length depends upon particle size and shear velocity and, in open ground, appears to be in the range 75–375 mm (Sørensen, 1985; Willetts and Rice, 1988). By establishing simple empirical relationships between mean transport length (h) and shear velocity, the sediment flux can be modelled. Using wind tunnel data from the studies of Sørensen (1985) yields:

$$h = 0.0275 + 0.2824\,u^*, \qquad r = 0.9765, \quad (6.10)$$

where h is in m and u^* in m/s. Since the intercept term is not significantly different from zero, the relationship may be simplified and the sediment flux or transport rate (qs) estimated as:

$$qs = DTW \cdot 0.28\,u^*. \qquad (6.11)$$

Vegetation will influence mean transport length in two ways. First, as seen above, it will reduce the shear velocity of the flow. Second, it will physically intercept saltating particles and cause them to fall to the ground without completing their trajectory. The effective jump length will thus be considerably reduced as a function of the spacing of the vegetation. No studies have been made of this second effect.

6.5 PREDICTION OF WIND EROSION

Despite considerable attention paid to the physics of the wind erosion process, research on prediction of wind erosion is rather slight and the state-of-art lags well behind that for water erosion. The only method which can take account of vegetation is the USDA Wind Erosion Equation (Woodruff and Siddoway, 1965), an empirical technique developed as a wind erosion counterpart to the Universal Soil Loss Equation used for prediction of water erosion. Like the water erosion equation, the method multiplies together values of a number of factors that influence the rate of erosion. These are soil erodibility, soil roughness, climate, width of open wind blow and vegetation cover. Factor relationships are more complex, however, because interactions between the factors are allowed for. Complicated charts and equations are required to make the prediction.

Vegetation is accounted for in the equation by using the graph shown in Figure 6.9 to convert soil loss predicted without a vegetation cover to that with a cover. In this approach, the above-ground biomass is the only property of the vegetation which is considered. This is clearly an inadequate treatment of the role of vegetation and does not provide a satisfactory starting point for developing process-based prediction models. The scheme presented in Figure 6.10, based on the material reviewed above and in Chapter 2, allows for the effects of projected area of the foliage facing the wind, leaf area density and average orientation of the leaves, in addition to that of biomass. The scheme is a modified version of that proposed by Morgan (1990) and uses many of the equations presented above. It simulates wind erosion over a succession of individual crop or plant rows aligned at right angles to the wind. Erosion is considered as a two-phase process of detachment of soil

particles from the soil mass and their transport downwind. The rate of soil loss is controlled either by the sediment flux (product of detachment rate and mean transport length) or by the transport capacity, whichever is the limiting factor. Transported sediment is routed downwind from one plant row to the next, where it is added to the sediment flux in that row to determine the amount of soil available for further transport.

Table 6.1 shows an example of the use of the model to predict soil loss from a 100 m long field of young (5 cm tall) sugar beet on a sandy loam soil tilled to a very fine seed-bed subjected to a wind velocity of 11 m/s at a height of 1 m above the ground. The wind is assumed to be non-turbulent (turbulence index, $TU = 0.17$, as defined in section 2.3.6) to give the greatest risk of wind erosion. Comparisons are made for bare soil, sugar beet alone, and sugar beet with an in-field shelter system of 0.15 m tall live barley strips. Since conditions stabilize downwind, results are tabulated only for the first 9 m of the field.

For the bare soil, erosion is found to be detachment-limited initially and increases with distance to reach 152.2 g/cm s after 9 m. Extrapolation of this rate means that transport capacity is reached after 40.9 m beyond which soil loss remains constant at 691.25 g/cm s. With the young sugar beet, erosion is transport-limited in the rows of beet and detachment-limited in the intervening rows of bare soil. Soil particles detached on the bare soil are therefore mostly deposited within the sugar beet. Assuming that the last row in the field is occupied by beet, the soil loss is 6.32 g/cm s. The combined barley strips and sugar beet system results in erosion being transport-limited in the barley strips at 0.82 g/cm s but detachment-limited in the sugar beet rows where an additional 0.2 g/cm s of sediment is added to the air flow, to give a sediment flux of 0.84 g/cm s, only to be deposited in the next barley strip. Thus, if the last row in the field is a barley strip, the erosion rate is 0.82 g/cm s.

If the soil loss rates are multiplied by 10 000

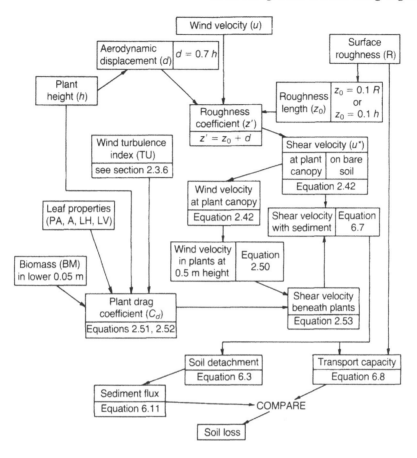

Figure 6.10 Flow chart of a simple wind erosion model incorporating the effects of vegetation (Modified from Morgan, 1990).

to convert values per centimetre-width to a hectare for the 100 m long field, and the wind velocity of 11 m/s is maintained continuously for 10 min, predicted erosion rates are 4140 t/ha on bare soil, 37.9 t/ha with the young sugar beet and 2.7 t/ha for the combined beet and barley. Although these rates do not seem unreasonable for such a strong wind over a fine seed-bed, it is not possible to validate them. The lack of suitable field data at present inhibits the validation of any wind-erosion prediction procedure. The results of the model demonstrate, however, that considerable reductions in erosion can be achieved, even on a highly erodible soil, with a plant cover.

6.6 CONTROLLING WIND EROSION USING VEGETATION

The earliest attempts to control wind erosion were based on erecting inert structures as windbreaks. In 1753, the Reverend Björn Halldórsson attempted to stop erosion and protect his land at Sauðlauksdalur in Iceland from drifting sand by building a large stone wall as boundary shelter (Arnalds, A., 1988b). The remains of this wall still exist with the sand built up against it on the windward side.

The first engineering work using a combination of inert structures and vegetation to control wind erosion dates to 1778 when the French

Table 6.1 Example of soil loss prediction using a simple wind erosion model (modified from Morgan, 1990)

Distance downwind (cm)	Soil detachment (g/cm s)	Sediment flux (g/cm s)	Transport capacity (g/cm s)	Soil loss (g/cm s)
Bare soil (smooth seed bed)				
30	6.57	5.07	691.10	5.07
60	6.57	5.07	691.25	10.15
90	6.57	5.07	691.25	15.22
120	6.57	5.07	691.25	20.29
810	6.57	5.07	691.25	136.98
840	6.57	5.07	691.25	142.05
870	6.57	5.07	691.25	147.13
900	6.57	5.07	691.25	152.20
Young sugar beet (0.05 m tall in rows perpendicular to wind)				
30	6.57	5.07	691.10	5.07
60	0.05	0.02	6.32	5.10
90	0.05	0.02	6.32	5.12
120	6.57	5.07	691.25	10.19
150	0.05	0.02	6.32	6.32
810	0.05	0.02	6.32	6.32
840	0.05	0.02	6.32	6.32
870	6.57	5.07	691.25	11.39
900	0.05	0.02	6.32	6.32
Young sugar beet with barley strips (sugar beet as above, barley strips 0.15 m tall grown in intervening rows)				
30	0.00	0.00	0.82	0.00
60	0.05	0.02	6.32	0.03
90	0.05	0.02	6.32	0.05
120	0.00	0.00	0.82	0.05
150	0.05	0.02	6.32	0.08
810	0.05	0.02	6.32	0.40
840	0.05	0.02	6.32	0.43
870	0.00	0.00	0.82	0.43
900	0.05	0.02	6.32	0.45

Parameter values used in the simulations: downwind length of crop row or bare soil segment (L) = 30 m; wind velocity (u) = 11m/s at 1 m height; soil surface roughness (R) = 0.000 15 m; wind turbulence index (TU) = 0.17; plant height (h) = 0.05 m for sugar beet and 0.15 m for barley strips; projected foliage area (PA) = 0.001 35 m^2 for sugar beet and 0.0044 m^2 for barley; leaf area density (A) = 2.7 m^2/m^3 for sugar beet and 8.8 m^2/m^3 for barley; across-wind leaf alignment (LV) = 45° for sugar beet and 50° for barley; downwind leaf alignment (LH) = 40° for sugar beet and 15° for barley; biomass (BM) = 0.399 kg DM/m^3 for sugar beet and 0.590 kg DM/m^3 for barley.

government sent an engineer, Baron de Villiers, to Gascogne to propose a method of stabilizing sand dunes. The system he developed was later perfected by François Bremontier and is still the basic system used today. First, fences, about 1 m high, are built on the windward side of the dunes by driving wooden stakes into the sand and linking them together with branches. Sand builds up around the fences and when the mound reaches 0.50–0.75 m high, a second fence is built on top of the first. When the height and slope of the dune are high enough to prevent sand from passing over it, the sands behind the barrier are fixed by planting grasses and pine trees.

The earliest organized programme using shelterbelts to reduce wind erosion was carried out in the latter part of the last century in western Jylland by the Danish Heath Society, following its foundation in 1866. Responsibility for this work has now passed to the Shelter Belt Division of Hedeselskabet (Danish Land Development Service) which helps to renovate and maintain some 55 000 km of shelterbelts throughout Denmark.

Despite much research and experience over the last 150–200 years, our knowledge of the properties of vegetation which are most important for wind erosion control is still very limited. The choice of vegetation species is obviously dominated by its ability to survive in the local climatic environment, its resistance to wind damage, tolerance to light and rapid growth rate. Less consideration is given to the plant morphology. Research on shelterbelts has emphasized that height and porosity are important design factors (Caborn, 1965; Seginer and Sagi, 1972; Skidmore and Hagen, 1977), whilst Marshall (1971) has shown the significance of the diameter-to-height ratio of the individual plant elements in a vegetation stand. Other properties have not been studied in detail (Heisler and De Walle, 1988). The above review, however, points to the importance of total biomass, the spatial uniformity of cover, projected area of foliage facing the wind, and the shape and orientation of the leaves. In addition, the ability of the vegetation to transpire will influence the moisture status of the soil and, therefore, its susceptibility to wind erosion. Thus, the E_t/E_o ratio is another salient property. Table 6.2 lists these various properties and describes their effects.

Table 6.2 Salient properties of vegetation and their role in controlling wind erosion

Vegetation property	Effect on factors affecting erosion				Methods of erosion control		
	Drag coefficient/ shear velocity	Soil moisture	Soil strength	Sedimentation	Boundary shelter	Infield shelter	Vegetation stands
Biomass	x			x	x	x	x
Projected area facing the wind	x			x	x	x	
Leaf area density/porosity	x			x	x	x	x
Leaf orientation	x						
Leaf shape and rigidity	x			x	x	x	x
Plant height					x	x	x
Plant diameter/height ratio							x
Uniformity of cover					x	x	x
Anchorage					x		
E_t/E_o ratio		x					x
Root density and strength			x				x

6.6.1   BOUNDARY SHELTERBELTS

The control of wind erosion by a shelterbelt can be appreciated by considering the effect of a barrier at right angles to the wind on the pattern of air flow. As shown in Figure 6.11 (Plate, 1971), the flow is broken up into a number of zones. Zone 1 represents the undisturbed boundary layer. In zone 2, the flow is displaced and distorted by the barrier; the lower boundary of this zone starts at the point of flow separation at the apex of the barrier, where wind velocity is locally increased, and ends at some distance downwind where the flow merges into zone 3 and the original air flow pattern is re-established. The upper boundary of zone 2 extends to a height of about four times the height of the barrier. Sometimes, a zone of backflow (zone 4) is established beneath zone 2 as a result of the setting-up of negative air pressure immediately downwind of the barrier (zone 5); this zone is most marked in the presence of solid barriers. Negative pressure may extend downwind for a distance of about three times the barrier height but then pressure rises rapidly to slightly above air stream pressure at the point where the boundary-layer air flow is re-established.

The effectiveness of a shelterbelt is measured by the distance downwind at which wind velocity remains less than 80% of the open wind speed at the same height. This distance is dependent upon the height, width, length and shape of the barrier and the resilience and porosity of the vegetation comprising it.

Wind tunnel studies of tree belts at right angles to the wind show that they should afford protection for a distance (L) equal to 17 times their height for open wind velocities, measured at 10 m above the ground, up to 44 km/h (Woodruff and Zingg, 1952). In practice, variability in growth and maintenance of the belt means that the distance protected downwind often does not exceed 12 times the belt height for trees, although it may reach 24 times belt height for 2–5 m tall hedges and thin solid fences. This latter distance may reduce to 17.5 times belt height in unstable air (Jacobs, 1984). The absolute distance protected is of course greater for trees than for hedges because of their greater height.

Bates (1924) showed that the distance protected is also related to the length of the barrier. This is because, with short barriers. the area protected is triangular in plan but becomes semi-circular with barrier lengths in excess of 12 times barrier height, if the barrier is at right angles to the wind, or 24 times the barrier height, if the wind direction is 45° to the belt.

The distance downwind over which protection is afforded is also affected by the degree of porosity or openness of the belt (Caborn, 1965; Figure 6.12). Optimum protection is achieved with porosity levels of 40–50%. More open belts

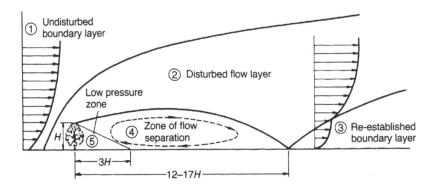

Figure 6.11 Patterns of air flow around a shelterbelt (after Plate, 1971).

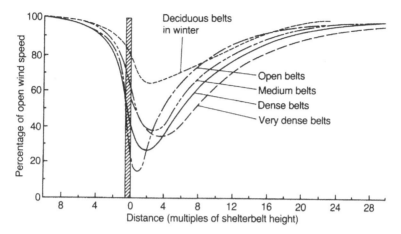

Figure 6.12 Percentage reduction in wind speed for shelterbelts of different densities (after Caborn, 1965).

give greater velocity reductions but these occur immediately downwind and last for only about nine times the barrier height. Dense belts give protection over a greater distance but with very limited reductions in wind velocity. They also give rise to eddying in the lee of the barrier, so overall they are not very effective.

Since between 50 and 72% of the soil moved by the wind is carried in saltation within 1 m of the ground surface, best protection is provided if the barrier gives vegetation cover on and close to the ground. In addition, the most effective belt is one that rises rapidly in height on the windward side. Generally, great width to a belt is not necessary and, indeed, may give rise to belts of too great a density. Tanner and Naegeli (1947) propose that belt width should vary with the height of the barrier from 2.5 m for 5 m tall belts to 10–15 m for belts that are 25 m tall.

The reduction in wind velocity brought about by a shelterbelt changes the micro-climate (Figure 6.13; Marshall, 1967) so that, in the lee of the belt, there is less evaporation, higher soil moisture, lower soil temperature in summer and higher soil temperature in winter. Except close to the belt, where shade occurs, these effects often increase vegetative growth and give higher crop yields. In warm and cool temperate climates, the higher soil

temperatures in winter serve to increase the length of the growing season but in cold climates, the lower soil temperatures in summer may shorten the growing season.

When selecting vegetation species for shelterbelts it is important to keep in mind both the engineering requirements and the opportunity that is provided to create a valuable ecological resource. The tree species selected must be capable of producing a barrier of appropriate porosity and straightness at all heights, tolerant of wind and resistant to windthrow (see Chapter 2), which means having pliable branches that do not break off in strong winds and deep widespread roots to provide firm anchorage. The trees must also have a long life and be easy to maintain.

The shelterbelt planting programme of the Danish Land Development Service (Olesen, 1979; Als, 1989) aims at meeting both engineering and ecological needs. A special type of three-row hedge is used comprising a mixture of tree and bush species adapted to local soil and climatic conditions and chosen to give fast growth and long life (Figure 6.14). The rows are 1.25–1.5 m apart, giving a total belt width of 3–4 m. When mature, the belt grows to about 10 m in height. Genetic variability produces a belt that is less vulnerable to pests and diseases.

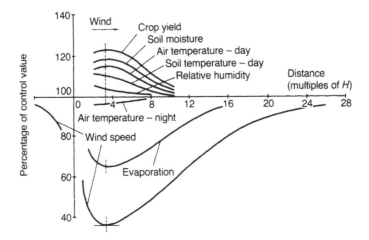

Figure 6.13 Effect of shelterbelts on microclimate (after Marshall, 1967).

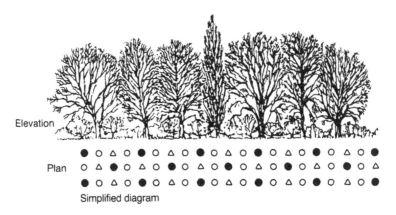

Figure 6.14 Danish shelterbelt system. ●, Permanent shelter trees; △, nurse trees; ○, bushes (after Olesen, 1979).

The hedge belt is made up of three groups of plants, designed to provide shelter in tiers and to replace each other in a plant succession:

1. Fast-growing nurse trees are used to give acceptable shelter within four to five years and thereby encourage the growth of the durable trees.
2. Durable trees are tall and long-living species which will grow up with, but more slowly than, the nurse trees and eventually replace them when they are either felled or die out. These trees will give the belt its life of 80–100 years or more.

3. Shade-tolerant and flowering bushes provide the undergrowth which will trap saltating grains and, through the litter they yield, provide a mulch to aid weed control.

On the sandy soils of western Denmark, grey alder (*Alnus incana*) is the most common nurse tree, and oaks (*Quercus robur* and *Quercus petraea*), Norway maple (*Acer platanoides*) and rowan (*Sorbus aucuparia*) are the most common durable trees. Bush species include cherry rum (*Prunus serotina*), snowy mespilus (*Amelanchier spicata*) and lilac (*Syringa vulgaris*). A mixture of deciduous species is used to give a belt that is less vulnerable to pests and disease and also more attractive. Coniferous species are not recommended because they are often too dense and are more subject to toppling when they become mature.

Successful establishment of the belt depends upon proper clearance of the old hedges, if any, followed by mechanical soil preparation and fertilization to give a suitable environment for plant growth. Plants are carefully selected to ensure disease-free stock. After planting, the ground must be kept weed-free, either mechanically or chemically, for at least two and a half growing seasons to encourage vigorous healthy growth of the nurse trees and bushes. Nurse trees are pollarded after a few years so that they do not inhibit the growth of the durable trees. The sides of the belt are cut mechanically every four or five years, to limit its width to about 5 m. The belt must be protected against spray-drift, fire, livestock and damage by machinery.

The effectiveness of a shelterbelt can be evaluated using a simple point scoring system relating its alignment to the frequency and direction of erosive winds. Available wind data should be examined for the 16 compass directions. A threshold value of wind speed should be selected as representing the minimum velocity of erosive winds. For wind speeds measured at a height of 10 m, a value of 34 km/h is suitable. Knowing the frequency of erosive winds from each direction, the effectiveness of the belt is scored as follows:

1. One point is awarded for every period of effective protection, i.e. wind direction at 90° to the belt.
2. A half-point is awarded for each period of substantial protection, i.e. wind direction between 60° and 90° to the belt.
3. No points are awarded for each period of indifferent protection, i.e. wind direction between 30° and 60° to the belt.
4. A half-point is subtracted for each period of poor protection, i.e. wind direction between 0 and 30° to the belt.

This scheme can be used to compare different layouts of shelterbelts. Where erosive winds come from several directions, maximum effectiveness may be achieved by providing substantial protection from all directions rather than full protection from any single direction. Often, however, the optimum layout cannot be obtained because of the constraints imposed by the need to place shelterbelts along existing field or property boundaries. Experience in Denmark, however, shows that 10-m-high hedge belts placed every 200–400 m over distances of 10–20 km can reduce regional wind velocities by 50% (Jensen, 1954). With 'effective' spacings of 20–40 instead of 17–24 times belt height, the combined effect of a coordinated layout of shelterbelts in the landscape is thus much greater than the summation of the effects of the individual belts.

Single-row shelter belts are recommended by the Agricultural Development and Advisory Service of the UK Ministry of Agriculture, Fisheries and Food, for controlling wind erosion and providing shelter on the peat soils of the Fens. The Bowle's hybrid variety of willow (*Salix vinimalis*), propagated from 75-cm-long rods pushed into the soil at 0.5 m spacing, produces an effective shelterbelt of 50% porosity when out of leaf, within five years, provided that the shoots are cut back to within 15 cm of their origin in either the second or third year to promote growth close to the ground (Wickens, 1981). Subsequent maintenance should be by coppicing to produce an A-shaped cross-section and to restrict the height of the belt to about 5 m. This restriction in height is necessary on these soils because the high groundwater table prevents deep root development and anchorage of the trees which are then subject to wind-throw. Other suitable species for single-row shelterbelts on these soils are the Italian alder (*Alnus cordata*), black alder (*Alnus glutinosa*) and grey alder (*Alnus incana*).

The planting of regional networks of shelter-belts is also recommended for wind erosion control in the lowlands of Iceland (Iceland Forestry Service, 1986) to improve conditions for livestock and hay production. Single-row belts of black willow (*Salix nigricans*) are used but experiments are now being carried out on double-row belts.

6.6.2 IN-FIELD SHELTER

In-field shelter systems are used in agriculture to control wind erosion whereby row crops and soil-protecting crops are grown in alternating strips aligned at right angles to the erosive winds. Similar systems could be used in non-agricultural situations. In-field shelter may either supplement or replace boundary shelter provided by trees and hedges. Based on studies of the effect of single-crop rows on wind velocities (Morgan and Finney, 1987; Morgan, Finney and Williams, 1988), barriers used for in-field shelter should satisfy the following design requirements (Morgan, 1989):

1. Spacing: the barriers should be spaced at distances of eight to ten times their height.
2. Biomass: each barrier should contain at least 0.65 kg DM/m³ in the lowest 5 cm.
3. Leaf shape: the plants should have small bladed leaves rather than ovate leaves. This minimizes streamlining of the foliage downwind in high wind speeds.
4. Leaf alignment: the leaves should be aligned full-face to the wind. This reduces the contact length between the foliage and the air and thereby minimizes the 'wall effect' (section 2.3.6).
5. Width: the barrier should be restricted to one or two plant rows. This also reduces the contact length between the foliage and the air flow.

These design requirements are best met in most agricultural situations by growing species of barley with an upright form, in alternating strips with the main crop. On the peat soils of the Fens, in England, barley strips are used by some farmers to protect both the soil from erosion and the young crops of sugar beet and onions from wind damage. The barley is sown in February or March, allowing time for it to emerge before the main crop is drilled in mid

to late April. Once the main crop has sufficient biomass to perform a protective role, the barley is killed with a selective herbicide. Care is required with the amount and timing of the herbicide applied to ensure that the barley is effectively removed without destroying or stunting the growth of the main crop.

Although barley strips can be used to prevent or reduce wind erosion in non-agricultural areas (e.g. on industrial or waste sites), grass species with relatively rigid leaves, an upright habit and which are capable of giving a uniform, non-tussocky cover up to 0.5 m tall, may be more suitable because they will provide longer-term protection.

6.6.3 VEGETATION STANDS

The best protection against wind erosion is provided by a dense uniform stand of ground vegetation covering at least 70% of the soil surface. Based on the work of Marshall (1971), the vegetation should mimic a dense network of equidistantly spaced cylinders. A simple empirical equation (Lyles, Schrandt and Schmeidler, 1974) may be used as a basis for design, setting a value of 5 for the critical friction velocity ratio below which no significant wind erosion will occur:

$$u^*/u_t^* = 1.32 + 3.52(H/L), \quad (6.12)$$

where $u_t^* =$ the threshold value of shear velocity for erosion to begin; $u^*/u_t^* =$ the critical friction velocity ratio; $H =$ average height of the vegetation elements; $L =$ average distance between the vegetation elements in a downwind direction.

Alternatively, design may be based on achieving a vegetation cover with a minimum value of 0.0104 for the bulk drag coefficient (*CD*), the value proposed by Lyles, Schrandt and Schmeidler (1974) above which no regional-scale wind erosion will take place.

The required vegetation conditions demand a uniformly distributed plant cover of relatively rigid standing stems or stalks. In practice, this can be difficult to achieve. Most agricultural crops are seldom uniformly distributed over the soil surface and many have either a prostrate habit or a tendency to lie flat in strong winds. Grasses that grow in semi-arid climates or on sand dunes, and which may therefore be suitable environmentally for areas prone to wind erosion, are often of bunch or tussocky type, or they spread horizontally and have limited vertical growth.

Cover crops are used in agriculture to produce temporary vegetation stands to protect the soil during the off-season. One method, developed in The Netherlands to control wind erosion on sandy and peaty soils, is known as the Dutch rye system. When either potatoes or sugar beet are grown, winter rye is sown in the previous November to provide cover over the winter and, in particular, the spring when the risk of wind erosion is greatest. The rye is either killed just before the sugar beet is sown, or potatoes are planted in the rye and the rye is killed as soon as the potato plants emerge. Barley would be a suitable alternative cover crop to rye and has the advantage of more rapid germination and emergence.

By careful selection of plant species for their engineering role and their ability to survive in the local environment, ecological successions can be planned and implemented to provide a permanent vegetation stand. Using these principles, Landgraeðsla Ríkisins Íslands (State Soil Conservation Service of Iceland)

has developed strategies to reclaim degraded land for productive use and control wind erosion.

Basis for an ecological succession

Guidance on determining a suitable ecological succession comes from two sources. First, the susceptibility of plant communities in Iceland to erosion was studied by Arnalds, O., (1984) in the Biskupstungnaafréttur area by estimating canopy cover in 0.25 m^2 ring frames. The land was assigned to one of six classes of susceptibility to erosion based on the following criteria:

S-1. No erosion; up to 100% vegetation cover.

S-2. 90–100% cover; some soil exposed; little erosion.

S-3. 80–90% cover; exposed soil areas easily seen; abrasion by airborne materials affects composition and vigour of the vegetation; some erosion.

S-4. 40–80% cover; rofabarði common; vegetation degraded in vigour and composition; considerable erosion.

S-5. 20–40% cover; rofabarði very common; vegetation is weak due to abrasion and burial; intensive erosion.

S-6. 0–20% cover; only isolated spots of soil and vegetation remain.

Table 6.3 shows the percentage distribution of

Table 6.3 Vegetation in the Biskupstungnaafréttur, Iceland, divided according to plant communities and erosion classes (%) (after Arnalds, O. 1984)

S-class	Moss heath	Dwarf shrub heath	Rush heath	Sedge heath	Grassland	Snow-patch	Primary succession	Semi-bog	Bog	Fen
1	17.8	1.9	–	–	0.4	3.3	0.8	1.8	0.2	0.8
2	5.1	4.9	1.5	0.1	1.2	4.7	0.8	0.9	11.9	2.8
3	0.5	14.2	1.4	1.1	1.9	1.9	2.2	0.2	1.8	0.9
4	–	6.5	0.8	0.6	0.1	0.2	–	–	0.2	–
5	–	3.0	–	–	0.7	–	–	–	–	–
6	–	0.7	–	–	<0.1	–	–	–	–	–
Total	23.4	31.2	3.7	1.8	4.3	10.1	3.8	3.0	14.1	4.5

each plant community by S-classes. The dwarf shrub heath community is the most susceptible to erosion with 78% of its occurrence falling in S-classes 3 to 6. The main species are *Betula nana*, *Empetrum hermaphoditum* and *Vaccinium uliginosum* and these are among the first species to disappear under abrasion by wind-blown material. Rush heaths and grasslands are also susceptible to erosion. Wetlands contain the most stable plant communities because the high groundwater table prevents wind erosion. The very shallow soils associated with snow-patches and moss heaths also resist erosion although *Rhacomitrium* moss withstands abrasion very poorly. The most resistant vegetation species are *Salix lanata*, *Elymus arenarius* (Lyme grass) and *Cerastium alpinum*, although the last of these does not have a widespread distribution (Arnalds, O., 1981).

The second source of information comes from observations of the natural plant succession on denuded lands. *Elymus arenarius* is generally the first plant to take hold, followed by mosses and then, in turn, by *Festuca rubra* (creeping red fescue), *Agrostis tenuis*, *Poa alpina* and *Poa glauca*. Ultimately, these grasses will be succeeded by willows (*Salix* spp.) and low growing birch (*Betula* spp.) but the rate of regeneration is highly variable and it may take decades or even centuries to reach this stage. However, as natural seed banks form and growing conditions improve, succession rates assume a logarithmic pattern (Gunnlaugsdóttir, 1985). This implies that if some assistance can be provided to enhance the early stages of succession, restoration may be achieved more rapidly. However, since the soils are deficient in organic matter, nitrogen, available phosphorus and clay particles, as well as having low water-holding capacity (Arnalds, O., Aradóttir and Thorsteinsson, 1987), it is generally not feasible to improve the soils fast enough to speed up succession to the late seral stages. The soil conservation work is therefore concentrated on accelerating the early stages of succession and then allowing the later stages to occur naturally.

Establishing the first phase

The first step in reclaiming eroded land is to sow *Elymus arenarius* to stabilize drifting sand. It performs the engineering functions of reducing wind velocity, trapping moving sand and increasing soil cohesion through root reinforcement. It is usually the first plant to take hold naturally on sand dunes in Iceland and it requires regular additions of loose sand around its roots to survive. *Elymus* does not produce a continuous vegetation cover but sand collects around the shoots of the individual plants to form small heaps or mounds. The shoots are renewed continually after burial and the mounds are bound together by the buried stems of the plant (Figure 6.15). Although erosion occurs in the craters between the tufts of *Elymus*, the material is not carried far and, in time, new mounds are formed. *Elymus arenarius* cannot withstand much pressure from grazing or cutting, so fencing is required to enclose areas being reclaimed. *Elymus* is usually seeded in strips at right angles to the erosive winds and fertilized annually until the most serious sand movement has been halted (Runólfsson, 1987). Where erosion is severe, boundary shelter needs to be provided in the form of timber or stone walls. *Elymus* then survives first in the lee of the walls and then spreads downwind in its own shelter, building and stabilizing its own dunes. Until most sand movement has ceased, it is not possible to establish the next stage in the plant succession.

Managing the second phase

The Landgraeðsla Ríkisins (State Soil Conservation Service) and the Rannsóknastofnun Landbúnaðarins (Agricultural Research Institute) have carried out numerous trials to determine the most suitable grass species for the next stage (Friðriksson, 1960, 1969a,b, 1971; Arnalds, A. *et al.*, 1978). The major engineering requirements of plant species for this second stage in the succession are to give dense uniform ground cover greater than 70% and with a biomass of

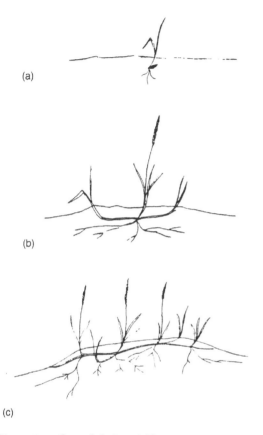

(a)

(b)

(c)

Figure 6.15 Growth habit of *Elymus arenarius*. The plant traps moving sand and so builds its own dune. The shoots grow from creeping stems and the roots reinforce and bind the sand (after Guðmundsson and Pálsson, 1990).

0.65 kg DM/m^3 or more in the lowest 5 cm, to maximize the roughness imparted to the air flow through rigid and erect foliage, and to produce a dense root mat which will increase water retention in the soil. An improved soil-water regime together with increasing organic content of the soil as the plant material decays will provide a suitable environment for native plant species to invade and allow the succession to progress to the next stage. These requirements mean that grasses with long-bladed leaves and laterally spreading mat-root systems are preferred. In addition, the species must be able to withstand the cold winter period and be able to complete their annual life cycle within a short

cool growing season; criteria which become important above 400 m altitude and usually limiting above 700 m. Other prerequisites may include the need to increase forage production, particularly below 400 m altitude where the grass is cut for hay, and to promote an aesthetically pleasing plant community. This last objective often creates a conflict because the seed bank of native species is very small.

Festuca rubra (red fescue) is one of the most widely used species for reseeding and it is commonly found in Icelandic hayfields. It is densely tufted with angular stems which grow to 50–55 cm tall and narrow-bladed leaves growing to 25–30 cm long (Figure 6.16). The plant spreads by creeping shoots or rhizomes and will produce a thick, dense cover. Trials with a native strain in central Iceland where the grass was fertilized annually for six years with the equivalent of 350 kg of ammonium nitrate, 300 kg of triple superphosphate and 100 kg of potassium chloride per hectare showed that 80% ground cover could be obtained in that time at altitudes up to 650 m (Figure 6.17; Friðriksson, 1971). Trials in east central Iceland at altitudes of 580–660 m produced 77% cover after two years and 90% cover after three (Friðriksson, 1969b).

Other commonly used species are *Poa pratensis* and *Phleum pratense* (Figure 6.16). *Poa pratensis* (smooth meadow grass) is a loosely-tufted grass with stems growing in the lowlands of Iceland to 40–45 cm tall and leaves to 25–35 cm long. It does not produce a uniform ground cover because of its tufty habit and suffers from winter kill at high altitudes. It does not seem as good as *Festuca rubra* because of a high variability in performance between different cultivars. *Phleum pratense* (timothy) is a tall upright plant with stems growing to 50–65 cm tall and leaves to 20–35 cm long. It is capable of producing dense leaf growth but covers the ground less rapidly than *Festuca rubra*. In the trials in east central Iceland it achieved 65% ground cover after three years compared with 90% for *Festuca rubra* (Friðriksson, 1969b). Again different cultivars vary in their hardiness but at low elevations it produces more biomass

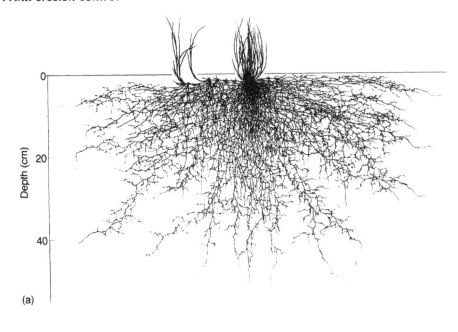

(a)

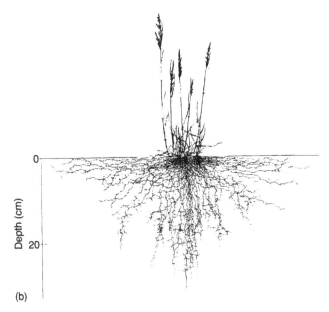

(b)

Figure 6.16 Growth habits of (a) *Festuca rubra*; (b) *Poa pratensis* and (c) *Phleum pratense* (after Kutschera and Lichtenegger, 1982).

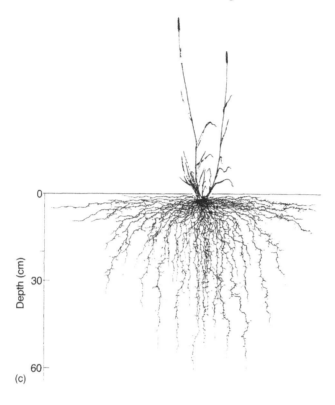

(c)

Figure 6.16 (cont.)

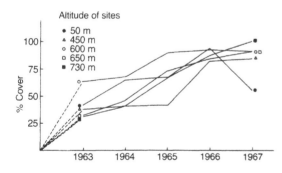

Figure 6.17 Trend in percentage cover of *Festuca rubra* at various sites in central Iceland following fertilization (after Friðriksson, 1971).

than other grasses. It performs poorly at high altitudes.

All three grass species meet the requirements of rigid-bladed leaves. *Festuca rubra* and *Phleum pratense* meet the requirement of uni-formity of cover but only *Festuca rubra* meets the percentage cover criterion. None of the grasses meets the required level of biomass. Typical hay yields for all three grass species with annual fertilizer application are about 30 kg DM/ha (Friðriksson, 1969b). Since this represents about a 15% take, total biomass production is around 200 kg DM/ha. Assuming that one-quarter of this is contained in the lowest 5 cm, this gives a biomass of about 0.1 kg DM/m^3 which is below the 0.65 kg DM/m^3 in the lowest 5 cm proposed by Morgan (1989) for an effective cover. For comparison, the equivalent figure for natural Icelandic range-land is about 0.45 kg DM/m^3 and for temperate grasslands is 2.5 kg DM/m^3 (Thorsteinsson, 1980). These figures emphasize the fragility of Icelandic ecosystems and indicate why slight disturbance of the vegetation cover can quickly lead to erosion.

Since seeds for all three species need to be imported, research is continuing for an alternative pioneer species. Trials are now being undertaken with *Lupinus nootkatensis* (Alaskan lupin). This legume seems to be well-adapted to Icelandic conditions and gives similar biomass yields to the grasses (Arnalds, A., 1988c). It fixes nitrogen which is then available to later species in the succession.

The seeding for this second stage of succession is carried out by air using one of two aircraft, a Piper Brave with an 800-kg payload, and a converted Douglas DC-3, a gift from Icelandic Air, with a 4000 kg payload. The pilots of Icelandic Air fly the aeroplanes free of charge, indicating their interest and commitment to the work of soil conservation. The procedure is to apply the seeds along with the fertilizer in the first year and then to provide fertilizer annually for two further years. Generally, annual fertilizer rates are 70 kg nitrogen and 70 kg phosphate per hectare (Runólfsson 1978) although for the Blanda revegetation programme the annual application was 400 kg/ha of 23-23 N-P205 (Arnalds, O., Aradóttir and Thorsteinsson, 1987).

The final phase

After three years, the land is left for the native species to invade. The seeded grasses will survive only a further two or three years without the regular supply of fertilizer but, in addition to the improvements they give to the growing conditions, they trap wind-dispersed seeds and therefore aid the development of a seed bank for local species. Friðriksson and Pálsson (1970) observed the changes in plant composition that occurred in central Iceland, east of the Thjórsa River, on fertilized and reseeded plots where the *Festuca rubra* failed. Native species quickly colonized the fertilized ground, however, so that after five years, on plots at 640 m altitude, total ground cover had increased from 4.8 to 70.3%, of which broad-leaved plants accounted for 46.0%, mosses and algae 12.8%, grasses (*Agrostis stolonifera* and *Festuca rubra*) 10.4%

and winter-killed plants 1.1%. Studies to the west of the Thjórsa River (Guðjónsson, 1980) showed that protection of the land from grazing led to decreases in the area under mosses, succession grasses (*Agrostis* spp.) and sedge heath. In areas where the land has been protected for more than 20 years, willows (*Salix* spp.) and low-growing birches (*Betula* spp.) are coming into the succession (Bjarnason, 1971), indicating that, in time, the land can be successfully restored to a climax botanical composition.

An alternative conclusion

With the recent decline in sheep numbers because of limited markets for the high-cost lamb production, it is now possible to look to afforestation as an alternative to restoration of the land for grazing. This means that instead of waiting for trees to invade the land naturally, saplings of native birch species (*Betula pubescens*) and Siberian larch (*Larix sibrica*) can be planted. This is more costly but it produces a tree cover more rapidly. Birch cuttings, 20-30 cm long, are notch-planted with 2-3 cm protruding above the soil (Guðmundsson and Pálsson, 1990). Afforestation is particularly valuable where recreational use of the land is important because of the aesthetic appeal of the traditional birch woodland.

Achievements

Between its formation in 1907 and 1986, the Landgraeðsla Ríkisins has carried out soil conservation and dune stabilization on 239 015 ha or 2.4% of the land area of the country (Figure 6.18; Arnalds, A., 1988d). Some 70% of this restoration work has taken place since 1952 and 38% since 1974, an increasing rate of activity which reflects the gradual development of the technology and ecological principles outlined above. Some 1260 km of fencing has been erected to enclose the protected land. Present-day fencing costs are about 200 000 Iskr/km (*c.* £1905 per km). The cost of the land reclamation varies with the complexity of the terrain but

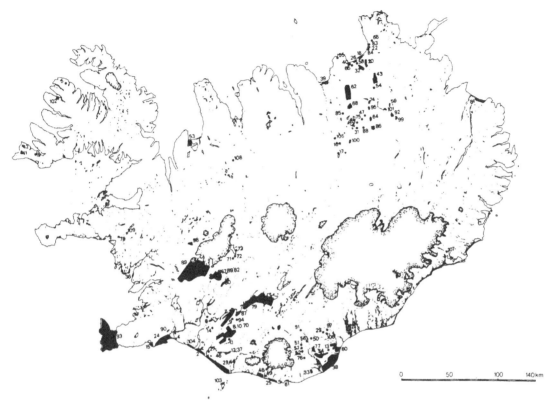

Figure 6.18 Location of soil conservation and dune stabilization sites in Iceland, fenced to exclude livestock from seedlings of *Elymus* and *Festuca* (after Arnalds, A. 1988d).

is typically between 30 000 and 45 000 Iskr/ha (£285 and £429 per ha) (Pálsson, 1989; Runólfsson and Sigurðsson, 1990). These costs are met entirely from the government which gives the Service an annual budget of about 160 million Iskr (£1.52 million).

REFERENCES

Als, C. (1989) How to succeed in planting 900 km of shelterbelts per year in a small country like Denmark? in *Soil Erosion Protection Measures in Europe* (eds U. Schwertmann, R. J. Rickson and K. Auerswald). Soil Technology Series 1. Catena Verlag, Cremlingen-Destedt, pp. 25–7.

Arnalds, A. (1987) Ecosystem disturbance in Iceland. *Arctic and Alpine Research*, 19(4), 508–13.

Arnalds, A. (1988a) Landgaeði á Íslandi fyrr og nú. *Árbók Landgraeðslu Ríkisins*, 1, 13–31.

Arnalds, A. (1988b) Brautin rudd – Saga Landgraeðslu á Íslandi fyrir 1907. *Árbók Landgraeðslu Ríkisins*, 1, 33–9.

Arnalds, A. (1988c) Lúpínan og landgraeðsland. *Árbók Landgraeðslu Ríkisins*, 1, 193–6.

Arnalds, A. (1988d) Friðuð Landgraeðslusvaeði. *Árbók Landgraeðslu Ríkisins*, 1, 139–55.

Arnalds, A., Árnasson, Th. O., Lawrence, Th. and Sigurbjörnsson, B. (1978) Grass variety trials for reclamation and erosion control. *Fjölrit RALA No. 37.*

Arnalds, O. (1981) Rannsókn á gróður- og jarðvegseyðingu á Biskupstungnaafrétti. *Fjölrit RALA No. 76.*

Arnalds, O. (1984) Eolian nature and erosion of some Icelandic soils. *Íslenzkar Landbúnaðarrannsóknir*, 16(1–2), 21–35.

Arnalds. O. (1989) Jarðvegseyðing. *Árbók Landgraeðslu Ríkisins*, 2, 47–68.

Arnalds, O., Aradóttir, A. L. and Thorsteinsson,

I. (1987) The nature and restoration of denuded areas in Iceland. *Arctic and Alpine Research*, 19(4), 518–25.

Bagnold, R. A. (1941) *The Physics of Blown Sand*. Methuen, London.

Bagnold, R. A. (1956) The flow of cohesionless grains in fluids. *Phil. Trans. Roy. Soc. London A*, 249, 235–97.

Bates, C. G. (1924) The windbreak as a farm asset. *USDA Farmers' Bull.*, 1405.

Bjarnason, H. (1971) Um friðun lands og frjósemi jarðvegs. *Ársrit Skógraektarfélags Íslands*, 4–19.

Bjarnason, H. (1974) Athugasemdir við sögu íslendinga í sambandi við eyðingu skóglendis. *Ársrit Skógraektarfélags Íslands*, 30–43

Caborn, J. M. (1965) *Shelterbelts and Windbreaks*. Faber, London.

Chepil, W. S. (1945) Dynamics of wind erosion. III. Transport capacity of the wind. *Soil Sci.*, 60, 475–80.

Chepil, W. S. (1950) Properties of soil which influence wind erosion. II. Dry aggregate structure as an index of erodibility. *Soil Sci.*, 69, 403–14.

De Ploey, J. (1977) Some experimental data on slopewash and wind action with reference to Quaternary morphogenesis in Belgium. *Earth Surf. Proc.*, 2, 101–15.

Einarsson, Th. (1963) Pollen-analytical studies on the vegetation and climatic history of Iceland in Late- and Post-Glacial times, in *North Atlantic Biota and Their History* (eds A. Löve and D. Löve). Pergamon, Oxford, pp. 355–65.

Finkel., H. J. (1959) The barchans in southern Peru. *J. Geol.*, 67, 614–47.

Finnsson. H. (1796) *Mannfaekkun af hallaerum*. Almenna bókafélagið, 1970.

Friðriksson, S. (1960) Uppgraeðsla og raektun affréttarlanda. *Árbók Landbúnaðarins*, 11, 201–18.

Friðriksson, S. (1969a) Uppgraeðslutilraun á Mosfellsheiði. *Íslenzkar Landbúnaðarrannsóknir*, 1, 28–37.

Friðriksson, S. (1969b) Uppgraeðslutilraun á Tungnaáröraefum. *Íslenzkar Landbúnaðarrannsóknir*, 1, 38–44.

Friðriksson, S. (1971) Raektunartilraunir á Kili. *Íslenzkar Landbúnaðarrannsóknir*, 3(1), 12–27.

Friðriksson, S. (1972) Grass and grass utilization in Iceland. *Ecology*, 53(5), 785–96.

Friðriksson, S. and Pálsson, J. (1970) Landgraeðslutilraun á Sprengisandi. *Íslenzkar Landbúnaðarrannsóknir*, 2(2), 34–49.

Guðjónsson, G. (1980) Gróðurbreytingar í Thór-

sárdal. *Íslenzkar Landbúnaðarrannsóknir*, 12(1), 27–59.

Guðmundsson, J. and Pálsson, D. (1990) Leiðbeiningar um Landgraeðslu. *Árbók Landgraeðslu Ríkisins*, 3, 137–46.

Gunnlaugsdóttir, E. (1985) Composition and dynamical status of heathland communities in Iceland in relation to recovery measures. *Acta Phytogeographica Suedica*, 75.

Heisler, G. M. and De Walle, D. R. (1988) Effects of wind break structure on wind flow. *Agriculture, Ecosystems and Environment*, 22/23, 41–69.

Iceland Forestry Service (1986) *Forestry in Iceland*. Skograek Ríkisins, Reykjavík.

Iversen, J. D. (1985) Aeolian threshold: effect of density ratio, in *Proceedings of International Workshop on the Physics of Blown Sand* (eds O. E. Barndof-Nielson, J. T. Møller, K. R. Rasmussen and B. B Willett). University of Aarhus, pp. 67–81.

Jacobs, A. F. G. (1984) Wind reduction near the surface behind a thin solid fence. *Agric. Fores. Met.*, 33, 157–62.

Jensen, M. (1954) *The Shelter Effect*. Danish Technical Press, København.

Jóhannesson, H. and Einarsson, S. (1990) Glefsur úr sögu hrauna og jarðvegs sunnan Heklu. *Árbók Landgraeðslu Ríkisins*, 3, 123–35.

Kutschera, L. and Lichtenegger, E. (1982) *Wurzelatlas Mitteleuropäischer Grünlandpflanzen. Band 1 Monocotyledoneae*. Gustav Fischer Verlag, Stuttgart.

Landsberg, J. J. and James, G. B. (1971) Wind profiles in plant canopies: studies on an analytical model. *J. Appl. Ecol.*, 8, 729–41.

Larsen, G. and Thórarinsson, S. (1977) H-4 and other acid Hekla tephra layers. *Jökull*, 27, 28–46.

Lettau, K. and Lettau, H. (1978) Exploring the world's driest climate. *Univ. Wisconsin-Madison, Inst. Env. Studies Report 101*.

Lyles, L. and Allison, B. E. (1981) Equivalent wind erosion protection from selected crop residues. *Trans. Am. Soc. Agric. Engnrs*, 24, 405–9.

Lyles, L., Schrandt, R. L. and Schmeidler, N. F. (1974) How aerodynamic roughness elements control sand movement. *Trans. Am. Soc. Agric. Engnrs*, 17, 134–9.

Magnússon, A. and Vídalín, P. (1913–1917) *Jarðabók*. Hið íslenska fraeðafélag, Reykjavík.

Marshall, J. K. (1967) The effect of shelter on the productivity of grasslands and field crops. *Field Crop Abstracts*, 20, 1–14.

Marshall, J. K. (1971) Drag measurements in roughness arrays of varying density and distribution. *Agric. Met.*, **8**, 269–92.

Morgan, R. P. C. (1989) Design of in-field shelter systems for wind erosion control, in *Soil Erosion Protection Measures in Europe* (eds U. Schwertmann, R. J. Rickson and K. Auerswald). Soil Technology Series 1, Catena Verlag, Cremlingen-Destedt, pp. 15–23.

Morgan, R. P. C. (1990) Modelling the effect of vegetation on air flow for application to wind erosion control, in *Vegetation and Erosion* (ed. J. B. Thornes). Wiley, Chichester, pp. 85–98.

Morgan, R. P. C. and Finney, H. J. (1987) Drag coefficients of single crop rows and their implications for wind erosion control, in *International Geomorphology 1986 Part II* (ed. V. Gardiner). Wiley, Chichester, pp. 449–58.

Morgan, R. P. C., Finney, H. J. and Williams, J. S. (1988) Leaf properties affecting crop drag coefficients: implications for wind erosion control, in *Land Conservation for Future Generations* (ed. S. Rimwanich). Department of Land Development, Bangkok, pp. 885–93.

Nickling, W. G. (1988) Prediction of soil loss by wind, in *Land Conservation for Future Generations* (ed. S. Rimwanich). Department of Land Development, Bangkok, pp. 75–94.

Olesen, F. (1979) *Collective Shelterbelt Planting*. Hedeselskabet, Viborg.

Pálsson, D. (1989) Afréttarnotkun í Biskupstungun. *Árbók Landgraeðslu Ríkisins*, **2**, 151–60.

Pálsson, H. and Stefánsson, O. E. (1968) *Farming in Iceland*. Búnaðarfélag Íslands, Reykjavík.

Plate, E. J. (1971) The aerodynamics of shelterbelts. *Agric. Met.*, **8**, 203–22.

Rasmussen, K. R., Sørensen, M. and Willetts, B. B. (1985) Measurement of saltation and wind strength on beaches, in *Proceedings of International Workshop on the Physics of Blown Sand* (eds O. E. Barndorff-Nielsen, J. T. Møller, K. R. Rasmussen and B. B. Willetts). University of Aarhus, pp. 301–25.

Runólfsson, S., (1978) Soil conservation in Iceland, in *The Breakdown and Restoration of Ecosystems* M. W. Holdgate and M. J. Woodman (eds). Plenum, New York, pp. 231–40.

Runólfsson, S. (1987) Land reclamation in Iceland. *Arctic and Alpine Research*, **19**(4), 514–17.

Runólfsson, S. and Sigurðsson, Th. (1990) Landgraeðsla í Austur-Skaftafellssýslu 1900–1989. *Árbók Landgraeðslu Ríkisins*, **3**, 85–97.

Savat, J. (1982) Common and uncommon selectivity in the process of fluid transportation: field observations and laboratory experiments on bare surfaces. *Catena Suppl.*, **1**, 139–60.

Seginer, I. and Sagi, R. (1972) Drag on a windbreak in two dimensional flow. *Agric. Met.*, **9**, 323–33.

Sigurðsson, S. (1977) *Birki á Íslandi*. Skógarmál, Reykjavík, pp. 146–72.

Shields, A. (1936) Anwendung der Ähnlichkeitsmechanik und der Turbulenzforschung auf die Geschiebebewegung. *Mitteilungen der Preussischen Versuchsanstalt für Wasserbau und Schiffbau*, **26**.

Skidmore, E. L. and Hagen, L. J. (1977) Reducing wind erosion with barriers. *Trans. Am. Soc. Agric. Engnrs*, **20**, 911–15.

Sørensen, M. (1985) Estimation of some aeolian saltation transport parameters from transport rate profiles, in *Proceedings of International Workshop on the Physics of Blown Sand* (eds) O. E. Barndorff-Nielsen, J. T. Møller, K. R. Rasmussen and B. B. Willetts). University of Aarhus, Aarhus, pp. 141–90.

Sveinsson, R. (1953) A memorandum on soil erosion and reclamation problems in Iceland. Landgraeðsla Ríkisins, Gunnarsholt. Cyclo.

Tanner, H. and Naegeli, W. (1947) Wetterbleobachtungen und Untersuchungen über die Windverhältnisse in Bereich von Laub- und Nadelholzstreifen. *Anhang Jahr. Melioraten der Rheinebene (Schweiz)*, 28–42.

Thom, A. S. (1975) Momentum, mass and heat exchange of plant communities, in *Vegetation and the Atmosphere*, Vol. 1 (ed. J. L. Monteith). Academic Press, London, pp. 57–109.

Thórarinsson, S. (1961) Uppblástur á Íslandi í ljósi öskulagarannsókna. *Ársrit Skógraektarfélags Íslands*, 17–54.

Thórarinsson, S. (1968) *Hekluedar*. Sögufélagið, Reykjavík.

Thórarinsson, Th. (1974) Thjóðin lifði en skógurinn dó. *Ársrit Skógraektarfélags Íslands*, 16–29.

Thorgeirsson, H. (1990) Raetur og lífsthróttur plantna. *Árbók Landgraeðslu Ríkisins*, **3**, 49–54.

Thorgilsson, A. (1068–1148) *Íslendingabók/Landnámabók*. Íslensk fornrit I (1986), Hið Íslenka Fornritafélag, Reykjavík.

Thóroddsen, Th. (1914) An account of the physical geography of Iceland with special reference to the plant life, in *The Botany of Iceland* (eds L. K. Rosenvinge and E. Warning). Ejnar Munksgaard, København, pp. 193–334.

Thorsteinsson, I. (1980) Nýting úthaga – beitarthungi. *Íslenzkar Landbúnaðarrannsóknir*, **12**(2), 113–22.

Thorsteinsson, I. and Blöndal, S. (1986) *Gróður-yðing og endurheimt landgaeða*. Námsgagnastof-nun, Reykjavík.

Tsoar, H. (1974) Desert dunes morphology and dynamics, El Arish (Northern Sinai). *Z.f.Geomorph. Suppl.*, 20, 41–61.

Wasson, R. J. and Nanninga, P. M. (1986) Estimating wind transport of sand on vegetated surfaces. *Earth Surf. Proc. Landf.*, 11, 505–14.

Wickens, R. (1981) Erosion control on peaty soils, in *Soil and Crop Loss: Developments in Erosion Control*. Soil and Water Management Association, Stoneleigh.

Willetts, B. B. and Rice, M. A. (1988) Particle dislodgement from a flat sand bed by wind. *Earth Surf. Proc. Landf.*, 13, 717–28.

Woodruff, N. P. and Siddoway, F. H. (1965) A wind erosion equation. *Soil Sci. Soc. Am. Proc.*, 29, 602–8.

Woodruff, N. P. and Zingg, A. W. (1952) Wind tunnel studies of fundamental problems related to windbreaks. *USDA-SCS Pub.*, SCS-TP-112.

Zingg, A. W., (1953) Wind tunnel studies of the movement of sedimentary material. *Proc. 5th Hydraulic Conference, Iowa State Univ. Bull.*, 34, 111–35.

SLOPE STABILIZATION 7

T. H. Wu

7.1 INTRODUCTION

The use of vegetation for slope stabilization started in ancient times. A fine historical review may be found in Greenway (1987). Krabel (1936) was among the first to use soil-biotechnical construction in the United States. In more recent times, the roles played by vegetation in some specific geotechnical processes have been recognized. Vegetation may affect slope stability in many ways. Comprehensive reviews may be found in Gray (1970), Gray and Leiser (1982), Greenway (1987), and Coppin and Richards (1990).

The stability of slopes is governed by the load, which is the driving force that causes failure, and the resistance, which is the strength of the soil–root system. The weight of trees growing on a slope adds to the load but the roots of trees serve as a soil reinforcement and increase the resistance. Vegetation influences slope stability indirectly through its effect on the soil moisture regime. Vegetation intercepts rainfall and draws water from the soil via evapotranspiration. This reduces soil moisture and pore pressure, increases the shear strength of the soil, and increases the resistance. Vegetation roots tend to increase soil permeability and increase infiltration and soil moisture, while the organic layer associated with vegetative cover tends to retard infiltration. The influences of these factors on stability are summarized in Table 7.1.

Several case studies have shown that slope failures may be attributed to the loss of tree roots as a soil reinforcement (O'Loughlin, 1974;

Wu, McKinnell and Swanston, 1979; Riestenberg, 1987). Field and laboratory studies have shown that vegetation reduces water content and increases soil-moisture suction in the soil (Gray, 1970; Gray and Brenner, 1970; Williams and Pidgeon, 1983; Greenway 1987). Greenway (1987) has given an extensive summary of observations on the effect of vegetation on slope stability, including several reports that vegetation tended to reduce stability. Considering the many factors that affect slope stability, it is unwise to make general conclusions. The approach adopted in this chapter is to emphasize the methods of stability analysis and the mechanisms by which vegetation affects slope stability. While methods of stability analysis are well known, analytical methods for evaluation of the contribution by roots are relatively new and still undergoing development. This chapter contains a review of the fundamental concepts, followed by a summary of simplified methods and available data that are applicable to design and construction problems. Examples are given to illustrate applications to specific cases. Equally important are construction methods for slope stabilization with vegetation. A summary of current methods is given in this chapter. More detailed treatment of construction methods may be found in Gray and Leiser (1982) and Schiechtl (1980).

7.2 STABILITY ANALYSIS

Stability analysis may be used to evaluate an existing condition or a proposed solution to

Slope Stabilization and Erosion Control: A Bioengineering Approach. Edited by R. P. C. Morgan and R. J. Rickson.
Published in 1995 by E & FN Spon, 2-6 Boundary Row, London, SE1 8HN. ISBN 0 419 15630 5.

Table 7.1 Effects of vegetation on slope stability

Process	Type	Effect on stability
1. Roots increase permeability, increase infiltration, and thereby increase pore pressure	Hydrologic	Negative
2. Vegetation increases interception and evapotranspiration, and thereby reduces pore pressure	Hydrologic	Positive
3. Vegetation increases weight or surcharge and thereby increases load on slope	Mechanical	Negative
4. Vegetation increases wind resistance, and thereby increases load on slope	Mechanical	Negative
5. Roots reinforce soil and thereby increase strength	Mechanical	Positive

determine if it meets the requirements of safety. The most common methods are based on limit equilibrium, in which the soil mass is considered to be at the verge of failure, and the shear strength of the soil is fully developed along a potential slip surface. The safety of the slope is generally expressed as a factor of safety

$$F_s = R/L, \qquad (7.1)$$

where F_s is the factor of safety, R is the resistance and L is the load. For slopes, the load consists of the weight of the soil mass and applied loads. The resistance is provided by the shear strength of the soil–root system along the slip surface.

7.2.1 SHEAR STRENGTH OF SOIL AND SOIL–ROOT SYSTEM

The Mohr–Coulomb criterion is commonly used to describe the shear strength of soils. The shear strength of the soil is expressed as (Coulomb, 1776),

$$s_s = c + \sigma \tan\phi, \qquad (7.2a)$$

where c is the cohesion, σ is the normal stress, and ϕ is the angle of internal friction. Where there is pore pressure in the soil, the shear strength is (Terzaghi, 1936)

$$s_s = c' + \sigma' \tan\phi' = c' + (\sigma - u) \tan\phi', \qquad (7.2b)$$

where the prime denotes effective stress and u is the pore pressure. A further refinement is to separate the pore pressure into pore-water pressure, u_w, and pore-air pressure, u_a, or (Bishop and Blight, 1963)

$$u = u_w + \chi(u_w - u_a), \qquad (7.2c)$$

where χ is a parameter. The shear strength can be expressed as (Fredlund, Morgenstern and Widger, 1978)

$$s_s = c' + (\sigma - u_w)\tan\phi' + (u_w - u_a)\tan\phi''. \qquad (7.2d)$$

This relationship is shown as a surface in Figure 7.1. Description of laboratory tests for the measurement of the strength parameters, c', ϕ' and ϕ'', may be found in Bishop and Henkel (1967) and Fredlund, Morgenstern and Widger (1978). For a saturated soil, $u_a = 0$, and equation 7.2d becomes equation 7.2b. For a dry soil, $u_w = 0$, and equation 7.2d becomes equation 7.2a.

The Mohr–Coulomb criterion is a simplification of soil behaviour and ignores the effects of several factors, the most important ones being the intermediate principal stress, rotation of principal axes, and soil anisotropy. However, for the type of applications described in this

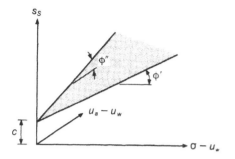

Figure 7.1 Failure envelope.

chapter, the Mohr–Coulomb criterion is considered to be adequate. A review of the influence of these factors on the shear strength may be found in Bishop (1966) and Symes, Gens and Hight (1984).

Where the soil contains roots, shear failure would involve failure of the soil–root system. A simple approach is to consider the root as a reinforcement which increases the shear strength by s_r. Then

$$s = s_s + s_r, \qquad (7.3)$$

where s_r is the contribution of the roots to the shear strength. This value of s may then be used in the conventional methods of stability analysis. Procedures for evaluation of s_r are presented in section 7.4.

7.2.2 EFFECTIVE STRESS AND TOTAL STRESS ANALYSES

To perform a stability analysis with effective stress, s_s is given by equations 7.2b or 7.2d. The pore pressure on the slip surface must be estimated. It is important to note that u should represent the value at the moment of failure. Hence,

$$u = u_s + u_e, \qquad (7.4)$$

where u_s is the hydrostatic pore pressure and u_e is the excess pore pressure caused by deformation of the soil during loading. For the undrained condition, the excess pore pressure is commonly expressed as a function of the applied stresses (Henkel, 1960)

$$u_{e,u} = \frac{B}{3}(\Delta\sigma_1 + \Delta\sigma_2 + \Delta\sigma_3) +$$
$$\alpha\left[(\Delta\sigma_1 - \Delta\sigma_2)^2 + (\Delta\sigma_2 - \Delta\sigma_3)^2 + (\Delta\sigma_3 - \Delta\sigma_1)\right]^{1/2}, \qquad (7.5)$$

where $\Delta\sigma_1$, $\Delta\sigma_2$, $\Delta\sigma_3$ are changes in principal stresses, and α and B is the pore pressure parameters.

For slope stability problems, the evaluation of u_e involves several difficulties. Prior to failure, a soil element on the potential failure surface is subjected to a substantial shear stress. Since this is long-term loading, the soil is in the drained state under this stress and $u_e = 0$. Failure may occur either when the shear stress is increased by loading, or when the hydrostatic pore pressure is increased during a period of heavy precipitation. In either case, the soil deformation prior to failure may cause an increase in u_e. This amount is likely to depend on the *in situ* stresses prior to failure and can be determined in the laboratory only by performing stress-path tests. In these tests, the perceived stress history *in situ* is duplicated as closely as possible. Where such tests are not practicable, the following simplified estimates may be made. If the soil is pervious and the loading is slow, or if u_e is small, it may be assumed that $u_e = 0$ and an effective stress analysis may be made with $u = u_s$. If the soil has low permeability and the soil is contractive, u_e may be important. When u_e cannot be estimated with confidence, an alternative is to assume that the shear occurs in the undrained condition. For saturated soils, $\phi = 0$ in equation 7.2a. Then the total stress analysis may be made with the undrained shear strength (Bishop and Bjerrum, 1960).

7.2.3 FAILURE MODES

A variety of failure modes is possible. The critical slip surface is determined by slope geometry, soil properties, and load. Where roots are not present in sufficient numbers or are not so large as to influence the shape of the slip surface, solutions developed for soil slopes may be adapted for use. The following is a review of

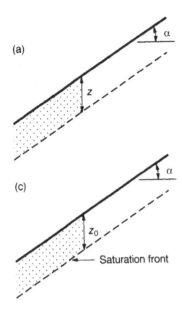

(a)

(c)

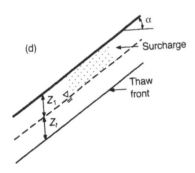

Saturation front

(d)

Surcharge

Thaw
front

Figure 7.2 Infinite slope: (a) no pore pressure; (b) seepage parallel to slope; (c) saturation from surface; (d) thaw from surface.

solutions for soil slopes. Where roots intersect the slip surface, s_r should be included in s, as given in equation 7.3.

Consider first, a slope of infinite extent. The slip surface is parallel to the ground surface (Figure 7.2(a)). The safety factor is

$$F_s = \frac{s}{\gamma z \sin\alpha \cos\alpha},\qquad(7.6)$$

where γ is the unit weight of the soil, z is the depth of the slip surface, and α is the slope angle. If a shallow soil layer lies over an imper-

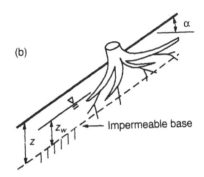

Impermeable base

meable base, seepage occurs parallel to the slope and the pore pressure is

$$u_s = \gamma_w z_w \cos^2\alpha.\qquad(7.7a)$$

In effective stress analysis, assuming $u_e = 0$,

$$F_s = \frac{c' + (\gamma z \cos^2\alpha - u_s)\tan\phi'}{\gamma z \sin\alpha \cos\alpha}.\qquad(7.7b)$$

For this case, the most critical slip surface is at the bottom of the soil layer (Figure 7.2(b)). If the soil has low permeability, saturation due to rainfall may proceed from the surface downward. Then a saturated zone forms below the surface as shown in Figure 7.2(c). The pore pressure in the saturated zone is 0_e then, for $u_e = 0$,

$$F_s = \frac{c' + \gamma z_0 \cos\alpha \tan\phi'}{\gamma z_0 \sin\alpha \cos\alpha}.\qquad(7.8a)$$

If $c' = 0$,

$$F_s = \frac{\tan\phi'}{\tan\alpha}.\qquad(7.8b)$$

For an infinite slope on permafrost (Figure 7.2(d)), excess pore pressure is generated during thaw. For $c' = 0$, the safety factor is (Morgenstern and Nixon, 1971; Pufahl and Morgenstern, 1979)

$$F_s = \frac{\gamma_1 z_1(1 - R'') + \gamma_1 z_t(t)(1 - R')}{\gamma_1 z_1 + \gamma z_t(t)}\frac{\tan\phi'}{\tan\alpha},\qquad(7.9a)$$

$$R = \frac{K_T T_s}{\Lambda c_v^{1/2}},\qquad(7.9b)$$

$$R' = \frac{1}{1 + \dfrac{1}{2R^2}}, \qquad (7.9c)$$

$$R'' = \frac{\text{erf}(R)}{\text{erf}(R) + \dfrac{e^{-R^2}}{R\sqrt{\pi}}}, \qquad (7.9d)$$

where c_v is the coefficient of consolidation, Λ is the latent heat of fusion, K_T is the thermal conductivity, T_s is the surface temperature, z_1 is the surcharge thickness, and z_t is the thickness of the thawed layer.

For a forested slope, a special case exists for an infinite slope over a firm base. If the number and diameter of roots intersecting the portion of the slip surface below the root mat is large, the tree and root mat may be anchored to the firm base. Then failure along the slip surface shown in Figure 7.2(b) cannot occur. However, the failure mode shown in Figure 7.3 should be considered. The trees may serve as buttresses and the soil between the root mats, whose diameter is D_r, may slide over the firm base. The force acting on the soil and root system under each tree is (Wang and Yen, 1974)

$$P = \frac{K_0}{2}\gamma z^2 (B + D_r) - pzB, \quad (7.10a)$$

where

$$p = \frac{\gamma z \cos\alpha \,(m\sin\alpha - m\cos\alpha\tan\phi_1 - K_0\tan\phi) - (c_1 m + 2c\cos\alpha)}{2K_0\cos\alpha\tan\phi}, \qquad (7.10b)$$

$$(1 - e^{-2K_0 n \cos\alpha \tan\phi}) + \frac{1}{2}\gamma z K_0 e^{-2K_0 n \cos\alpha \tan\phi},$$

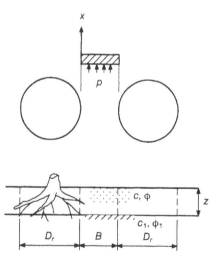

Figure 7.3 Root buttressing.

$$m = B/z, \qquad n = x/B, \qquad (7.10c)$$

K_0 = coefficient of earth pressure at rest, B = width, and c_1, ϕ_1 = shear strength parameters of material at the base of the sliding mass. It can be seen that for a single row of trees, p increases

with x, the distance along the slope. The buttress is no longer effective when $B \geqslant B_{cr}$

$$B_{cr} = \frac{zK_0(K_0 + 1) + 2c/\gamma}{\cos\alpha\,(\tan\alpha - \tan\phi) - c_1\gamma z\cos\alpha}. \quad (7.10d)$$

The value of B_{cr} increases rapidly with $c_1 > 0$.

For a finite slope on uniform soil, the slip surface may be approximated by a circular arc (Figure 7.4(a)). The safety factor is

$$F_s = M_R/M_L = sLR/Wx, \qquad (7.11a)$$

where M_R is the resisting moment, M_L is the moment due to load, L is the length of arc ab, W is the weight of soil mass, and R is the radius of arc ab. In most problems, the soil properties and pore pressure are not uniform along the slip surface. Then, the slip surface is divided into slices (Figure 7.4(b)) and the moments are summed over all slices. Equation 7.11a is written as

$$F_s = \sum_i \frac{[c'l_i + (w_i\cos\alpha_i - u_i)\tan\phi_i']}{\sum_i w_i\sin\alpha_i} \quad (7.11b)$$

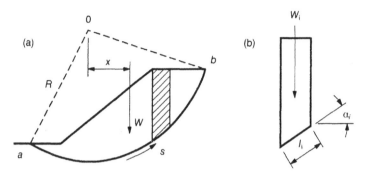

Figure 7.4 Circular slip surface. (a) Scheme for complete slip; (b) method of slices.

This is the Fellenius (1936) solution and is based on the assumption that the effect of the interslice forces is negligible. Where the nonuniformity in soil properties becomes important, the slip surface may be quite different from a circular arc. Solutions for general slip surfaces are available (Morgenstern and Price, 1965; Janbu, 1973). These methods include the interslice forces in the computation of the safety factor and may also be used for the circular arc.

Where the soil near the surface contains roots to a depth z_r, a modified circular arc analysis may be made to account for the resistance s_r in the area abcd on the ends and along ab and cd of the cylindrical surface (Figure 7.5(a)). The three-dimensional slip surface is shown in Figure 7.5(b). M_R in equation 7.11a becomes (Wu, 1984a):

$$M_R = 2s_s BR^2\theta_0 + s_s R^3 (2\theta_0 - \cos\theta_0 \sin\theta_0) + 4s_r z_r R^2\theta_0 \cos^2\theta_0 + 2s_r z_r BR/\sin\theta_0.$$

$$(7.12)$$

The same approach may be used to modify the method of slices to account for s_r.

Equations 7.7 and 7.8 may be used to illustrate the influence of the various factors on stability. To emphasize the key issues, the results of a simple parametric study are given in Figure 7.6 for two sets of soil strength, $\phi' = 27°$, $c' = 0$ and $\phi' = 40°$, $c' = 0$. The former represents a plastic cohesive soil in the drained condition and the latter, a cohesionless soil. The figure shows the influence of s_r on the safety

factor for various z. It is important to note that the effect of s_r is largest at small z, i.e. when the failure surface is at a shallow depth. Comparison of Figures 7.6(a) and (b) shows the important role of the seepage force. Vegetation also affects stability indirectly through its effect on seepage. If it can reduce the vertical infiltration, it will reduce z_0, the depth of the saturated zone in Figure 7.2(c). For seepage parallel to the slope, reduced infiltration will reduce z_w in Figure 7.2(b). Figure 7.6(c) shows the influence of the surcharge p due to the weight of the trees and of w, the shear force due to wind; p is added to the weight γz in equations 7.7 and 7.8 and w is added to the shear stress in the numerator of equations 7.7 and 7.8 (Brown and Sheu, 1975). The value of $w = 1\text{kPa}$ corresponds to the drag force created by a wind velocity of $90\,\text{km/h}$, directed parallel to the slope (Wu, McKinnell and Swanston, 1979).

7.3 HYDROSTATIC PORE PRESSURE AND SUCTION

Calculation of pore pressure and suction is based on the conservation of mass. For a differential element, this is expressed as (Philip, 1957, 1969):

$$\frac{\partial}{\partial x}\left[K_x(\psi)\frac{\partial\psi}{\partial x}\right] + \frac{\partial}{\partial y}\left[K_y(\psi)\frac{\partial\psi}{\partial y}\right] +$$

$$\frac{\partial}{\partial z}\left[K_z(\psi)\frac{\partial\psi}{\partial z}\right] + \frac{\partial K}{\partial\psi}\frac{\partial\psi}{\partial z} = C(\psi)\frac{\partial\psi}{\partial t} \quad (7.13a)$$

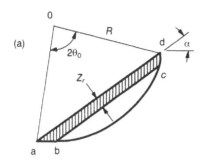

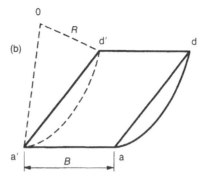

Figure 7.5 Simplified three-dimensional slip surface. (a) Section; (b) three-dimensional view.

or

$$\frac{\partial}{\partial x}\left[D_x(\Theta)\frac{\partial \Theta}{\partial x}\right] + \frac{\partial}{\partial y}\left[D_y(\Theta)\frac{\partial \Theta}{\partial y}\right] +$$

$$\frac{\partial}{\partial z}\left[D_z(\Theta)\frac{\partial \Theta}{\partial z}\right] - \frac{\partial K}{\partial z} = \frac{\partial \Theta}{\partial t}, \quad (7.13b)$$

where C is the storage coefficient; D_x, D_y, D_z represent diffusivity and K_x, K_y, K_z represent permeability in X, Y and Z directions, Θ is the volumetric water content, Ψ is the potential, and t is time. Pore pressure and suction are obtained by solving the equation for the appropriate initial and boundary conditions. Complications in boundary conditions and in the relations $\partial K/\partial \Psi$ and $C(\Psi)$ often require numerical solutions. Freeze (1971) has presented a finite difference solution of saturated–unsaturated flow in a groundwater basin. The solution accounts for variable $\partial K/\partial \Psi$ and $C(\Psi)$ and hysteresis in wetting and drying. Simplified solutions that are useful in some situations are described below.

7.3.1 UNSATURATED FLOW

Problems involving infiltration and evapotranspiration require solution of the equation for unsaturated flow. Simplified solutions for infiltration may be obtained by using the assumption that the flow is vertical. The closed form solution for one-dimensional infiltration into a homogeneous soil of infinite depth, with constant moisture, Θ_1, at the surface, is (Philip, 1957)

$$Q(t) = St^{1/2} + [A_2 + K(\Theta_0)]t + A_3 t^{2/3} + \dots . \quad (7.14)$$

For large times, an approximate solution is

$$Q(t) = St^{1/2} + Kt, \quad (7.15a)$$

where

$$S = 2(\Theta_1 - \Theta_0)\frac{D}{\sqrt{\pi}} \quad (7.15b)$$

is called sorptivity, and Θ_0 is the initial value of Θ.

For flow into a soil layer with a water table at some depth below the surface, the unsaturated zone may be subdivided into discrete zones (Figure 7.7), each representing a dominant force that controls the moisture movement (Eagleson, 1978). Conservation of mass is expressed as

$$\frac{\Delta \Theta_j}{\Delta t}h_j = (v_{j-1} - v_j - E_j), \quad (7.16)$$

where j denotes the zone, E is evapotranspiration, h is the thickness of a zone, and v is velocity and is governed by Darcy's Law,

$$v_j = \frac{1}{2}K(\Theta)\left[\frac{\psi_j - \psi_{j-1}}{h_j} + 1 + \frac{\psi_{j+1} - \psi_j}{h_{j+1}} + 1\right]. \quad (7.17)$$

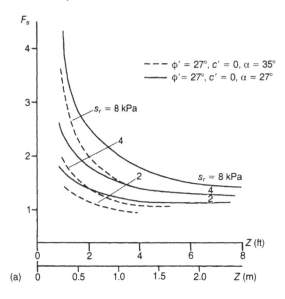

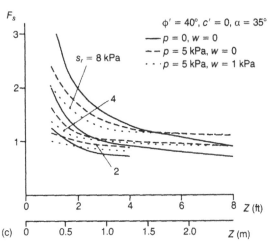

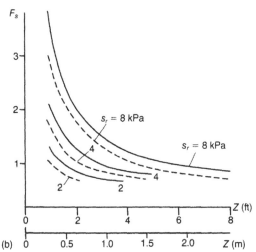

Figure 7.6 Computed factors of safety. (a) Vertical seepage; (b) seepage parallel to the slope; (c) seepage parallel to slope (with surcharge and wind).

The rainfall may be applied as a boundary condition at the ground surface, subjected to the limitation that it cannot exceed the saturated permeability. Then

$$z = 0, \quad K_z d\psi/dz = q, \; q \leqslant K_{z,s} \quad (7.18)$$
$$K_z d\psi/dz = K_{z,s}, \; q > K_{z,s},$$

where q is the rainfall, and $K_{z,s,}$ is the saturated permeability in the z direction. Numerical solution is used to obtain Θ_j and Ψ_j as functions of time.

7.3.2 SATURATED FLOW

For saturated flow in an incompressible soil, the right-hand side of equation 7.13 is 0. The solution for steady state flow is conveniently done by construction of the flow net, composed of a system of equipotential lines and flow lines (Figure 7.8). Along the equipotential lines, Ψ is constant and along the flow lines, $Kd\Psi/dl = v$, where v = velocity and l = distance along the flow line. The equipotential and flow lines are perpendicular to each other for isotropic soils. Given the flow net, the hydrostatic pore pressure at a point (point a in Figure 7.8) is

$$u_s = \gamma_w z_w, \quad (7.19)$$

where γ_w is the unit weight of water. Numerical solutions using finite difference (Remson, Hornberger and Molz, 1971), finite element (Pinder and Gray, 1977) or boundary integral (Ligget and Liu, 1983) methods are necessary for solving problems of transient flow. For unconfined transient flow over an impermeable base, simplified solutions have been proposed by

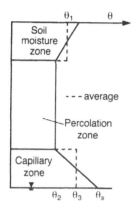

Figure 7.7 Lumped parameter model, simplified moisture profile.

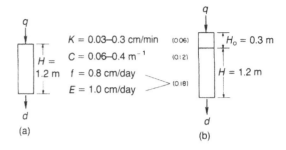

Figure 7.9 Mass balance model. K = permeability; C = storage coefficient; d = drainage; E = evapotranspiration.

7.3.3 COMBINED FLOW

Because of the complexities associated with solutions for combined saturated and unsaturated flows, simplified models based on mass balance may be used instead, as shown in Figure 7.9(a). Precipitation (q) enters at the top and d is the discharge, which is the sum of evapotranspiration (E) and drainage by gravity flow (f). A variety of models has been proposed for calculation of the flow through the partially saturated zone and the drainage (Beven, 1982; Sangrey, Harrop-Williams and Klaiber, 1984; Greenway, Anderson and Brian-Boys, 1984; Reddi and Wu, 1988).

Beven (1981), Sloan, Moore and Eigel (1983) and Sangrey, Harrop-Williams and Klaiber (1984).

7.3.4 EFFECTS OF VEGETATION

Vegetation influences the pore pressure and suction largely through its contribution to evapotranspiration and interception. A fraction of precipitation is intercepted by the leaves of vegetation (section 2.2.2). It is retained there and a part of it is evaporated. Moisture is removed from soil by evaporation. It is also withdrawn from soil by roots of vegetation and evaporated through leaves (section 2.2.1). Models of interception include those of Rutter *et al.* (1972) and Massman (1980). Simplified methods for estimating interception may be used for many engineering problems. Interception may be expressed as (Zinke, 1967).

$$I = a_1 + a_2 P \qquad (7.20a)$$

or (Jackson, 1975)

$$I = b_1 + b_2 \ln t + b_3 \ln p, \qquad (7.20b)$$

in which P is gross precipitation (mm), p is the rainfall intensity (mm/min) for a storm duration of t (min). Values of a, and a_2, b_1, b_2, and b_3 are of the order of 1 and 0.1, 0.9, 0.5 and 0.5, respectively, for forests. Average seasonal values of interception range between $0.1P$ and $0.35P$ for forests, grass and crops (Lull, 1964).

Several methods may be used for estimating

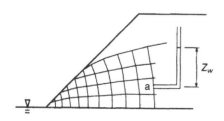

Figure 7.8 Flow net.

the maximum rate of evapotranspiration. The Penman (1948) equation is

$$E_{max} = \frac{1}{\Lambda_v(\zeta + \rho)} \left[\zeta H_r + \rho f(u)(p_a - p_s) \right].$$

$$(7.21a)$$

A simple model for maximum evapotranspiration is (Priestley and Taylor, 1979)

$$E_{max} = \frac{\lambda \zeta}{\Lambda_v(\zeta + \rho)} (H_r - H_s - H_c), \quad (7.21b)$$

where ζ is the slope of saturation vapour pressure curve, ρ is the psychrometric constant, Λ_v is the latent heat of vapourization, H_r is the net radiation flux density, H_s is the soil heat flux density, H_c is the rate of latent, sensible and photosynthetic energy storage in the canopy, $f(u)$ is the wind function, and p_a, p_s are vapour pressures at air temperature and at dewpoint. Values of λ range between 0.6 and 1.1, depending on the cover (McNaughton and Black 1973; Black 1979). Below some critical value of water content, evapotranspiration becomes limited by available moisture, and

$$E_s = \beta\Theta, \quad (7.21c)$$

where β is a parameter for the site. The evapotranspiration may be taken as E_{max} during rainstorms, and the lesser of E_{max} or E_s during dry intervals. A more complex model is PROSPER (Goldstein, Mankin and Luxmoore, 1974). Calibration studies and forecasting with PROSPER are described in Huff and Swank (1985). Simplified equations for maximum evapotranspiration use air temperature and solar radiation as parameters (Thornthwaite and Mather, 1957; Jensen and Haise, 1963). Mean evapotranspiration rates of crops and forests are summarized in Kozlowski (1968).

Vegetation also affects pore pressure and suction in important but indirect ways. Soil permeability may be significantly altered by vegetation. Vegetation roots may increase permeability while reduced moisture changes would reduce the amount of shrinkage cracks, which would in turn reduce permeability. Organic materials contributed by vegetation would increase the storage capacity of soils.

An example illustrates the effect of various parameters on the piezometric level in a slope shown in Figure 7.10(a). The infiltration and drainage were computed separately (Wu and Swanston, 1980). The values of E and f are 1.0 and 0.8 cm/day, respectively. The range in the parameters is shown in Figure 7.9(a) and the values that are considered to be the best estimates are given in parentheses. Curve A in Figure 7.10(b) shows the piezometric levels calculated with the best estimates of the parameters. To study the sensitivity of the model to the parameters, C, K and e were changed. The results are shown as curves B, C and D. Only the changed parameters are shown in Figure 7.10(b). The effect of evapotranspiration on the maximum piezometric level is obvious when curves B and D are compared. In a fifth trial, a surface layer with $C = 2\,\text{m}^{-1}$ is added to simulate the organic layer near the ground surface (Figure 7.9(b)). This is shown as curve E. Comparison of curves E and B shows the effect of destruction of the organic layer, as may occur during a fire or after clear-cutting. Reduction in pore pressure by evapotranspiration has also been demonstrated in laboratory experiments (Brenner, 1973) and *in situ* measurements on forested and clear-cut slopes (Gray, 1970). Somewhat different results were obtained by Terwilliger (1990), who measured soil moisture and pore pressure on slopes covered by chaparral and burned grassland. After a dry season, the slopes with chaparral had a lower moisture than the burned slopes but chaparral slopes absorbed more moisture following the beginning of heavy rains in the autumn. The pore pressure, immediately after the beginning of heavy rains, was not determined but there was no significant difference between the subsequent pore pressures under the two conditions.

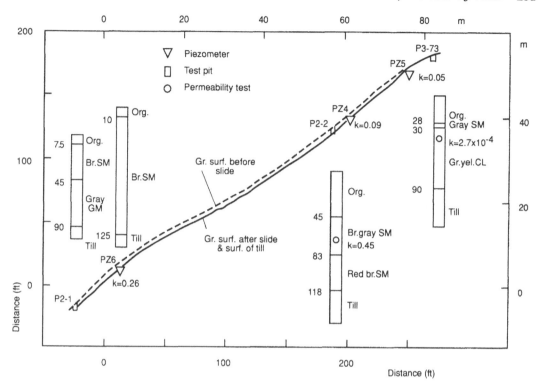

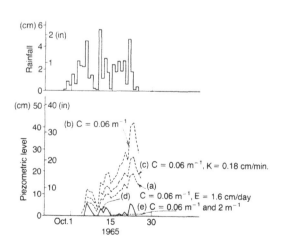

Figure 7.10 (a) Slope in Maybeso Valley. k in cm/min; depth in cm. (b) Computed piezometric levels (Wu, 1984).

7.4 SOIL REINFORCEMENT BY ROOTS

To evaluate the roots' contribution, s_r, to the shear strength, s, it is necessary to consider the soil–root interaction. Following the assumptions in limit equilibrium methods, the shear deformation along the slip surface is assumed to be restricted to a narrow zone with thickness, t (Figure 7.11). The root is represented in Figure 7.11 by a bar which is initially straight (dashed line) and is displaced into the position shown by the solid line. The displacement introduces an axial force as well as a shear and a bending moment in the bar. The addition to the shear strength is

$$s_r = \frac{T_z}{A}\tan\phi + T_x, \qquad (7.22a)$$

where A is the area of the section, and T_z and T_x are the z and x components of the force carried by the bar. For a given shear

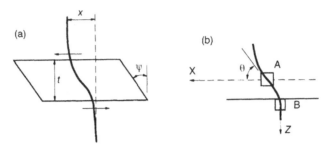

Figure 7.11 Root displacement (a) shear zone; (b) detail of root displacement.

displacement, x, the deformation of the bar, θ, the axial force, the shear and the moment will depend on the relative stiffness of the bar with respect to the soil and on the initial position and length of the bar. A complete solution requires treatment of the soil–root interaction as a three-dimensional problem in continuum mechanics. To date, only two-dimensional solutions have been obtained for large displacements and in elastic soil behaviour (Juran *et al.*, 1981). However, simplified decoupled solutions are available for several special cases. These are described below.

To evaluate the force T in equation 7.22a, it is necessary to consider the soil–root interaction. The simplest interaction model is to assume that the soil deforms in simple shear and the root deforms with the soil through the angle, ψ (Figure 7.12(a)). This would be true for a perfectly flexible root. Then,

$$\theta = \tan^{-1}\left(\frac{1}{\dfrac{x}{t} + \dfrac{1}{\tan\alpha}}\right), \tag{7.22b}$$

$$s_r = \frac{1}{A}\left[T\cos\left(\frac{\pi}{2} - \theta\right)\tan\phi + T\sin\left(\frac{\pi}{2} - \theta\right)\right]. \tag{7.22c}$$

Where there is more than one root,

$$s_r = \frac{1}{A}\left[\sum_{i=1}^{n} T_i\cos\left(\frac{\pi}{2} - \theta_1\right)\right.$$
$$\left. \tan\phi + T_1\sin\left(\frac{\pi}{2} - \theta_1\right)\right], \tag{7.22d}$$

where n = number of roots in an area A, and ϕ may be in terms of either total or effective stress, but should correspond to the method used in stability analysis.

For values of $(\frac{\pi}{2} - \theta)$ between 48° and 72°, and ϕ between 30° and 40°, s_r in equation 7.22b is insensitive to θ and equation 7.22d may be approximated by (Wu, McKinnell and Swanston, 1979)

$$s_r = 1.2\sum_{i=1}^{n} T_i/A. \tag{7.23a}$$

If it is assumed that the different roots are loaded to approximately the same stress, σ_r, equation 7.23a becomes

$$s_r = 1.2\frac{\sigma_r A_r}{A}, \tag{7.23b}$$

where A_r is the sum of cross-sectional areas of all roots in area A, and $a_r = A_r/A$.

Model tests by Shewbridge and Sitar (1989) show that the shape of the displaced reinforcement depends on the stiffness of the reinforcements. If it is assumed that the soil displacement is equal to the reinforcement displacement, then an approximate relation between t and the reinforcement stiffness $E_r I_r$ can be obtained with Shewbridge's data. This is shown in Figure 7.13. For comparison, t was also estimated from the measured soil displacements outside the shear box in *in situ* tests (Wu, Beal and Lan, 1988a). These are also shown in Figure 7.13. The boundary conditions and E_s for the *in situ* test are not the same as those in Shewbridge's tests.

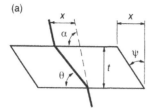
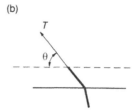

Figure 7.12 Root–soil interaction model for flexible root: (a) shear zone; (b) force in root.

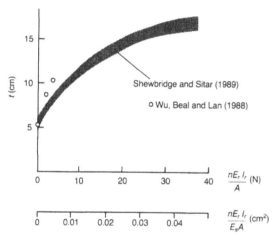

Figure 7.13 Relation between thickness of shear zone and reinforcement stiffness $E_S = 7800\,\text{kPa}$ (Durgunoglu and Mitchell, 1973). $n =$ no. of reinforcements.

However, the general agreement is encouraging. Thus, for a given stiffness and shear displacement, x, t and θ can be estimated.

To account for the stiffness of the root, we need to consider the relative displacement between soil and root. At A (Figure 7.11(b)), the root and soil are assumed to move together, while at B, below A, the root moves ahead of the adjacent soil. In this zone the soil–root interaction may be represented approximately as a beam on elastic–plastic support. Available solutions (Hetenyi, 1946; Scott, 1981) may be used to calculate the force in the root at small displacements. At large relative displacements the soil at B is in a plastic state. The soil reaction on the root may then be evaluated with the equa-

tions for bearing capacity. For a constant pressure along the root, the root force may be evaluated with the solution for a tie (Oden and Ripperger, 1981). If the stiffness of the root is small relative to T, the solution degenerates to that for a flexible cable (Figure 7.14(a)):

$$T(0) = \frac{P_y l}{\cos\theta}, \qquad (7.24a)$$

$$T(l) = T(0)\sin\theta. \qquad (7.24b)$$

The displacement is

$$x(z) = \frac{T(0)l\cos\theta - P_y z(2l - z)}{2T(l)}, \qquad (7.24c)$$

where l is the length of yield zone, and P_y is the soil resistance at yield (force/length) along z. For a given force $T(0)$ and angle θ, equation 7.24 can be used to calculate l and x. The cable equation is applicable for soil–root systems with a stiffness ratio

$$\eta = \left(\frac{T\sin\theta}{E_r I_r}\right)^{1/2} L \qquad (7.24d)$$

greater than 2.5, where E_r, $I_r =$ Young's modulus and moment of inertia of the root, respectively. The beam solution and the cable solution represent two extremes of large and small displacements, respectively, and may be used to estimate the upper limit of the root force due to displacement. Examples are given in Wu, Bettandapura and Beal (1988b). The force T also cannot exceed the tensile and compressive strengths of the root. An alternative is to estimate the deformed shape of the reinforcement

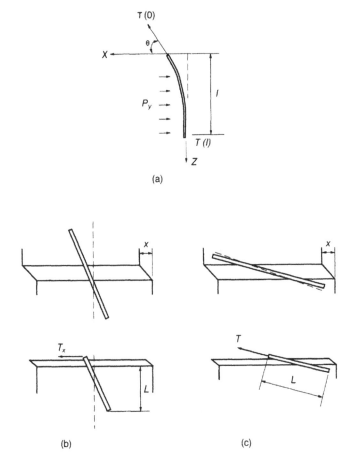

Figure 7.14 Simplified solutions for (a) tie, (b) stiff bar nearly perpendicular to shear zone; (c) stiff bar almost parallel to shear zone.

from the results of Shewbridge's tests and derive the work required to produce the deformation. The axial and shear forces in the reinforcement are then calculated from the work (Shewbridge and Sitar, 1990).

Stiff bars that are nearly perpendicular to the shear zone ($\alpha \approx \frac{\pi}{2}$) may be analysed as laterally loaded piles (Figure 7.14(b)). The ultimate resistance is (Broms, 1964a)

$$T_{x,f} = \frac{cN_cDL}{2.4} \qquad (7.25a)$$

for cohesive soils ($\phi = 0$), and (Broms, 1964b)

$$T_{x,f} = \frac{1}{2}\gamma DL^2 \tan^2\left(\frac{\pi}{4} + \frac{\phi}{2}\right) \qquad (7.25b)$$

for cohesionless soils ($c = 0$), where $N_c \approx 7$ (Wu *et al.*, 1988b). If the bar is long, bending failure could occur. The solutions may be found in Broms (1964a, b). If the bar is oriented almost parallel to the shear zone ($\alpha \to 0$), the ultimate resistance is close to the ultimate tensile resistance (Figure 7.14(c)). The tensile force that can be developed in a root is limited by the tensile strength of the root, or

$$T_f = \sigma_t A_r, \qquad (7.26a)$$

where σ_t = tensile strength. If the root consists of a rod, then T_f cannot exceed the shearing resistance along the soil–root interface

$$T_f = \pi DfL, \qquad (7.26b)$$

where f is the interface resistance.

Most roots contain branch roots such as the root which passes through the shear box in Figure 7.15(a). An idealized representation of the root system that extended beyond the shear box is shown in Figure 7.15(b), in which the segments are denoted by $R_{i,j}$ and the branch points, or joints, by $J_{i,j}$. J_0 denotes the point where the root intersects the shear surface. The force T, applied at J_0, is transmitted to the branch point and then to the branches. The forces at J_0 and J_1 may be computed with equation 7.24. The loads in the root branches, $R_{1,1}$ and R_2 may be found by considering the displacement compatibility and force equilibrium at the branch point J_1. Equations for computation of loads and displacements have been presented by Wu *et al.* (1988b). This procedure may be used to compute displacement and branch forces at successive branch points and to construct the load displacement relation for the root. Since the diameter of a branch root may be smaller than that of the main root, there is the possibility that a branch may fail first. Progressive failure occurs with successive failure of other branches. The system fails when all branches that support the main root fail. For a shear displacement x (Figure 7.11), the root displacement at the middle of the shear zone is $x/2$. The force that corresponds to this displacement can be calculated as described above and used in equations 7.3 and 7.22 to obtain the shear strength of the soil–root system.

7.5   ROOT PROPERTIES

The root properties that are needed for the computation of soil–root interaction include the root geometry and the strength properties. While data are available for several species, these are limited to the sites from which the data were obtained. It is known that root properties depend on site conditions but the relations are not well established. Hence, extrapolation of data from one site to another

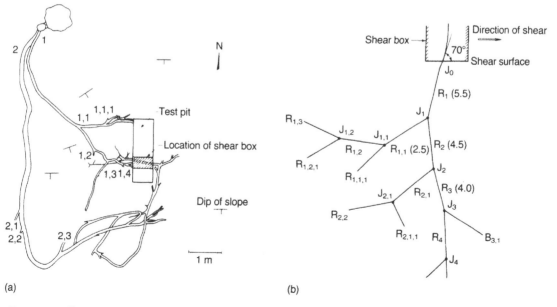

(a)

(b)

Figure 7.15 Root system: (a) root system of a western hemlock (numerals denote root and branch numbers); (b) idealized root system. Numerals in parentheses denote diameters in mm (Wu *et al.*, 1988a).

involves uncertainties. However, available data are sufficient for approximate calculations in a number of cases. These should be verified by *in situ* tests whenever possible. Most of the data collected pertain to tree roots. Data on roots of shrubs and grasses are much more limited.

7.5.1 ROOT GEOMETRY

In the comprehensive sense, root geometry denotes all the properties that are necessary to define the positions and dimensions of the roots in the system. Figure 7.16 shows root systems of a variety of plants and serves to illustrate the wide range in root geometry. Comprehensive descriptions of root morphology of trees have been given by Stout (1956), Sutton (1969), Kozlowski (1971) and others. The major parts of a root system are shown in Figure 7.17. The root crown or root stock includes the bases of the lateral roots and the concentration of small roots beneath the root crown. It may be spherical or heart-shaped. The diameter of lateral roots decreases rapidly with distance from the root crown. This is the zone of rapid taper. The third part consists of lateral roots beyond the zone of rapid taper. In this zone, the lateral roots are oriented nearly parallel to the ground surface. An example of a lateral root system is shown in Figure 7.15(a). Sinker roots may extend vertically downward from lateral roots. In some species, a tap root with branches

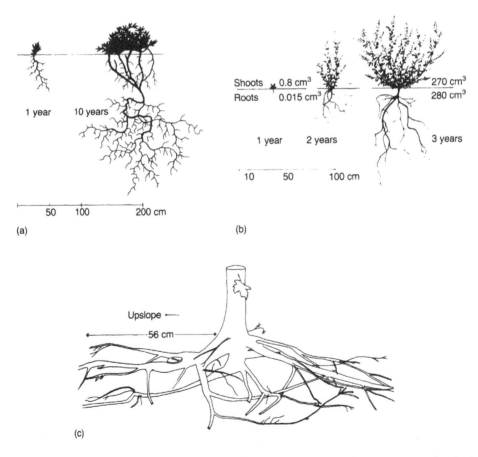

Figure 7.16 Examples of root systems. (a) *Lathyrus sylvestris*, wild pea, (b) *Artemesia vulgaris*, (c). *Acer saccharum*, sugar maple. (a) and (b) from Schiechtl (1980), (c) from Riestenberg (1987).

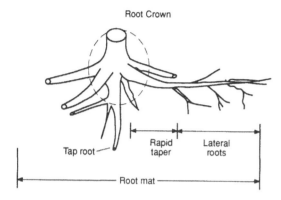

Root Crown

Tap root

Rapid taper

Lateral roots

Root mat

Figure 7.17 Parts of a root system.

may extend to some depth. The mass that contains most of the lateral roots is sometimes called the root mat, which is often exposed after trees have been toppled by wind.

Data from excavated roots have been used to establish the dimensions of components of a root system and their relations with environmental factors. It is known that the depth of a root system is strongly influenced by soil moisture and soil profile. Most of the root system is found within the zone of aeration and few roots extend into impermeable soils or bedrock. Where not limited by the above factors, the root crown and lateral roots of deciduous trees in eastern US usually extend to depths up to 0.5 m (Stout, 1956). Sinker and tap roots may extent to greater depths. In arid regions, some trees and shrubs may grow tap roots as long as 27 m to reach water (Williams and Pidgeon, 1983; Coatsworth and Evans, 1984). Relations between dimensions of different components have also been studied. The diameter of the root mat can be related to the crown and to the stem diameter as shown in Table 7.2.

More detailed correlation between different dimensions of a root system can be developed and used in computer simulations to generate distribution of root diameters at various distances from the stem (Wu *et al.*, 1988c). In the meantime, a more practical approach is to use available data on root density and distribution of root diameters in some of the simplified computations. Data available from excavated roots are summarized in Table 7.3. Figure 7.18 shows the values of area ratio, $a_r = A_r/A$, for a Sitka spruce. Figure 7.18(a) shows that lateral roots with diameters larger than 2 cm are concentrated in a zone with a radius of about 2 m from the stem. Figure 7.18(b) shows the area ratio of roots passing through a plane parallel to the ground surface. The area ratio of roots with diameter less than 2 cm does not change very much with distance. Roots passing through the bottom of the weathered soil, or B-horizon, are limited to those with small diameters. The range in a_r obtained from three trees (Sitka spruce, western hemlock and Alaska yellow cedar) in the same region is between 3.4 and 10.0×10^{-4} (Table 7.3). Figure 7.19 shows the changes of a_r of two sugar maples and a white ash with depth. The soil, at a depth of 40 cm, was sometimes saturated and this may be a factor that limits the depth of the maple roots. The area ratio of sugar maple roots passing through a slip surface in the same area is given in Table 7.3. The growth of roots from cuttings used in soil bioengineering systems (section 7.6) may be expected to be different from that of natural seedlings. Some examples are shown in Figure 7.20. Data on the rate of growth is scarce. Studies were made on roots of cuttings of willows and poplars, one year after planting, by

Table 7.2 Diameter of root mat (D_r)

Diameter of root mat		Source
$D_r = H$ to $D_r = 2H$	Estimated from damage survey	Greenway (1987)
$D_r = 1.5D_c$ to $D_r = 3.0D_c$	Fruit trees	Kozlowski (1971)
$D_r(m) = 2 + 5D_t(m)$, D_r from 4–16 m	Western hemlock and Sitka spruce	Wu (1984a)
$D_t = 18D_t$, D_r from 1.2–1.6 m	Sugar maple and white ash	Riestenberg (1987)

H = height of tree; D_c = diameter of crown; D_t = diameter of stem.

Table 7.3 Characteristics of excavated roots

Site	Soil	Species	Depth	Diameter (cm)	$\dfrac{A_r}{A} = a_r$	Roots s_r (kPa)	θ	$\dfrac{s_r}{a_r}$ (kPa)	Reference
Estimated from a									
Cincinnati, OH	Colluvium, Eden silty clay loam	Sugar maple	Slip surface, above boundary between B&C horizons, 0.5 m	<2.5	1.4×10^{-4}	5.70 4.30[b]	90°	2.8×10^4	Riestenberg and Sovonick-Dunford (1983)
Maybeso Valley, AK	Till, colluvium	Sitka spruce, Western hemlock, Alaska cedar	Slip surface, boundary between B&C horizons, 0.5–1 m	<1.3	$3.7\text{–}10.0\times10^{-4}$ 3.7×10^{-4}	4.3–12.6 5.0[b] 3.4–4.4[b]	90° 90°	1.4×10^4	Wu (1984a) Wu et al. (1979) Swanston (1970)
Hong Kong	Decomposed granite		0–1.5 m	<1.0	$0.5\text{–}15\times10^{-4}$	0.5–10.0[d]		10^4	Greenway (1987)
New Zealand		Willow, poplar cuttings (1 year old)		<1.0	$A_r=5.2$ cm^2				Hathaway (1973)
Netherlands		Marram Grass	15 cm	<0.3	$1.5\text{–}15\times10^{-4c}$	1.5–15[d]	−90°–90°		Wu (1984a)
Alps		Grass	25–75 cm	<1	$2\text{–}8\times10^{-4c}$	2–8[d]	−90°–90°		Schiechtl (1980)
		Willow	1 m	<2	6×10^{-4c}	6[d]			Schiechtl (1980)
		Poplar	0.5 m	<4	2×10^{-4c}	2[d]			Schiechtl (1980)
In situ shear tests									
Oregon	Slickrock Preacher Loam	Hemlock	0.3–0.6 m	<3.0	$10\text{–}80\text{–}\times10^{-4}$	1–8		0.1×10^4	Wu et al. (1988)
California		Pinus contorta						0.1×10^4	Ziemer (1981)
Japan	Loam	Alder			$0\text{–}2\times10^{-4}$	0–1		0.05×10^4	Endo and Tsuruta (1969)
Laboratory shear tests									
California Temescal Ranch, CA	Castaic silty clay loam	Barley Chaparral[a] Grassland[a]	30 cm 20–45 cm 20–45 cm	<.05	$0.2\text{–}0.8\times10^{-4}$	0.6–2.6 −0.6–3.0 −0.9–2.4		3×10^4	Waldron (1977) Terwilliger (1988) Terwilliger (1988)

[a] See Terwilliger (1988) for species inventory.
[b] Calculated from slope failures.
[c] Area is estimated from drawings and photographs of excavated roots.
[d] s_r is calculated with $\sigma_r = 10$ MPa.

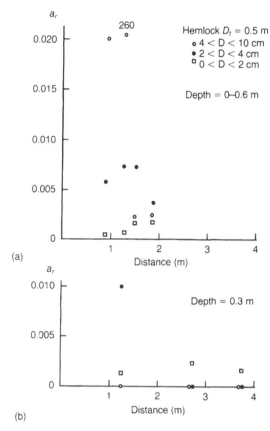

(a)

(b)

Figure 7.18 Root area ratio $a_r = A_r/A$ for a Sitka spruce ($D_t = 0.5\,\text{m}$), Maybeso Valley, Alaska: (a) roots intersecting vertical planes in the organic-weathered soil; (b) roots intersecting a plane parallel to the ground surface and at a depth of 0.3 m in the organic-weathered soil. Bottom boundary of the weathered soil is at a depth of 0.6 m.

Hathaway (1973), and the results are included in Table 7.3.

The maximum depth of roots of grass and forbs in temperate zones is usually less than 0.5 m. Root area ratios of grass and forbs on horizontal planes estimated from drawings are also given in Table 7.3.

7.5.2 STRENGTH PROPERTIES OF ROOTS

Consider first the strength of roots in tension. The strength of the root material may be measured by performing the simple tension test

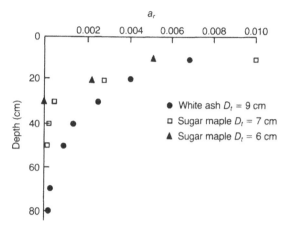

Figure 7.19 Root area ratio of sugar maples and white ash, Cincinnati, Ohio. (Source: Riestenberg, 1987.)

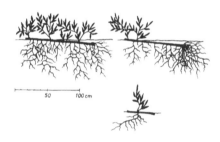

Figure 7.20 Growth of roots from cuttings. (a) Horizontally placed cuttings of *Salix purpurea*; (b) shoot and root development of a cutting of *Salix elaeagnos*, approximately 10 cm in diameter (from Schiechtl, 1980).

on a root segment. Available data on tensile strength are summarized in Table 7.4. The detailed summary by Greenway (1987) shows

Table 7.4 Tensile strength of roots

		Tensile strength (MPa)	Young's modulus (MPa)	Reference
Salix	Willows	9–36	200–300	Hathaway and Penny (1975)
Populus	Poplars	5–38	200–300[a]	Hathaway and Penny (1975)
Alnus	Alders	4–74		
Pseudotsuga	Douglas fir	19–61		
Acer sacharinum	Silver maple	15–30	600[b]	Beal (1987)
Tsuga heterophylla	Western hemlock	27	170[b]	Beal (1987)
Vaccinium	Huckleberry	16		
Hordeum vulgare	Barley	15–31	40–90	Waldron & Dakessian (1981)
	Grass, forbes	2–20		
	Moss	2–7 kPa		Wu (1984b)

Data are from Schiechtl (1980), except where otherwise noted.
[a] Values were estimated from stress–strain curves.
[b] Roots were tested without removing bark; cross-sectional area includes area of bark.

that differences between root strengths of species of the same family can be very large. There also may be large differences between root strengths of one species growing in different locations. Also, for a given species, the tensile strength decreases with diameter (Burroughs and Thomas, 1977). Considering the wide range of strengths, the values given in Table 7.4 may only serve as a rough guide in the estimation of root strength. Data on roots of shrubs are scarce, but the available data indicate that the range is not significantly different from that of trees. A distinction should be made on whether the root was tested with or without the bark and whether or not the diameter includes the thickness of the bark. The Young's modulus is less frequently used than the tensile strength. Limited available data are also given in Table 7.4.

It is necessary to distinguish between the tensile strength of a root segment from the tensile resistance of a root system. Consider the idealized root system shown in Figure 7.15(b). If the tensile force applied at J_0 is increased, tension failure may occur anywhere in the segment R_1. Also, because of the load distribution between branches R_2 and $R_{1,1}$, failure may occur first in $R_{1,1}$ and then in R_2 at which point the root system fails. The tensile resistance and the load–displacement relation of root systems may be measured by *in situ* pullout tests. Consider the case where the applied load is in the direction of the root. Test methods have been described by Wu, McKinnell and Swanston (1979) and Riestenberg (1987). In addition to the strength properties of the soil and root, the load–displacement relation is strongly dependent on the number and orientation of branch roots with respect to the direction of displacement. Examples of typical load–displacement curves are shown in Figure 7.21 together with the respective root geometries. The white ash root (Figure 7.21(a)) consists of a nearly straight main root with two branches. The peak resistance is reached when the main root fails near the branch point at the top of the segment. This case approaches the tensile test on a root segment and the strength, σ_p, in the pullout test would be close to the tensile strength of the root σ_t. In contrast, the sugar maple root is not straight and has many branches. The many peaks in the load–displacement curve (Figure 7.21b) reflect points at which different branch roots fail. Eventually, all the branches fail. The resistance of such a system may be considerably less than that of the main root segment. From the load–displacement curve, peak and ultimate resistances can be determined. The ultimate

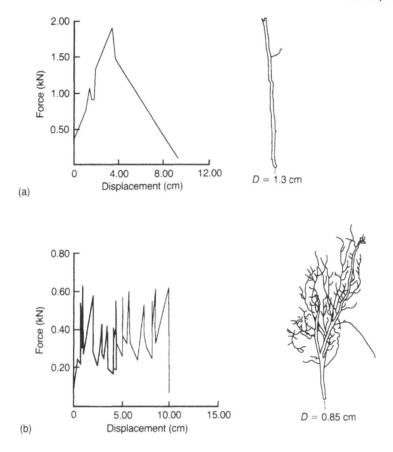

(a)

(b)

Figure 7.21 Typical load–displacement relations: (a) white ash; (b) sugar maple (from Riestenberg, 1987).

resistance is that just before the root system fails, which occurs at axial displacements of 10 cm or greater (Riestenberg, 1987).

Figure 7.22 shows plots of peak and ultimate resistances against the diameter of the root at the pulled end. Also shown are the curves for equation 7.26a with σ_t = tensile strength determined in tension tests on root segments. This relation is approximately the upper limit of the test data. It gives a reasonable prediction of the resistance of the root system for diameters less than 1 cm. This is because the smaller roots have fewer branches than the larger roots and failure of the smaller roots resembles that of the white ash root shown in Figure 7.21(a). Similar conclusions may be drawn from the results of pull-out tests by Wu, McKinnell and Swanston

(1979) on Sitka spruce, western hemlock and Alaska yellow-cedar.

7.5.3 CONTRIBUTION TO SHEAR STRENGTH

If the root is flexible and the tensile force in the root can be estimated, equation 7.23 may be used to calculate s_r. One procedure is to obtain a_r from measurements. If the root diameters are less than 1 cm, use $\sigma_r = \sigma_t$ in equation 7.23b or use equation 7.26a to get T and then equation 7.22c. Values of s_r, estimated from a_r are summarized in Table 7.3a. The values of s_r for Cincinnati and Maybeso Valley were calculated with σ_t and σ_p respectively, but were also checked against the values of $s = s_s + s_r$ at failure of the slopes. Table 7.3 shows the values

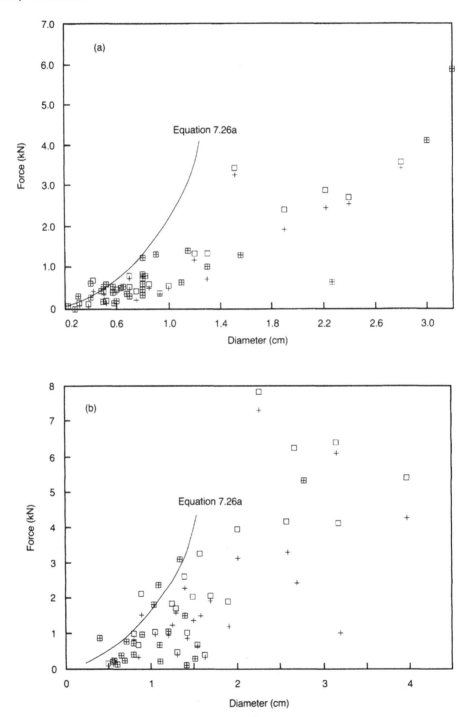

Figure 7.22 Tensile resistance versus diameter: (a) sugar maple; (b) white ash. □, Peak resistance; +, ultimate resistance (from Riestenberg, 1987).

obtained from *in situ* shear tests and laboratory shear tests on soil–root systems, respectively. These results are plotted in Figure 7.23 for comparison. The wide spread in the data is reflected in the large differences between values of s_r/a_r in Table 7.3. The difference between slope failures and *in situ* shear tests should be noted. Values obtained from slope failures are substantially higher than those from *in situ* shear tests. Observations of slip surfaces at Cincinnati and Maybeso Valley showed that most of the roots, including those with diameter up to 2.5 cm, failed by tension. On the other hand, the *in situ* tests at Oregon contained many roots that were cut off by the shear box and many roots, especially those with diameters larger than 1 cm, did not fail by tension. These conditions probably also hold for other *in situ* shear tests. Equation 7.26a is not applicable to roots that did not fail in tension. As an example, T and s_r were calculated with equations 7.23 and 7.24 for two roots with diameters of 0.5 and 2 cm. The results are shown in Table 7.5. The computed values agree with the trend shown in

Table 7.5 Computed values of s_r for two roots

Parameter	Root 1	Root 2
Diameter (D)	0.5 cm	2.0 cm
Area (A)	400 cm^2	400 cm^2
a_r	0.0005	0.008
E_r	27.5 kN/cm^2	27.5 kN/cm^2
$E_r I_r$	0.008 Nm^2	5.27 Nm^2
$\dfrac{E_r I_r}{A}$	0.02N	5 N
t (Figure 7.13	7 cm	12 cm
x	20 cm	20 cm
θ	70°	40°
P_y	7 N/cm	28 N/cm
$T(1)$ (Equation 24)	140 N	560 N
s_r (Equation 23)	4 kPa	15 kPa

Figure 7.23 for *in situ* tests. On the other hand, equation 7.26a, which assumes failure of the roots, overestimates s_r for the root with $d = 2$ cm by an order of magnitude. Thus, the results of *in situ* shear tests are likely to underestimate s_r and represent a conservative estimate of s_r at the failure of a slope.

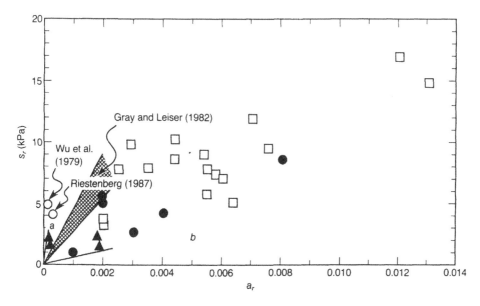

Figure 7.23 Root strength versus area ratio (Wu, 1991): Line a = barley (Waldron, 1977); line b = alder (Endo and Tsuruta, 1969). □, *Pinus contorta* (Ziemer, 1981); ●, hemlock, Oregon (Wu *et al.*, 1988b); ▲, maple, Ohio (Wu *et al.*, 1988b).

Field observations of failures of roots with diameters much larger than 2.5 cm are scarce. For such roots, equation 7.26a is likely to overestimate the resistance of the root system because of the extensive branching often associated with roots of larger diameters. Progressive failure of the root system should be considered. Available data are inadequate for analysis of root systems. Hence, the resistance needs to be determined by *in situ* tests. Tests of the type described by Wu, Bettandapura and Beal (1988c) may be used to determine the relation between T and the shear displacement x. The measured T at a given x may be used in equation 7.22 to compute s_r at that x. If *in situ* tests cannot be performed it may be necessary to estimate T from available test data, such as those shown in Figure 7.11.

Roots may also be subjected to compression. For roots smaller than 1 cm in diameter, failure is usually by buckling. The buckling load is small when compared to the tensile resistance and may be ignored. Experience with compression of larger roots is lacking. However, trial computations can be made to determine the likelihood of buckling failure.

7.6 RELIABILITY

The prediction of stability involves many uncertainties and the computed safety factor represents only an estimate based on the engineer's best choice of input for the computations. The major contributions to uncertainty come from uncertainty about future events or loads, uncertainty about material properties, and uncertainty about analytical models. Our estimate of the material properties may be in error because of the variability of natural materials and inaccuracies in the strength model used to interpret test results. The methods for estimating u (section 7.3) and for stability analysis (section 7.2) all involve simplifications about the geometry of the slope and about the physical processes. The combination constitutes the error in the analytical model. Reliability methods can be used to estimate the effect of the uncertainties on the output or the computed safety factor.

In reliability analysis, the uncertain quantity is treated as a random variable with a mean and a variance. The mean should represent the engineer's best estimate, without conservative assumptions, and the variance should represent the engineer's estimate of the uncertainty. The mean and variance of the input can be used to estimate the uncertainty about the output. A simple procedure is the first-order second-moment procedure (Ang and Tang, 1975). The mean and variance of the output Y, which is a function, g of the random variables $X_1, X_2 \ldots$, is

$$E[Y] = E[g(X_1, X_2 \ldots)] \qquad (7.27a)$$

$$\text{Var}\ [Y] = \sum_i \sum_j \frac{\partial g}{\partial X_i} \frac{\partial g}{\partial X_j}\bigg|_m \text{Cov}(X_i, X_j). \qquad (7.27b)$$

The errors associated with estimation of soil strength have been recognized by engineers. The uncertainties are represented as (Tang, Yuceman and Ang, 1976)

$$Y = \prod_i^n N_i X, \qquad (7.28)$$

where Y is the average *in situ* strength over the area of the slip surface. N_i is the correction factor for the ith source of error, and X is the estimated soil strength.

When used in a stability analysis, the error in the analytical model can be included as N_i. Uncertainties are expressed as coefficients of variation of the N_is. According to the first-order second-moment method, if the N_is are independent, the coefficient of variation of the resistance is

$$\Omega^2(R) = \sum_i \Omega^2(N_i). \qquad (7.29)$$

Estimated uncertainties encountered in geotechnical design and construction of soil slopes are summarized in Table 7.6.

Uncertainties about s_r are more difficult to estimate because of lack of data. Temascal

Table 7.6 Uncertainties in geotechnical design of slopes (after Tang, Yuceman and Ang, 1976)

	Coefficient of variation
Soil variability	0.10–0.15
Strength model	0.10–0.20
Analytical model	0.08–0.10
Resistance	0.16–0.27

Ranch is one of the few sites where systematic sampling was done over a fairly large area (16 ha approximately). The distributions of s_r in chaparral and grassland (Terwilliger, 1988) have coefficients of variation of 1 or larger. This is larger than what may be expected from the other data shown in Table 7.6. Nevertheless, it is obvious that the errors associated with estimating s_r are much larger than those associated with estimating s_s. To get an idea of the uncertainty about s, consider a soil with $c' = 0$, $\phi' = 30°$, $\gamma = 18\,\text{kN/m}^3$. At a depth of 0.6 m, $\sigma \approx 6\,\text{kPa}$ if the soil layer is saturated, and $s_s \approx 3\,\text{kPa}$. If $E[s_s] = 3\,\text{kPa}$, $\Omega[s_s] = 0.2$, $E[s_r] = 3\,\text{kPa}$, $\Omega[s_r] = 0.6$, we obtain, via equation 7.27, $\Omega[s] = 0.3$. This shows that at shallow depths, or low s_s, uncertainty about s_r can have a significant influence on the uncertainty about s.

In the estimate of pore pressure, precipitation is one of the key inputs. Since future storms cannot be predicted with certainty, the precipitation to be used in the pore pressure computations is a random variable. One approach is to use the largest rainfall for the design life of the structure, which may be 20–50 years. The mean and variance of the largest storm may be determined using extreme value statistics.

For a given design life, the reliability is

$$P_r = 1 - P_f, \quad (7.30)$$

where P_f is the failure probability. Failure probability is the probability that the random variables $X_1, X_2 \ldots$ will take on such values so the $F_s \leqslant 1$. If F_s is assumed to be a log-normal distribution, the failure probability is

$$P_f = \Phi\left[-\overline{\ln F_s}/\text{Var}^{1/2}(\ln F_s)\right]. \quad (7.31)$$

Using the first-order second-moment approach (equation 7.27), the mean and variance of $\ln F_s$ can be computed. The reliability index, β, is a measure of the safety factor after accounting for the uncertainties,

$$\beta = (\overline{F}_s - 1)/\text{Var}^{1/2}[F_s]. \quad (7.32)$$

For a given coefficient of variation of F_s, the mean safety factor required to achieve a level of safety can be calculated with equations 7.31 and 7.32. The results are given in Figure 7.24. The historic failure probability of earthworks is between 5×10^{-3} and 10^{-3} with design safety factors of 1.5 (Meyerhof, 1970). For P_f between 10^{-2} and 10^{-3}, or $\beta \approx 2$, the ranges in $\Omega(F_s)$ and $\overline{F}_s$ are indicated by the area shaded in Figure 7.24. Consider the preceding example, in which the uncertainty about s_r may be large. If we assume that the load is deterministic and $\Omega(R) \approx \Omega(s)$, the value of $\Omega(R)$ for a slope stabilized with vegetation may be expected to be near the upper limit of the shaded area. The safety factors to be used in design should be increased accordingly.

7.7 SOIL BIOENGINEERING SYSTEMS

Vegetation for slope stabilization ranges from grasses to shrubs and trees. These may be

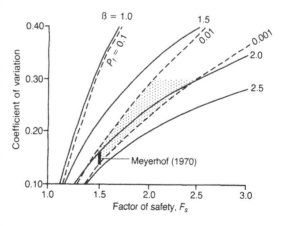

Figure 7.24 Relationship between safety, mean factor of safety and coefficient of variation.

established by conventional seeding or live planting (section 3.44–3.46). Specialized methods have been developed for establishing vegetation on slopes and these are called soil biotechnology or soil bioengineering systems. These construction methods use mainly unrooted cuttings, which are taken from live plants and installed in the ground by various means and in various configurations. The plant cuttings take root and become established on the slope. Soil–bioengineering systems and construction methods have been described by Gray and Leiser (1982), Schiechtl (1980), Coppin and Richards (1990). Some common systems are summarized in Table 7.7 and Figure 7.25. Most of the systems serve the dual purpose of reducing surface erosion and reinforcing the soil. Phreatophytes such as willows are effective at increasing evapotranspiration. A system's effectiveness as soil reinforcement depends on the depth at which the cuttings can be placed and the depth to which the roots will penetrate. The growth rate of roots is related to the volume of the cutting and some guides on choice and preparation of cuttings have been given by Gray and Leiser (1982) and Schiechtl (1980).

As shown in section 7.5, the root properties of vegetation can vary over a wide range. Hence, slope stabilization by vegetation requires judicious choice of the type of vegetation. For stability, the species should have a root system that extends to a sufficient depth. In humid regions plants with high transpiration will reduce soil moisture and pore pressure. Plant characteristics that should be considered in the choice of species are summarized in Table 7.8. Wherever feasible, native vegetation is preferred and the succession from pioneer to climax vegetation in the site environment, primarily climate and soil type and moisture, should be considered. This subject is treated in section 3.3.2 and in Gray and Leiser (1982) and Schiechtl (1980).

Live plantings and/or soil bioengineering systems may provide adequate resistance in many cases. They may also be used in combination with conventional retaining structures (section 3.4.7). In this case they serve as

Table 7.7 Summary of soil bioengineering systems

Name	Construction	Primary function(s)
1. Live stake	Sticks are cut from rootable plant stock and tamped directly into ground	Live plants reduce erosion and remove water by evapotranspiration. Plant roots reinforce soil
2. Live facine (wattling)	Sticks of live plant material are bound together and placed in a trench. They are tied to the ground by live stakes (Figure 7.25(a))	Same as 1
3. Brush mattress	Live branches are placed close together on the surface to form a mattress (Figure 7.25b))	Same as 1. In addition, it provides immediate protection against erosion
4. Brushlayer, branchpacking	Live branches are placed in trenches or between layers of compacted fill (Figure 7.25(c) and (d))	Same as 1
5. Vegetated geogrid	Live branches are placed in layers between compacted soil wrapped in geogrid (Figure 7.25(e))	The geogrid provides immediate stability. The plants serve the same functions as in 1
6. Rooted plants	Rooted plants grown in a nursery or in the wild are planted	Same as 1. In addition, roots provide buttressing

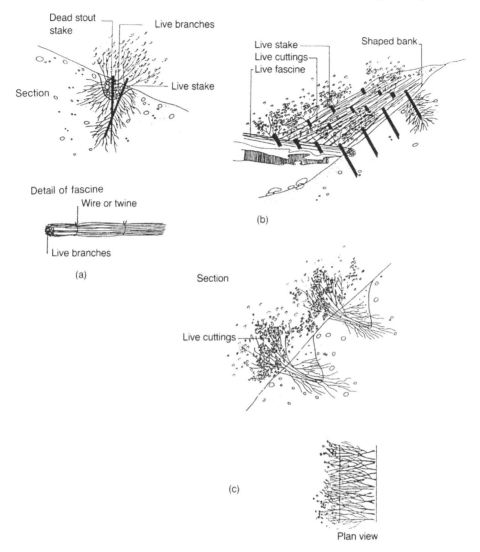

Figure 7.25 Soil bioengineering systems: (a) live fascine; (b) brush mattress; (c) brushlayer; (d) branch packing; (e) live soft gabion. Leaves and roots are not representative of condition at time of installation (Robbin B. Sotir and Assoc.).

Table 7.8 Characteristics of plant groups

1. Ecological criteria	Resistance to drought, salt, and temperature extremes
2. Growth characteristics	Ease of propagation, growth rate
	Requires consideration of cutting material, humidity, temperature, light, soil type and time of propagation
3. Engineering properties	Root strength, depth and diameter of root systems, water use

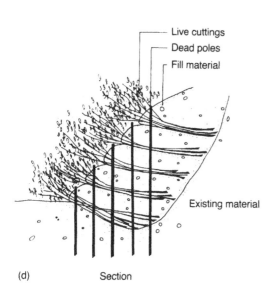

Live cuttings
Dead poles
Fill material

Existing material

(d)   Section

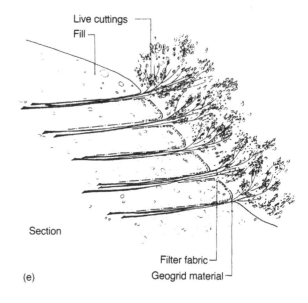

Live cuttings
Fill

Section

Filter fabric
Geogrid material

(e)

Figure 7.25 (cont.)

supplementary measures. A common example is illustrated in Figure 7.26. The retaining structure is necessary to provide stability against a deep slip surface (a in Figure 7.26) whereas vegetation is used to prevent erosion and shallow slips (b in Figure 7.26) on the slope above the structure. Such a combination allows the use of a smaller retaining structure. Vegetation may also be grown in openings of structures such as crib walls and grids and in interstices of rip-rap, revetments, and gabions to reinforce the soil behind these structures (section 3.4.7). Detailed descriptions of many vegetation–structure combinations may be found in Gray and Leiser (1982).

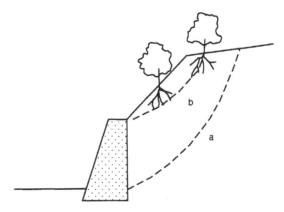

Figure 7.26 Combination of a retaining wall with vegetation for slope stabilization. a = deep slip surface; b = shallow slip surface.

7.8 EXAMPLES OF SLOPE STABILIZATION

Vegetation has been used for slope stabilization under a variety of conditions. The following examples illustrate some recent applications. It is important to note that the choice of vegetation is not always based exclusively on slope stability and plant survival but often requires consideration of ecology. This lies beyond the

scope of this chapter but some of the considerations are mentioned in the examples.

7.8.1 HILLSIDE SLOPES

Steep hillside slopes with shallow soil cover are vulnerable to failures. A common problem is slope failures following removal of vegetation.

Logging on slopes has been shown to increase the frequency of slope failures. The solution is to restore vegetation where necessary or refrain from clearing potentially unstable slopes. Stability analysis can be used to estimate the safety of slopes before and after clearing and to identify zones of high risk.

An example is the effect of logging on the slopes in Maybeso Valley, Alaska. Figure 7.10(a) shows a typical slope that was clear-cut and the slide that took place several years after clearing. The original forest was composed of Sitka spruce (*Picea sitchensis*), western hemlock (*Tsuga heterophylla*), and Alaska yellow cedar (*Chamaecyparis nootkatensis*). Investigation of the site showed that roots of trees decayed after the trees were cut. Because the slope was nearly uniform it was analysed as an infinite slope using equation 7.7. The root area ratio measured in excavations was 3.7×10^{-4} (Table 7.3). Using $\sigma_r = 10$–$12 \, \text{MPa}$, equation 7.23b gives $s_r = 4.2$–$5.5 \, \text{kPa}$. From tests of dead roots, it was found that four years after the trees were cut, the strength was about one-sixth of that of live roots. This was used to calculate s_r four years after clear-cutting. The maximum measured piezometric levels were used to calculate u_s in equation 7.7. Results of stability analysis are given in Table 7.9. It shows that the reinforcement provided by roots was largely responsible for the stability of the forested slope. In addition, the measured z_w in the forested slope was considerably lower than that in the clear-cut slope. This also contributed to the larger safety factor of the forested slope. Such analysis may be used to plan timber

harvests. Clear-cutting should be avoided where the failure probability exceeds the acceptable level or where the expected cost of failure probability exceeds the projected income (Wu and Swanston, 1980).

When investigating potential failures, measured piezometric levels are often not available. Then z_w can be computed as described in section 7.3, Figures 7.9 and 7.10. The surface flux was taken to be equal to the rainfall minus the evapotranspiration, which was computed by Thornthwaite and Mather's (1957) method. The measured and computed piezometric levels are compared in Figure 7.27. The wide scatter in the measured piezometric levels is noteworthy. This can only partly be attributed to local topography and illustrates the difficulty encountered in estimating u. However, the effect of trees on pore pressure is clear when one compares the pore pressures for 1965 and 1974. The pore pressures for 1965 were measured four years after clear-cutting and the pore-pressures for 1974 were measured in these same slopes covered with regrowth. Although the precipitation was much greater in 1974 than in 1965, the pore pressure was about the same. Calculations show that increased evapotranspiration and increased storage capacity could account for the difference. Pore pressure changes caused by infiltration of snow-melt have also been found to be different in forested and clear-cut slopes (Megahan, 1984).

A similar study by Gray and Megahan (1981) covered a large area called the Idaho Batholith in central Idaho. Slope failures were found to be related to road building, logging and fires. The

Table 7.9 Stability analysis of slopes in the Maybeso Valley, Alaska (after Wu, McKinnell and Swanston, 1979)

Case	z (cm)	$z_w \cos \alpha$ (cm)	α (degrees)	s_r (kPa)	F_s	Condition and time
1	120	75	39	1	0.9	Clear-cut slope, 1961–1965
2	120	50	39	1	1.0	Clear-cut slope, 1961–1965
3a	100	38	50	6	1.3	Forested slope, 1974
3b	100	>38	50	6	<1.3	Forested slope, 1969
4	100	38	35	6	1.8	Forested slope, 1974

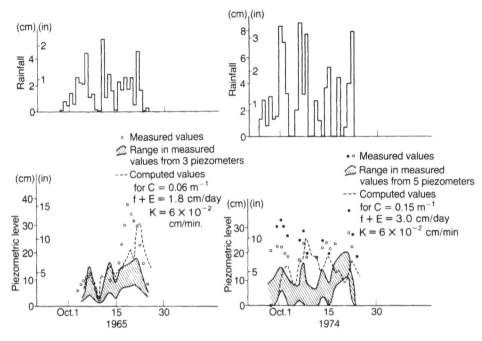

Figure 7.27 Measured and computed piezometric levels, Maybeso Valley, Alaska. (a) 1965, (b) 1974, 4 years after clear-cutting. d = drainage; e = evapotranspiration; k = permeability; C = storage coefficient.

study was made to develop methods for identifying critical areas where trees and vegetation should be left undisturbed in order to avoid failures. Some of the results are summarized here. The soil consisted of a sand (SW–SM according to the Unified Classification) whose shear strength varied over a considerable range: $\phi = 29°–37°$, $c = 0–6$ kPa. Surveys have shown that the average opening between root mats was around 2 m. Calculations were made with equation 7.10 and $\phi_1 = 0.5\phi$, $c_1 = 0.12c$, $\phi = 35°$, $c = 2.8$ kPa, $z = 0.9$ m. This yielded $B_{cr} = 6.3$ m. It should be noted that B_{cr} is very sensitive to c and c_1 which can vary over a considerable range. Nevertheless, the above figures suggest that buttressing would be effective. Then the soil and root mat would move as a unit and the stability was evaluated with equation 7.7. Using $\phi = 35°$, $\alpha = 35°$, $z = z_z = 0.9$ m, it was found that in order to obtain $F_s > 1$, c' should be greater than 4.5 kPa. Considering the range in c, there a likely to be areas where c' exceeds 4.5 kPa, whereas in other areas, c'

would be less. A value of $a_r = 4.5 \times 10^{-4}$ was considered a lower limit for roots across the potential slip surface in this region. From equations 7.26a and 7.23a and σ_t from Burroughs and Thomas (1977), $s_r = 1$ kPa was obtained. Then $c' + s_r$ is larger than the cohesion required for stability. Thus, tree roots can be effective in preventing failure where c' is less than that required for stability. This emphasizes the importance of 'vegetation leave areas' where trees and woody vegetation should be left undisturbed (Gonsier and Gardner, 1971).

Another example is the contribution of trees to the stability of colluvial slopes in the Cincinnati area (Riestenberg and Sovonick-Dunford, 1983). The residual shear strength of the soil was estimated to be $c' = 0$, $\phi' = 12°$. If the soil is saturated with seepage parallel to the slope, calculations with equation 7.7 show that slopes with angles of 10° or larger are unstable. Hillside slopes in this area commonly have slopes from 15° to 35°. Root areas determined from excavations at the Rapid Run slide are given in

Table 7.3. Calculations with equation 7.7 using $s_r = 4.3\,\text{kPa}$ show that a slope with $\alpha = 33°$ would be fairly stable. Thus, clearing of slopes with angles greater than $10°$ will likely lead to failures and should be avoided.

7.8.2 EMBANKMENT AND CUT SLOPES

Slopes of embankments and cuts are subjected to infiltration of precipitation. If the soil has a low permeability and the groundwater is located at some depth below the surface, saturation during periods of prolonged rainfall progresses from the surface downward. Because the suction is reduced to zero in the saturated zone, equation 7.8b shows that a shallow slide would occur if ϕ' is equal to or less than the slope angle

α for soils with $c' = 0$. For many clays of low plasticity, the softened strength (Skempton, 1964) is around $\phi' = 30°$ and $c' = 0$. Hence, it is not surprising that shallow slips are frequently found on clay slopes with α close to $30°$. Vegetation roots increase the shear strength by s_r, which can significantly increase the safety factor if the roots extend below the slip surface.

An example is the embankment failure on route I-77, near Caldwell, Ohio, shown in Figure 7.28 (Wu, Randolph and Huang, 1993). During wet seasons, the slope was found to be saturated to a depth of up to 0.6 m. Shallow slips and their dates of occurrence are also shown. The embankment material consisted of a clay derived from weathering of a red Conemaugh shale. The shear strength of the clay

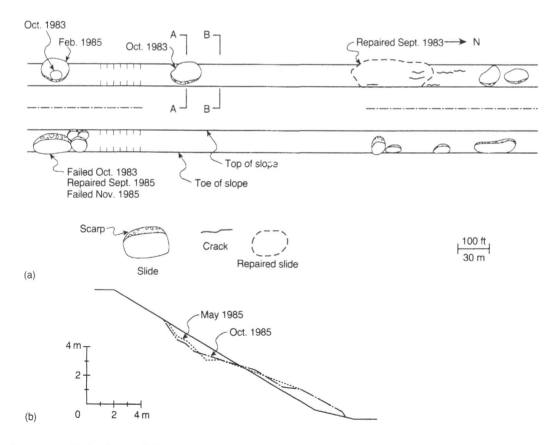

Figure 7.28 Embankment failure, I-77, Ohio: (a) plan; (b) section A–A (Wu *et al.*, 1992).

was approximately $\phi' = 30°$, $c' \approx 0$. The slope of the embankment was approximately 27°. Thus, if the slope is saturated by infiltration from the surface, the safety factor would be close to 1, according to equation 7.8. If seepage parallel to the slope occurs the safety factor, according to equation 7.7b, is less than 1. The measured suctions at section B (Figure 7.28) on 30 July, 15 October and 12 November 1985 are shown in Figure 7.29. The rain began on 1 November. On 12 November, saturation or $\Psi = 0$, was found at a depth of 0.6 m. This was accompanied by an increase in volumetric water content Θ at the same depth. The period of 1–15 November was one of the wettest in history. Therefore, the saturation depth shown in Figure 7.29(a) can be considered as the maximum depth of the saturated zone. Comparison of the measured suction with the results of calculations using equation 7.14 and measured permeability indicate that the soil near the surface could have a much higher permeability than the soil below.

Then, seepage parallel to the slope could occur through the surface layer and failure would result.

In the same region, shallow surface slips are also common on cut slopes in red Cavenaugh shales. The failure mechanism is similar to that of the embankment because the weathered shale near the surface has a shear strength close to that of the embankment material.

One of the measures for repair of the embankment was to use live planting on the slopes. Black locust (*Robinia pseudoacacia*) seedlings were planted at 1.2 m spacings. Black locust is a pioneer species and is expected to establish quickly. Other native species, such as oaks and maples, may also become established voluntarily as a process in plant succession. The contribution of the tree roots to the shear strength can only be estimated. The data given in Table 7.3 show that a_r for most species is greater than 10^{-4}. Using $a_r = 2 \times 10^{-4}$ and $\sigma_r = 20\,\text{MPa}$ in equation 7.23 gives $s_r = 4\,\text{kPa}$. Using this in

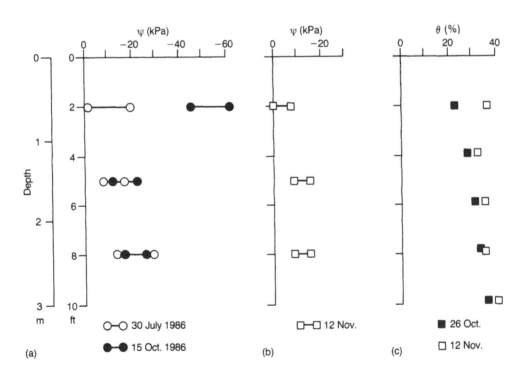

Figure 7.29 Measured and computed suctions, I-77, Ohio (Wu *et al.*, 1992).

equation 7.8a gives an adequate safety factor.

A combination of soil bioengineering systems and mechanical reinforcement was used in the reconstruction of an embankment slope on State Highway 126, near Marion, North Carolina (Figure 7.30) (Sotir and Gray, 1989). Most of the slides consisted of shallow slips that extended to depths of 1–1.5 m. The embankment material consisted primarily of clayey to sandy silt, ML and SM, according to the Unified Classification System. The shear strength, as measured in triaxial tests, was variable; $\phi' = 30°$ and $c' = 0$ were considered representative of the average. Since the embankment slope was also 30°, the failure mechanism was believed to be similar to that in the preceding example. Live brushlayers

consisting of stems 1.25–5 cm in diameter, were placed in three directions as shown in Figure 7.30, with about 15 stems/m in each direction. These layers were placed at 0.75 m spacing vertically. Tensar SS2 geogrids were placed at 1.5 m spacing vertically, in addition to the brush-layers. Where new material was placed over the old slope, as shown in the cross-sections, roots from the brushlayers were expected to grow into the old slope and tie the two parts together. Live staking was used over the entire area to tie the various systems together and prevent shallow slips. Plant species used included both woody and herbaceous vegetation. The woody species used in brushlayer construction were: black willow (*Salix nigra*), willow (*Salix* sp.), redtwig

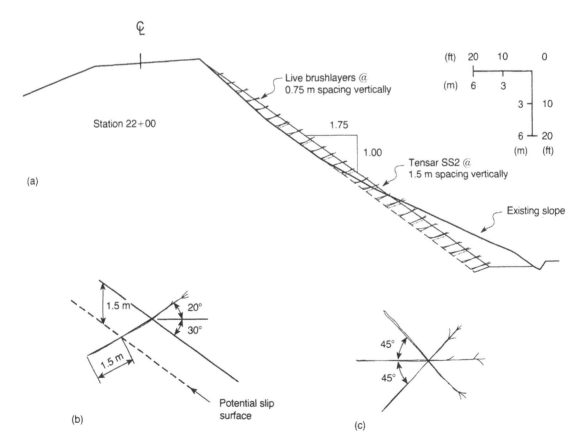

Figure 7.30 Embankment on Route NC126: (a) cross-section; (b) side view of brushlayer; (c) top view of brushlayer (Soil Bioengineering Corporation, 1986; Robbin B. Sotir and Assoc.).

dogwood (*Cornus* sp.), river birch (*Betula* sp.) and privet (*Ligustum* sp.). These were chosen because their cuttings root readily and because they are abundant in this locality. A mixture of grasses and legumes was used for the herbaceous cover.

The following calculations illustrate the methods that may be used to evaluate the effectiveness of soil reinforcement by soil bioengineering systems. The infinite slope analysis is considered appropriate for shallow slips. Immediately after installation, the stems would have no roots and the maximum tension is equal to what can be developed by bond between soil and stem. Assume that the suction is zero, then the interface resistance in equation 7.26b is

$$f = \gamma z \tan\delta, \qquad (7.33)$$

where δ is the friction between soil and stem. A conservative estimate is $\tan \delta = \frac{1}{2} \tan \phi'$. The portion of the stems that extends below the potential slip surface (Figure 7.30b) gives $L = 1.5$ m. For $D = 1.27$ cm and $T = 596$ N, equation 7.23a give $s_r = 2.67$ kN/m of slip surface and equation 7.8 gives $F_s = 1.68$. For comparison purposes, the contribution of the geogrid is computed in the same way. The creep strength of Tensar SS2 is about 7.41 kN/m (Tensar Corp., 1987). When the geogrids are placed at 0.75 m spacings vertically, $s_r = 2.37$ kN/m and $F_s = 1.4$. When the geogrids are placed at 1.5 m spacings in combination with the live brushlayers at 0.75 m spacings, $s_r = 3.56$ kN/m and $F_s = 1.9$. The long-term stability will depend on the roots that grow from the stem. This cannot be predicted because of inadequate knowledge. One assumption is that the root system would ultimately approach that of a tree. Table 7.3 shows that there is little information on roots at depths greater than 1 m. Hence the long-term strength cannot be estimated. The safety factors in this and the preceding cases are larger than that normally used for conventional highway slopes. This is consistent with the larger uncertainties associated with slopes reinforced with vegetation as shown in Figure 7.24.

The use of vegetation to prevent shallow slips on cut slopes is illustrated with the slope at Kananaskis County, near Canmore, Alberta. Figure 7.31 shows the shallow slips and erosion on the slope before reconstruction. The average slope angle, α, was approximately 45°. The soil bioengineering system used in the reconstruction is shown in Figure 7.32. The vegetated geogrids were used to provide stability to the steepest part of the slope and were designed to function in the same way as a conventional

Table 7.10 Live plant materials and method used at Kananaskis

Common name	Botanical name	Typed used
Alder	*Alnus crispa*	Cuttings
Balsam poplar	*Populus balsamifera*	Cuttings
Bog birch	*Betula pumila*	Cuttings
Buffaloberry	*Shepherdia conadensis*	Cuttings and rooted
Bush cinquefoil	*Potentilla fruiticosa*	Rooted
Gooseberry	*Ribes oxyacanthoides*	Rooted
Lodgepole pine	*Pinus contorta latifolia*	Cuttings
Prickly rose	*Rosa acicularis*	Rooted
Redtwig dogwood	*Cornus canadensis*	Cuttings
Trembling aspen	*Populus tremuloides*	Cuttings
Willow	*Salix maccalliana*	Cuttings
Willow	*Salix scouleriana*	Cuttings
Wolf willow	*Elaeagnus commutata*	Cuttings and rooted
Indian paintbrush	*Castilleja miniata*	Seed
Legume	*Hedysarum sulphurescens*	Seed

Figure 7.31 Kananaskis slope, Alberta, before reconstruction (Soil Bioengineering Corporation, 1987a; Robbin B. Sotir and Assoc.).

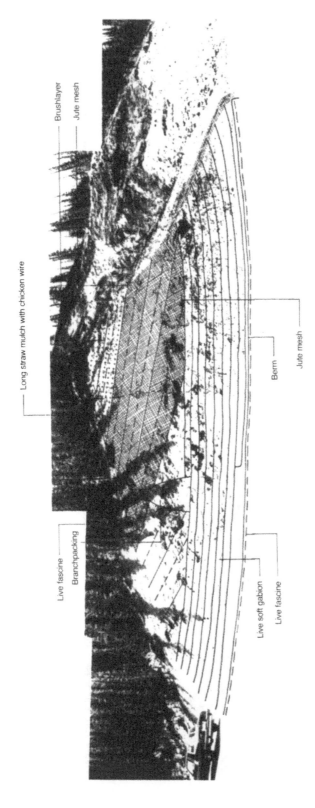

Figure 7.32 Soil bioengineering system for Kananaskis slope, Alberta (Soil Bioengineering Corporation, 1987a; Robbin B. Sotir and Assoc.). Live stakes were installed over the entire site except in the live soft gabion area and the branchpacking area. Rooted plants were installed over the upper slope area and above the live soft gabions, except in the branchpacking area.

gravity wall. Brushlayer and the live fascines used in the upper part of the slope provide soil reinforcement against shallow slips. Table 7.10 lists the species used in the construction. The soil was a sand derived from an alkaline sandy till. Analysis similar to those for the two preceding examples shows that the brushlayer should provide adequate reinforcement. As in all such constructions, a requirement is that the plant material should develop an adequate root system that extends below the potential slip surface. A point worth noting is that the soil at the site had a pH of 9.0. Mixing with soil from a nearby borrow pit and addition of potash and fertilizer lowered the pH to 7.8. A year after construction, plant growth was found to be medium to poor, largely because of the alkalinity of the soil. No shallow slips and no erosion rills were found on the slope at that time.

A slope in Hong Kong has been described in detail by Greenway (1987) and is summarized here to illustrate the effect of root reinforcement after trees have become established. The embankment, which was 9 m high and had a slope angle of 34°, was built in 1958. Acacia (*Acacia confusa*), Chinese banyan (*Ficus microcarpa*) and candlenut (*Aleurites moluccana*) were planted on the slope. In 1984, the trees were 25–54 cm in diameter, spaced at 5–10 m. The decomposed granite in the embankment was partially saturated. The suction was measured and its value used in equation 7.2d to obtain the shear strength of the soil. Roots of trees were excavated, the area ratios and strengths were determined (Table 7.3) and s_r was calculated from equations 7.26a and 7.22. Several potential slip surfaces were investigated. Stability analysis using Janbu's (1973) method showed that the roots would increase the safety factor from 1.2 to between 1.4 and 2.0, depending on the root area ratio used. For the three-dimensional slip surface (Figure 7.5) and using equation 7.12, the computed safety factors were 2.4 and 3.3, with and without roots respectively. In this case, the slope was stable initially and the trees have increased significantly the factor of safety.

7.8.3 RIVER BANKS

Erosion of river banks is a common phenomenon. The stability of the bank can be analysed by the methods described in section 7.2. The pore pressure is estimated from the flow net as shown in Figure 7.7. If necessary, the suction above the water table may be estimated by the methods outlined in section 7.3. For low banks, where the slip surface is likely to be shallow, vegetation roots are likely to intersect the potential slip surface and become an effective reinforcement. Then the constructions described in the preceding section may be applied. A separate factor is the undercutting of the toe of the slope by streamflow. Figures 7.33 and 7.34 show a combination of soil bioengineering systems proposed for bank stabilization at Winkler Creek, Virginia. The soil is a sandy loam of the Potomac Soil Group. Live fascine is proposed for soil reinforcement to prevent erosion on the banks. Vegetated geogrid will be used on steep banks on outside bends of the creek to prevent undercutting. Brush mattress will be used where the slope is less steep. Woody species will include willow (*Salix* sp.), redtwig dogwood (*Cornus* sp.), button bush (*Cephalanthus occidentalis*), alder (*Alnus* sp.) and aspen (*Populus tremuloides*). Cuttings of willow and redtwig dogwood root readily. The survival rate of the other species is expected to be lower but they are included to provide a diverse plant community that will adapt to locally variable growing conditions. All these species are found in the locality. The average soil pH is 4.7 and available phosphorous, nitrogen and calcium levels are low. Addition of lime and fertilizer is recommended to assure success of the soil bioengineering systems. Natural plant succession on the site is likely to bring in red maple, tulip, ash and sweet gum in the intermediate stage. The climax vegetation is expected to be the oak–hickory forest.

7.8.4 SLOPES ON PERMAFROST

Several types of flow landslides, including skin flows and solifluction, that occur on low angle

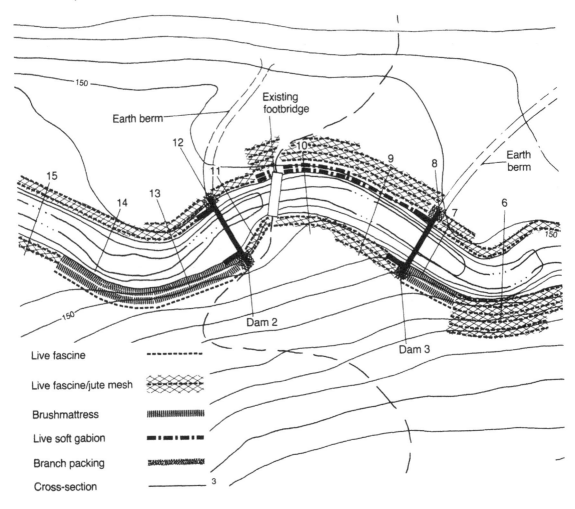

Live fascine - - - - - - - - -

Live fascine/jute mesh ✕✕✕✕✕✕✕

Brushmattress |||||||||||||||||||||||||

Live soft gabion ▬ ▪ ▬ ▪ ▪ ▬

Branch packing ▨▨▨▨▨▨▨

Cross-section ———— 3

Figure 7.33 Soil bioengineering systems for Winkler Botanical Preserve (Soil Bioengineering Corporation, 1987b; Robbin B. Sotir and Assoc.).

slopes are attributed to high pore pressures during thaw (McRoberts and Morgenstern, 1974). As shown in equation 7.9, the safety factor depends on the pore pressure parameter R, which in turn depends on the rate of thaw and rates of generation and dissipation of pore pressure. Natural organic cover, such as moss, serves as an insulating material that retards thaw and as a reinforcement that restrains movements in the thawed material. When the organic cover is destroyed, catastrophic movements can take place. Where the safety factor

of a slope is inadequate, the stability can be improved by using a surcharge to increase the effective stress. Also, an insulating layer may be placed on the ground surface to reduce the rate of thaw. When an insulating layer is present, a numerical solution is used to obtain F_s and design charts have been presented by Pufahl and Morgenstern (1979). These may be used to compute the amount of insulation and surcharge required to produce an adequate safety factor. Peat may be used for insulation, although it may not be readily available in northern

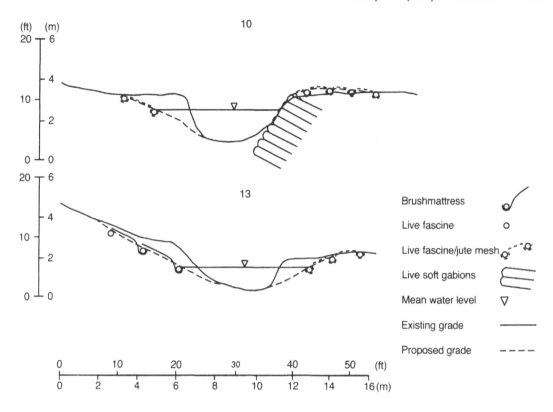

Figure 7.34 Cross-sections of river bank, Winkler Botanical Preserve (Soil Bioengineering Corporation, 1987b; Robbin B. Sotir and Assoc.).

regions. Where synthetic insulation is used, vegetation grown on top of the surcharge contributes additional insulation, although its thermal properties are not well known.

One type of low-angle flow slide is the bimodal flow which is characterized by a low-angle tongue and a steep headwall. Bimodal flows may be initiated by skin flows or other slope failures which expose ice-rich soil in the headwall or scarp (McRoberts and Morgenstern, 1974). The exposed permafrost in the scarp thaws rapidly. The melt is removed and the headscarp retreats – a process called ablation. The heat flux required for continuous melting and retreat has been studied by Pufahl and Morgenstern (1980). The temperature boundary conditions can be changed by covering the surface with organic vegetation or synthetic insulation. Natural organic vegetation that is draped

over the headscarp may effectively retard the retreat. However, this process is only effective if the height of the headscarp is less than 3 m. Wire mesh may be used to strengthen the organic vegetation and increase the height. The bimodal flow is also a good representation of instability of many cut slopes. Small cuts less than 3 m high are capable of stabilizing themselves by the above process. Figure 7.35 shows the evolution of a cut slope. Use of synthetic materials for stabilization has been described by Phukan (1985).

7.8.5 CONCLUDING REMARKS

The above examples show diverse ways that vegetation can be used for slope stabilization. In the near future, rapid growth in new construction methods is expected, especially in the

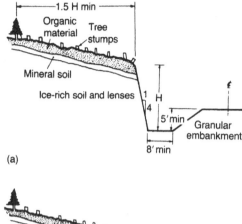

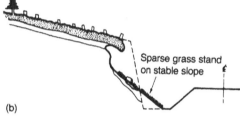

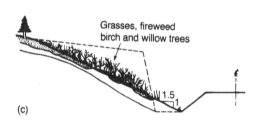

Figure 7.35 Idealized development of stability in ice-rich cut (from Berg and Smith, 1976). (a) Initial conditions of cut in frozen soil. (b) End of first thaw season. Slope is mostly unstable and very unsightly; the ditch will require cleaning if massive ice is present. (c) End of fifth or sixth thaw season. Slope stabilizes with reduced thaw and vegetation established. Free water from minimal thawing is used by plants whose root systems develop new organic material.

combined use of vegetation with synthetic reinforcement systems. The mechanisms of soil reinforcement and their effectiveness will require evaluation. It is hoped that the principles outlined in this chapter will be useful in evaluation of new construction techniques as well as in design of construction using current techniques.

ACKNOWLEDGEMENTS

I am grateful to several individuals that have made significant contributions to the writing of this chapter. Robbin B. Sotir provided several of the examples in section 7.8 and helped in the writing of sections 7.7 and 7.8. Information on root and/or soil properties was made available to me by Mary M. Riestenberg, E. David Penny, Robert R. Ziemer, and W. D. Bingham. Donald H. Gray and Lakshmi N. Reddi reviewed the manuscript and made many meaningful comments.

REFERENCES

Ang, A. H.-S., and Tang, W. H. (1975) *Probability Concepts in Engineering Planning and Design*, Vol 1, John Wiley and Sons, New York.

Beal, P. E. (1987) Estimation of the shear strength of root reinforced soils. MS Thesis, Ohio State University, Columbus, OH.

Berg, R. and Smith, N. (1976) Observations along the pipeline haul road between Livengood and the Yukon River, *Cold Regions Research and Engineering Laboratory, Special Report* 76–11, Hanover, NH.

Beven, K. (1981) Kinematic subsurface storm flow. *Water Resources Research*, 17, 1419–24.

Beven, K. (1982) On subsurface storm flow: predictions with simple kinematic theory for saturated and unsaturated flows. *Water Resources Research*, 18, 1627–33.

Bishop, A. W. and Bjerrum, L. (1960) The relevance of the triaxial test to the solution of stability problems, *Proceedings of Research Conference on Shear Strength of Cohesive Soils*, ASCE, pp. 437–501.

Bishop, A. W. and Blight, G. B. (1963) Some aspects of effective stress in saturated and partly saturated soils, *Géotechnique*, 13, 177–97.

Bishop, A. W. (1966) The strength of soils as engineering materials, 6th Rankine Lecture. *Géotechnique*, 16, 91–128.

Bishop, A. W. and Henkel, D. J. (1967) *The Measurement of Soil Properties in the Triaxial Test*, 2nd edn. Edward Arnold, London.

Black, T. A. (1979) Evapotranspiration parameters. *Water Resources Research*, 15, 164.

Brenner, R. P. (1973) A hydrological model study

of a forested and a cutover slope. *Hydrological Sciences Bulletin*, 18, 125–44.

Broms, B. (1964a) Lateral resistance of piles in cohesive soils. *J. Soil Mechanics and Foundations Division, ASCE*, 90, 27–63.

Broms B. (1964b). Lateral resistance of piles in cohesionless soils. *J. Soil Mechanics and Foundations Division, ASCE*, 90, 123–56.

Brown, C. B. and Sheu, M. S. (1975) Effects of deforestation on slopes. *J. Geotechnical Division, ASCE*, 101, 147–65.

Burroughs, E. R. and Thomas, B. R. (1977) Declining root strength in Douglas-fir after felling as a factor in slope stability. *Intermountain Forest and Range Experiment Station, Research Paper* INT-90, US Forest Service, Ogden, UT.

Coatsworth, A. and Evans, J. (1984) Discussion on Influence of vegetation on shrinking and swelling of clays. *Géotechnique*, 34, 154–5.

Coppin, N. J. and Richards, I. G. (1990) *Use of Vegetation in Civil Engineering*. Butterworths, London.

Coulomb, C. A. (1776) Essai sur une application des règles de maximis et minimis a quelques problèmes des statiques relatifs à l'architecture, *Memoirs Academie Royale des Sciences, Paris*, 3, 36.

Durgunoglu, H. T. and Mitchell, J. K. (1973) Static penetration resistance of soils, *Space Science Laboratory, Univ. of California, Series 14, Issue 24*, Berkeley, CA.

Eagleson, P. S. (1978) Climate, soil, vegetation, Parts 3 and 4. *Water Resources Research*, 14, 731–9, 741–8.

Endo, T. and Tsuruta T. (1969) The effect of tree roots upon the shearing strength of soil. *Annual Report, Hokkaido Branch, Tokyo Forest Experiment Station*, 18, 168–79.

Fellenius, W. (1936) Calculation of the stability of earthdams, *Trans. 2nd Congress on Large Dams*, 4, 445–60.

Fredlund, D. G., Morgenstern, N. R. and Widger, R. A. (1978) Shear strength of unsaturated soils. *Canadian Geotechnical Journal*, 15, 313–21.

Freeze, R. A. (1971) Three-dimensional, transient, saturated–unsaturated flow in a groundwater basin. *Water Resources Research*, 7, 347–66.

Goldstein, R. A., Mankin, J. B. and Luxmoore, R. J. (1974) *Documentation of PROSPER: A model of atmosphere-soil-plant water flow*. EDFB/IBP-73-9. Oak Ridge National Laboratory, Oak Ridge, TN.

Gonsier, M. J. and Gardner, R. B. (1971) Investigation of slope failures in the Idaho batholith. *Inter-*

mountain Forest and Range Experiment Station, Research Paper INT-97 US Forest Service, Ogden, UT.

Gray, D. H. (1970) Effect of forest clear-cutting on the stability of natural slopes. *Bulletin Association of Engineering Geology*, 7, 45–66.

Gray, D. H. and Brenner, R. P. (1970) The hydrology and stability of cutover slopes. *Proc. Interdisciplinary Aspects of Watershed Management, ASCE*, New York, pp. 295–326.

Gray, D. H. and Leiser, A. T. (1982) *Biotechnical Slope Protection and Erosion Control*. Van Nostrand Reinhold Co., New York.

Gray, D. H. and Megahan, W. F. (1981) Forest vegetation removal and slope stability in the Idaho batholith, *Intermountain Forest and Range Experiment Station, Research Paper* INT-271, US Forest Service, Ogden, UT.

Greenway, D. T. (1987) Vegetation and slope stability, in *Slope Stability* (eds M. G. Anderson and K. S. Richards). John Wiley, New York, pp. 187–230.

Greenway, D. R., Anderson, M. G. and Brian-Boys, K. C. (1984) Influence of vegetation on slope stability in Hong Kong. *Proceedings of 4th Int. Symp. Landslides*, Vol. 1, Toronto, Canada, pp. 399–404.

Hathaway, R. L. (1973) Factors affecting the soil binding capacity of the root systems of some *Populus* and *Salix* clones. MS, Thesis, Massey University, Palmerston North.

Hathaway, R. L. and Penny, D. (1975) Root strength in some *Populus* and *Salix* clones. *New Zealand J. Botany*, 13, 333–44.

Henkel, D. J. (1960) The shear strength of saturated remolded days. *Proceedings Research Conference on Shear Strength of Cohesive Soils, ASCE*, New York, 533–54.

Hetenyi, M. (1946) *Beams on Elastic Foundations*. University of Michigan Press, Ann Arbor, MI.

Huff, D. D. and Swank, W. T. (1985) Modelling changes in forest evapotranspiration, in *Hydrological Forecasting* (eds M. G. Anderson and T. P. Burt). John Wiley, New York, pp. 125–51.

Jackson, L. J. (1975) Relationships between rainfall parameters and interception by tropical forest. *J. Hydrology*, 24, 215–38.

Janbu, N. (1973) Slope stability computations, in *Embankment Dam Engineering* (eds R. C. Hirschfeld and S. J. Poulos). John Wiley, New York, pp. 47–86.

Jensen, M. E. and Haise, H. R. (1963) Estimating

evapotranspiration from solar radiation. *J. Irrigation and Drainage Division, ASCE*, **89**, 15–41.

Juran, I., Schlosser, F., Louis, C., Kernos, M. and Eckmann, B. (1981) Le reinforcement des sols par barrés passives. *Proc. 10th International Conference on Soil Mechanics and Foundation Engineering*, Vol. 3, pp. 713–16, Balkema, Rotterdam.

Kozlowski, T. T. (ed.) (1968) *Water Deficits and Plant Growth*. Academic Press, New York.

Kozlowski, T. T. (1971) *Growth and Development of Trees*. Academic Press, New York.

Krabel, C. J. (1936) Erosion control on mountain roads. *Dept. of US Agriculture, Circular* 380, Washington, DC.

Ligget, J.A. and Liu, P. L.-F. (1983) *Boundary Integral Equation Method for Porous Media Flow*. George Allen and Unwin, London.

Lull, H. W. (1964) Ecological and silvicultural aspects, in *Handbook of Applied Hydrology* (ed. V. T. Chow). McGraw-Hill, New York, pp. 6.1–6.30.

Massman, W. J. (1980) Water storage on forest foliage: A general model. *Water Resources Research*, **16**, 210–16.

McNaughton, K. G. and Black, T. A. (1973) A study of evapotranspiration by means of the equilibrium evaporation concept. *Water Resources Research*, **21**, 383–91.

McRoberts, E. C. and Morgenstern, N. R. (1974) The stability of thawing slopes. *Canadian Geotechnical J.*, **11**, 447–69.

Megahan, W. F. (1984) Snowmelt and logging influence on piezometric levels in steep forested watersheds in Idaho. *Transportation Research Board, Transportation Research Record* 965, Washington, DC, pp. 1–8.

Meyerhof, G. G. (1970) Safety factors in soil mechanics. *Canadian Geotechnical J.*, **7**, 349–55.

Morgenstern, N. R. and Nixon, J. F. (1971) One dimensional consolidation of thawing soils. *Canadian Geotechnical J.*, **8**, 558–65.

Morgenstern, N. R. and Price, V. E. (1965) Analysis of the stability of general slip surfaces. *Géotechnique*, **15**, 79–83.

O'Loughlin, C. L. (1974) The effects of timber removal on stability of forest soils. *J. Hydrology, New Zealand*, **13**, 121–34.

Oden J. T. and Ripperger, E. A. (1981) *Mechanics of Elastic Structures*, 2nd edn. McGraw-Hill, New York.

Penman, H. L. (1948) Natural evaporation from open water, bare soil, and grass. *Proceedings Royal Society A*, **193**, 120–45.

Philip, J. R. (1957) The theory of infiltration: 1. The infiltration equation and its solution. *Soil Science*, **83**, 345–57.

Philip, J. R. (1969) Theory of infiltration. *Advances in Hydroscience*, **5**, 215–95.

Phukan, A. (1985) *Frozen Ground Engineering*. Prentice-Hall, Englewood Cliffs, NJ.

Pinder, G. F. and Gray, W. G. (1977) *Finite Element Simulation in Surface and Subsurface Hydrology*. Academic Press, New York.

Priestley, C. H. B. and Taylor, R. J. (1979) On the assessment of surface heat flux and evaporation using large scale parameters. *Monthly Weather Review, US Dept. of Commerce*, **100**, 81–92.

Pufahl, D. E. and Morgenstern, N. R. (1979) Stabilization of planar landslides in permafrost. *Canadian Geotechnical J.*, **16**, 734–47.

Pufahl, D. E. A. and Morgenstern, N. R. (1980) The energetics of an ablating headscarp in permafrost. *Canadian Geotechnical J.*, **17**, 487–97.

Reddi, L. N. and Wu, T. H. (1988) Probabilistic analysis of groundwater levels in hillside slopes. *J. Geotechnical Engineering, ASCE*, **117**, 872–90.

Riestenberg, M. M. (1987) Anchoring of thin colluvium on hillslopes by roots of sugar maple and white ash. PhD Dissertation, University of Cincinnati, Cincinnati, OH.

Riestenberg, M. M. and Sovonick-Dunford, S. (1983) The role of woody vegetation in stabilizing slopes in the Cincinnati area, Ohio. *Geol. Soc. Am. Bull.*, **94**, 506–18.

Remson, I., Hornberger, G. M. and Molz, F. J. (1971) *Numerical Methods in Subsurface Hydrology*. John Wiley, New York.

Rutter, A. J., Kershaw, K. A., Robins, P. C. and Morton, A. J. (1972) A predictive model of rainfall interception in forests, 1. Derivation of the model from observations in a plantation of Corsican pine. *Agricultural Meteorology*, **9**, 367–84.

Sangrey, D. A., Harrop-Williams, K. O. and Klaiber, J. A. (1984) Predicting groundwater response to precipitation. *J. Geotechnical Engineering Division, ASCE*, **110**, 957–75.

Schiechtl, H. M. (1980) *Bioengineering for Land Reclamation and Conservation*. University of Alberta Press, Edmonton.

Scott, R. F. (1981) *Foundation Analysis*. Prentice Hall, Englewood Cliffs, NJ.

Shewbridge, S. E. and Sitar, N. (1989) Deformation characteristics of reinforced sand in direct shear. *J. Geotechnical Engineering, ASCE*, **115**, 1134–47.

Shewbridge, S. E. and Sitar, N. (1990) Deformation-based model for reinforced sand. *J. Geotechnical Engineering, ASCE*, **116**, 1153–70.

Skempton, A. W. (1964) Long-term stability of clay slopes, *Géotechnique*, **14**, 75–101.

Sloan, P. G., Moore, G. B. and Eigel, J. D. (1983) Modeling surface and subsurface storm flow on steeply-sloped forested watersheds, *Water Resources Institute, University of Kentucky, Report 142*, Lexington, KY.

Soil Bioengineering Corporation (1986) *Failure Fill Slope Stabilization using Soil Bioengineering Technology on State Highway NC 126*, McDowell Co., NC. Now Robbin B. Sotir Assoc., Marietta, GA.

Soil Bioengineering Corporation (1987a) *Case History Studies of Three Installed Jute Mesh Fabric/Geojute Systems in North America*. Now Robbin B. Sotir Assoc., Marietta, GA.

Soil Bioengineering Corporation (1987b) *Streambank Stabilization along Winkler Creek, I-395 Tributary, in the Winkler Botanical Preserve, Alexandria, VA*. Now Robbin B. Sotir Assoc., Marietta, GA.

Sotir, R. B. and Gray, D. H. (1989) Fill slope repair using soil bioengineering systems. *Public Works*, December 1989, 37–40, 77.

Stout, B. B. (1956) Studies of the root systems of deciduous trees. *Black Rock Forest Bulletin No. 15*, Harvard University, Cambridge, MA.

Sutton, R. F. (1969) Form and development of conifer root systems, *Commonwealth Forestry Bureau, Technical Communication No. 7*, Oxford.

Swanston, D. N. (1970) Mechanics of debris avalanching in shallow till soils of southeastern Alaska. *Forest Service, Research Paper PNW 103*, US Dept. of Agriculture.

Symes, M. J. P. R., Gens, A. and Hight, D. W. (1984) Undrained anisotropy and principal stress rotation in saturated sand. *Géotechnique*, **34**, 11–27.

Tang, W. H., Yuceman, M. S. and Ang, A. H-S. (1976) Probability-based short term design of soil slopes. *Canadian Geotechnical J.* **13**, 201–15.

Tensar Corp. (1987) Slope reinforcement with Tensar geogrids, Design and Construction Guideline. *Tensar Corp., Technical Note* TTN: SR1, Morrow, GA.

Terwilliger, V. J. (1988) Mechanical effects of chaparral disturbances on soil slip patterns in the Transverse Ranges of Southern California. PhD Dissertation, University of California, Los Angeles, CA.

Terwilliger, V. J. (1990) Effects of vegetation on soil slippage by pore pressure modification *Earth Surface Processes and Landforms*, **15**, 553–70.

Terzaghi, K. (1936) The shearing resistance of saturated soils and the angle between planes of shear. *Proc. 1st International Conference on Soil Mechanics and Foundation Engineering*, **1**, 54 Harvard University, Cambridge MS.

Thornthwaite, C. W. and Mather, J. R. (1957) Instructions and tables for computing potential evapotranspiration and water balance. *Drexel Inst. Technology, Publications in Climatology*, Vol. 10 (3), Centerton, NJ.

Waldron, L. J. (1977) The shear resistance of root-permeated homogeneous and stratified soil. *J. Soil Sci. Soc. Am.*, **41**, 843–9.

Waldron, L. J. and Dakessian, S. (1981) Soil reinforcement by roots: calculation of increased shear resistance from root properties. *Soil Science*, **132**, 427–35.

Wang, W. L. and Yen, B. C. (1974) Soil arching in slopes. *J. Geotechnical Engineering Division, ASCE*, **100**, 61–78.

Williams, A. A. B. and Pidgeon, J. T. (1983) Evapotranspiration and heaving clays in South Africa. *Géotechnique*, **33**, 141–50.

Wu, T. H. (1984a) Effect of vegetation on slope stability, in *Soil Reinforcement and Moisture Effects on Stability*, Transportation Research Record 965, Transportation Research Board, Washington, DC, pp. 37–46.

Wu, T. H. (1984b) Soil movements on permafrost slopes near Fairbanks, Alaska. *Canadian Geotechnical J.*, **21**, 699–709.

Wu, T. H. (1991) Soil stabilization using vegetation. *Proceedings of a Biotechnical Stabilization Workshop*, University of Michigan, Ann Arbor.

Wu, T. H. and Swanston, D. N. (1980) Risk of landslides in shallow soils and its relation to clear-cutting in southeastern Alaska. *Forest Science*, **26**, 495–510.

Wu, T. H., McKinnell, W. P., III and Swanston, D. N. (1979) Strength of tree roots and landslides on Prince of Wales Island, Alaska. *Canadian Geotechnical J.*, **16**, 19–33.

Wu, T. H., Beal, P. E. and Lan, C. (1988a) In situ shear tests of soil root systems. *J. Geotechnical Engineering, ASCE*, **114**, 1376–94.

Wu, T. H., McOmber, R. M., Erb, R. T. and Beal, B. E. (1988b) A study of soil root interaction. *J. Geotechnical Engineering, ASCE*, **114**, 1351–75.

Wu, T. H., Bettandapura, D. P. and Beal, P. E. (1988c) A statistical model of root geometry. *Forest Science*, **34**, 980–97.

Wu, T. H., Randolph, B. and Huang, C. S. (1993) Stability of shale embankments *J. Geotechnical Engineering, ASCE* **119**, 127–46.

Ziemer, R. R. (1981) Roots and the stability of

forested slopes, in *Erosion and Sediment Transport in Pacific Rim Steeplands*, Publication 132, International Association of Hydrological Sciences, London, pp. 343–61.

Zinke, P. J. (1967) Forest interception studies in the United States, in *International Symposium on Forest Hydrology* (eds. W. E Sopper and H. N. Lull). Pergamon Press, Oxford, pp. 137–62.

CONCLUSIONS 8

R. P. C. Morgan and R. J. Rickson

8.1 OVERCOMING UNCERTAINTY

Vegetation is an integral part of most landscapes and has the potential to play a major engineering role in stabilizing slopes and protecting the soil from erosion. The effects of vegetation on the geomorphological processes of water erosion, wind erosion and mass movement of soil, and on the properties of the soil material are, as seen in the previous chapters, reasonably well understood in a qualitative way. The ability to quantity the effects, however, is often limited. Without this quantification, vegetation cannot be included even in a simple way in engineering design procedures. Although the outcome of removing or introducing vegetation on a slope can be forecast in terms of increasing or decreasing either erosion or slope instability, the magnitude of the effect cannot be precisely predicted. This means that engineers are faced with a degree of uncertainty when using vegetation in engineering practice. The uncertainty is enhanced when allowing for variability in the performance of vegetation spatially and temporally, both seasonally and from year to year. Wu, in Chapter 7, demonstrates how this uncertainty can be analysed with respect to slope stability and thereby accounted for in engineering work. It can also be allowed for by designing vegetated slopes with higher factors of safety than the 1.2–1.5 commonly used for unvegetated slopes with conventional engineering support structures. Present attempts to evaluate the factor of safety on vegetated slopes (Chapter 7) show that F values of 1.7–1.8 are common. What is not clear from these studies, however,

is whether still higher factors of safety cannot be achieved or whether the present methods of evaluation underestimate the factor of safety because our knowledge does not allow the effects of vegetation to be properly taken into account.

Using vegetation to design slopes for higher factors of safety may be a valuable insurance against long-term declines in soil strength. This can be particularly beneficial on slopes in over-consolidated clays where the reduction in over-burden pressure following excavation causes the clays to expand, but water is transmitted so slowly through the impermeable material that equilibrium pressures are not attained quickly enough and pore-water pressure falls. Over a period of 10–15 years, however, water is able to move into the whole soil mass, equilibrium pressures are restored and the strength of the material is reduced (Skempton, 1970; Greenwood, Holt and Herrick, 1985).

As shown in Chapter 7, one of the major contributions of vegetation in slope stabilization is the increase in shear strength of the soil arising from the root system. The magnitude of this effect is, however, difficult to quantify. First, there is considerable spatial variability in the effect because of the variations in root density. As shown by Wu (Chapter 7), measured increases in the shear strength of the soil due to roots may have coefficients of variation as high as 30%. Second, there is uncertainty on whether field measurements of the increase in strength are representative because the small size of the shear boxes used means that shallow soil failures can be induced before the tensile

Slope Stabilization and Erosion Control: A Bioengineering Approach. Edited by R. P. C. Morgan and R. J. Rickson. Published in 1995 by E & FN Spon, 2-6 Boundary Row, London, SE1 8HN. ISBN 0 419 15630 5.

strength of the whole root system is fully mobilized. Third and working contrary to the last point, it is unlikely that full mobilization of the tensile strength of the roots will occur uniformly and simultaneously over a whole hillslope. As a result, it may well be that although the present *in situ* tests of shear strength underestimate the maximum reinforcement effect, they can give suitable 'safe' values of the increase in shear for design purposes.

In reality, of course, the increase in shear strength is only one aspect of the effect of vegetation. Important interactions exist between the various effects described in detail in Chapters 2 and 7 which, at present, cannot be easily analysed and quantified. For instance, it is extremely difficult with our present knowledge to separate out all the effects of increases in soil suction from those of root-reinforcement of the soil. Whilst increases in suction due to evapotranspiration lowering the groundwater level can be accounted for in slope stability analysis by decreases in pore-water pressure, suction effects in the unsaturated zone above the water table are either ignored or combined with the root-reinforcement effect and allowed for as an increase in the effective cohesion of the soil (Coppin and Richards, 1990). It is not known by how much our assessments of the factor of safety may be in error because of the failure to account properly for soil suction but studies by Fredlund (1987) in Hong Kong suggest that matric suction may increase the factor of safety in unsaturated soils by about 20%.

Research is also needed to improve our understanding of the hydrological effects of vegetation, particularly with respect to infiltration. Simply knowing that infiltration generally increases with percentage vegetation cover (Chapter 2) is not enough to forecast the effects of changes in vegetation on the magnitude of runoff and erosion. Account needs to be taken of differences in the root depths and densities, the spatial uniformity of the root distribution, and the nature and extent of any litter layer. These factors help to explain why the spatial variability in terminal infiltration rates is often

extremely high with vegetation covers of 80–100% but decreases as the extent of bare ground increases (Faulkner, 1990) even though considerable variability will still exist as a function of inherent soil properties such as structure and porosity. Also, little is known on the response time for infiltration to change as vegetation cover increases; for example, whether increases in infiltration are immediate or whether there is a time-lag before any effect is observed.

With imperfections in our understanding of both the mechanical and hydrological effects of vegetation on a slope, it is difficult to evaluate the interactions between different slope processes. Without this overview, we cannot easily identify the true nature of the processes operating on a slope and this inhibits our ability to make a proper assessment of the risk of slope failure. In particular, we often need to evaluate the trade-off between using vegetation to increase infiltration and reduce surface erosion by water, and the effect of the increased infiltration on soil moisture and therefore on pore-water pressure and slope instability. The trade-off problem is enhanced by the need to determine the extent to which the potential instability arising from higher pore-water pressure is counteracted by the strengthening effect of plant roots on the soil. The relative magnitudes of these effects will condition whether or not the slope remains stable and may also influence decisions on whether measures on a particular slope should be introduced to control primarily surface erosion or mass soil failure. Under extreme situations, as on steep slopes in tropical monsoon climates, where both processes are active, it may be necessary to introduce measures to control one and live with the effects of the other, for example by continuous repair and maintenance.

Despite these imperfections in knowledge and understanding, the examples cited in Chapters 4, 5, 6 and 7 show that both natural and simulated vegetation can be used successfully to control surface erosion and stabilize slopes. Although the use of vegetation in practice relies heavily on engineering experience rather than application of design formulae, this experience can often be analysed to develop procedures for selecting suitable bioengineering techniques according to local site conditions. Figure 8.1 shows an example of this state-of-art approach proposed by Clark and Howell (1992) for use in eastern Nepal. Where bioengineering techniques are adopted, the opportunity should be taken to conduct research to quantify their effects and thereby demonstrate whether their use is justified. As emphasized above, particular attention should be given to studying the spatial and temporal variability of bioengineering systems. The data collected in such research should be used to develop physically-based models which can analyse this variability and thereby help to explain and, possibly, to reduce the degree of uncertainty involved in choosing a bioengineering option.

8.2 VEGETATION SELECTION AND MANAGEMENT

Criteria for selecting vegetation have been discussed with reference to specific applications in Chapters 5, 6 and 7. In general terms, the vegetation must be capable of establishing and growing on the site concerned and must be able to perform the required engineering role. The latter point needs to be emphasized given that, as also seen in the earlier chapters, certain types of vegetation may not be beneficial and may, in fact, enhance erosion and slope instability. Plant species need to be carefully scrutinized before selection to ensure that they possess the right qualities for the task in hand. The qualities required for various types of bioengineering work are given in Table 8.1 (Coppin and Richards, 1990). Since the best species for bioengineering are not necessarily the dominant species at a given site, selection must also take account of ecological principles in order to evaluate the chances of suitable species establishing themselves successfully or the type of management necessary for this to happen. In some parts of the world, plant species have been evaluated for their engineering role and lists are

available of suitable species for specific applications in New Zealand (van Kraayenoord and Hathaway, 1986), Nepal (Howell *et al.*, 1991), the USA (United States Department of Agriculture, 1984) and the United Kingdom (Coppin and Richards, 1990).

The effectiveness of vegetation varies over time due to seasonality, climatic extremes and long-term changes in plant succession. Seasonal variability arises from the die-back and dormancy of the vegetation during either cold or dry parts of the year (Chapter 3). This variability must be considered when designing bioengineering systems to ensure that the vegetation will function satisfactorily at the times of the year when the risks of erosion and soil failure are highest.

Extremes of climate, such as a sequence of cold winters or dry summers, can cause lasting damage to a vegetation cover which will reduce its bioengineering value. Species selection should always be made after a long-term analysis of climatic data to establish the risk of extreme conditions occurring. If the risk is high, only species that can withstand such conditions should be chosen. Ecological analysis of the plant communities in an area may sometimes reveal local, resilient species that are indicative of extreme conditions and which are adapted to survive them.

The best way of ensuring against the possible impairment of the engineering function of vegetation is to select a diverse plant community rather than relying on a limited number of species. If, for any reason, one species fails, the community can respond by another species taking over its role. The management of a diverse community, however, must allow for the fact that vegetation evolves naturally over a long period of time as a result of ecological succession (Chapter 3). In some cases, as seen in the rangeland reclamation work described in Chapter 6, allowing the succession to take its course can be desirable but, in other situations, as seen in Chapter 5, the plants that perform the important engineering role can get shaded out. In these latter instances, long-term management

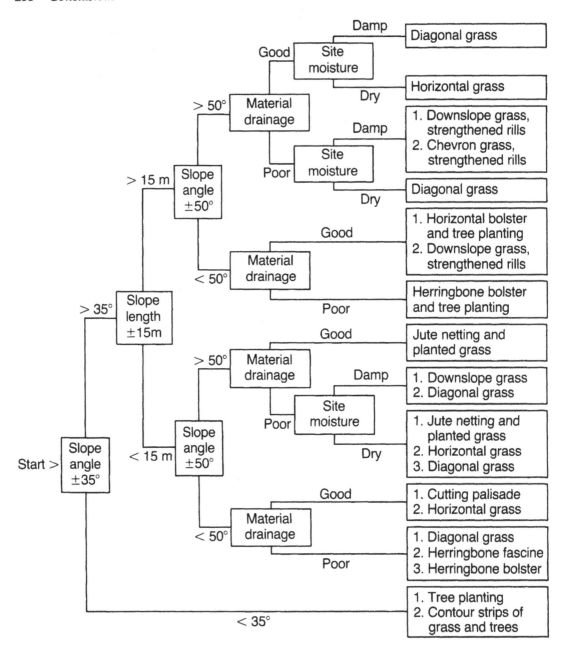

Figure 8.1 Flow chart for selection of bioengineering techniques according to site properties for slope stabilization in eastern Nepal (after Clark and Howell, 1992).

Table 8.1 The functional requirements of vegetation (after Coppin and Richards, 1990)

Function	Qualities required	Principal considerations
Soil reinforcement and enhancement of soil strength	Maximum root development to the required depth (i.e. below the slip surface)	Deep-rooting species Anchorage Suitable soil profile conditions
Soil water removal	Vigorous root development throughout soil volume Large transpiration area	Vigorous rooting species Substantial top growth which transpires throughout year Soil water balance
Surface protection against traffic	Vigorous development at soil surface of both roots and shoots Ability to self-repair quickly	Species selection, short growth habit Management Soil fertility Inherent soil trafficability Use of reinforcement
Surface protection against erosion by wind and water	Vigorous development at soil Surface of both roots and shoots Resistance to damage under high flow conditions Rapid establishment Uniform density of cover	Erosion risk Behaviour of vegetation under high flow conditions Soil surface conditions Species selection Use of reinforcement with geotextiles
Bank and channel reinforcement	Ability to grow in wet conditions perhaps with variable water levels Rapid effectiveness Root reinforcement Top growth absorbs wave impact Low hydraulic roughness under high flow conditions Ability to self-repair	Species selection with respect to ecological preference Growth habit Management Reinforcement with geotextiles
Shelter or screening	Top growth of suitable height and/or density Rapid development	Species selection Density of foliage Structural arrangement

will be needed to control or even halt the succession. Suitable management strategies may include cutting of the vegetation, e.g. mowing grasses or pruning and coppicing trees and shrubs, or controlled levels of grazing. Overall, however, the long-term management of a diverse community is probably easier and cheaper than that of a monoculture or a limited number of species. Managing a monoculture generally requires regular inputs of fertilizers, control over soil moisture, removal of all other invasive species using some form of weed con-

trol, and protection against pests and diseases which may wipe out all the vegetation cover completely.

The management of a vegetation cover for bioengineering work has three phases. First, there is the design and implementation phase in which the ground is prepared and the vegetation established by planting, turfing or seeding (Chapter 3). In this phase, use may be made of mulches and geotextiles to aid plant establishment (Chapter 4). Second, there is a period of after-care in which the growth of the vegetation is monitored, programmes of fertilizer application, mulching and weed control carried out, and any deficiencies in the plant cover made good by reseeding or replanting. At the end of this period, which may last two to three years, the proposed initial vegetation cover should be reasonably well established and the third phase, that of long-term management, can commence.

8.3   COSTS

The cost of implementing a programme of controlling erosion or slope instability using vegetation depends on the condition of the ground at the time of intervention. As an example, Figure 8.2 shows, qualitatively, the costs involved in reclaiming rangeland by revegetation in Iceland (Aradóttir, Arnalds and Archer, 1992). As long as the vegetation cover remains above 50–60%, the costs of restoring land to its original condition are relatively small. They increase as the vegetation cover deteriorates and the erosion rate rises dramatically until, when the cover falls below 5–10% and almost all the soil has been removed, the costs become extremely

high. This pattern of costs has general validity and demonstrates the importance of recognizing the engineering role of vegetation and planning a suitable programme of erosion control and slope stabilization before any vegetation removal takes place. Such an approach is feasible where land is being logged or cleared for agricultural development, but is less easy to adopt where the vegetation cover and soil need to be completely removed, for example in road construction or open-cast mining.

Even if complete removal of the vegetation cover has occurred, control of erosion and slope instability may be cheaper using vegetation than with conventional engineering solutions. Coppin and Richards (1990) compare the likely expenditure profiles for bioengineering works and inert structures (Figure 8.3). The initial costs are higher with structures but these may be offset by lower maintenance and monitoring costs. The real advantages of vegetation, however, are in the long term. Whereas inert structures have a design life and have to be replaced, vegetation is effective for an indefinite period and, subject to the constraints mentioned

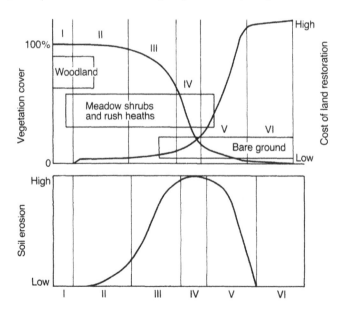

Figure 8.2 Cost profile for restoring Icelandic rangeland to its original condition as a function of the degree of land degradation (Phases I–VI) and the rate of soil erosion (after Aradóttir, Arnalds and Archer, 1992).

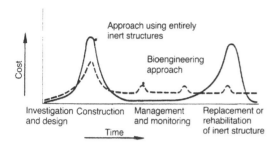

Figure 8.3 Cost profiles for use of bioengineering and inert structures for slope stabilization (after Coppin and Richards, 1990).

in section 8.1, requires only occasional low-cost repair and refurbishment. In the short term, it may be necessary to use simulated vegetation, such as mulches and geotextiles (Chapter 4), to protect an area until the proposed vegetation has become established. Although this will increase the cost, the increase may not be additional because similar protection is frequently required pending the building of inert structures, particularly if, for any reason, a project is delayed.

8.4 THE FUTURE

Even though bioengineering and biotechnical engineering are considered as relatively new subjects and are generally not covered formally as part of civil engineering degree courses, there is, as this book has endeavoured to show, considerable experience in the use of vegetation for erosion control and slope stabilization. Much of this experience comes from agricultural engineering which, together with geomorphology, also provides the basis for our general understanding of how vegetation performs its engineering role. In bioengineering, this experience is combined with that of the civil engineer and the geotechnical engineer and underpinned with the necessary botanical, biological and ecological skills. Bioengineering thus crosses several disciplines. Its future depends on drawing these together as a basis for improving our theoretical understanding of the engineering functions of vegetation and for analysing previous and present practical experience. From these developments there needs to emerge a better quantification of the various effects of vegetation and the uncertainties involved together with a set of numerical design procedures.

REFERENCES

Aradóttir, A. L., Arnalds, O. and Archer, S. (1992) Hnignun gróðurs og jarðvegs. *Árbók Landgraeðslu Ríkisins*, **IV**, 73–82.

Clark, J. E. and Howell, J. H. (1992) The application of bioengineering in the developing world, in *The Environment is Our Future*. Proceedings of Conference XXIII of the International Erosion Control Association, Reno, Nevada, pp. 275–84.

Coppin, N. J. and Richards, I. G. (1990) *Use of Vegetation in Civil Engineering*. Butterworths/CIRIA, London.

Faulkner, H. (1990) Vegetation cover density variations and infiltration patterns on piped alkali sodic soils: implications for the modelling of overland flow in semi-arid areas, in *Vegetation and Erosion* (ed. J. B. Thornes). Wiley, Chichester, pp. 317–46.

Fredlund, D. G. (1987) Slope stability analysis incorporating the effect of soil suction, in *Slope Stability* (eds M. G. Anderson and K. S. Richards). Wiley, Chichester, pp. 113–44.

Greenwood, J., Holt, D. A. and Herrick, G. W. (1985) Shallow slips in highway embankments constructed of overconsolidated clay. *Proceedings, Symposium on Failures in Earthworks*. Institution of Civil Engineers, London, pp. 79–92.

Howell, J. H., Clark, J. E., Lawrance, C. J. and Sunwar, I. (1991) *Vegetation Structures for Stabilising Highway Slopes. A Manual for Nepal*. UK/Nepal Eastern Region Interim Project, Kathmandu.

Skempton, A. W. (1970) First-time slides in overconsolidated clays. *Géotechnique*, **20**, 320-4.

United States Department of Agriculture (1984) *National Plant Materials Handbook*. United States Soil Conservation Service, Washington, DC.

Van Kraayenoord, C. W. S. and Hathaway, R. L. (1986) *Plant Materials Handbook for Soil Conservation. Vol. 1: Principles and Practices*. National Water and Soil Conservation Authority, Water and Soil Miscellaneous Publication No. 94, Wellington, New Zealand.

INDEX

Printed and bound by CPI Group (UK) Ltd, Croydon, CR0 4YY

01/11/2024

01782610-0011